Force

Multiply By ↑ / From →	To → Ounces	Pounds	Kips	Tons (short)
dynes	1.405×10^{-7}	2.248×10^{-6}	2.248×10^{-9}	1.124×10^{-9}
grams	3.527×10^{-2}	2.205×10^{-3}	2.205×10^{-6}	1.102×10^{-6}
kilograms	3.527×10^{1}	2.205	2.205×10^{-3}	1.102×10^{-3}
newtons	3.597	2.248×10^{-1}	2.248×10^{-4}	1.124×10^{-4}
kilonewtons	3.597×10^{3}	2.248×10^{2}	2.248×10^{-1}	1.124×10^{-1}
tons (metric)	3.527×10^{4}	2.205×10^{3}	2.205	1.102

Pressure (or Stress)

Multiply By ↑ / From →	To → lb/in²	lb/ft²	Kips/ft²	Tons (short)/ft²	Feet of Water (39.2°F = 4°C)	Atmosphere
gm/cm²	1.422×10^{-2}	2.048	2.048×10^{-3}	1.024×10^{-3}	3.281×10^{-2}	9.678×10^{-4}
kg/cm²	1.422×10^{1}	2.048×10^{3}	2.048	1.024	3.281×10^{1}	9.678×10^{-1}
kN/m²	1.450×10^{-1}	2.090×10^{1}	2.088×10^{-2}	1.044×10^{-2}	3.346×10^{-1}	9.869×10^{-3}
ton (metric)/m²	1.422	2.048×10^{2}	2.048×10^{-1}	1.024×10^{-1}	3.281	9.678×10^{-2}

Torque (or Moment)

Multiply By ↑ / From →	To → lb-in	lb-ft	Kip-ft
gm-cm	8.677×10^{-4}	7.233×10^{-5}	7.233×10^{-8}
kg-m	8.6777	7.233	7.233×10^{-3}
kN-m	9.195×10^{3}	7.663×10^{2}	7.663×10^{-1}

Velocity

Multiply By ↑ / From →	To → ft/s	ft/min	mi/h
cm/s	3.281×10^{-2}	1.9685	2.236×10^{-2}
km/min	5.467×10^{1}	3.281×10^{3}	3.728×10^{1}
km/h	9.116×10^{-1}	5.467×10^{1}	6.214×10^{-1}

1 mile = 1610 meters = 5282.152 ft

Unit Weight

Multiply By ↑ / From →	To → lb/in³	lb/ft³
gm/cm³	3.613×10^{-2}	6.248×10^{1}
kg/m³	3.613×10^{-5}	6.248×10^{-2}
kN/m³	3.685×10^{-3}	6.368
tons (metric)/m³	3.613×10^{-2}	6.428×10^{1}

Geotechnical Engineering:
Foundation Design

Geotechnical Engineering:
Foundation Design

Professor John N. Cernica, P.E., Ph.D.
Department of Civil Engineering
Youngstown State University

John Wiley & Sons, Inc.
New York   Chichester   Brisbane   Toronto   Singapore

Acquisitions Editor	Cliff Robichaud
Marketing Manager	Susan Elbe
Senior Production Editor	Savoula Amanatidis
Text and Cover Designer	Pedro A. Noa
Manufacturing Manager	Lori Bulwin
Illustration Coordinator	Anna Melhorn

This book was typeset in New Baskerville by Bi-Comp, Inc., and printed and bound by Hamilton Printing Company.

Library of Congress Cataloging in Publication Data:
Cernica, John N.
    Geotechnical engineering : foundation design / John N. Cernica.
        p.   cm.
    Includes index.
    ISBN 0-471-30887-0 (cloth)
    1. Foundations—Design and construction.   2. Soil mechanics.
I. Title.
TA775.C39   1995
624.1′5—dc20

94-3494
CIP

Printed in the United States of America.

10 9 8 7 6 5 4 3 2 1

To my beautiful and most precious family:
My wonderful wife Patricia and our six daughters and their families

Preface

Foundations transfer the loads from the superstructure to the underlying soil formation or rock. *Foundation Design,* in a broad sense, is a two step process: (1) the tentative selection of the foundation scheme based on building site conditions and structural constraints, and (2) the sizing, reinforcing, and detailing of the structural element. *Foundation Engineering* reaches beyond, encompassing a third step: the construction of the structural elements developed in step 2; typically, construction is done by a contractor and the supervision by the engineer.

Intimately related to foundation engineering is the interaction of the foundation components and soil or rock that support these components. To assess this interaction, the student generally relies on design procedures learned in class, but for real-life situations, the final design is normally the product of analytical procedures, experience, judgment, and a common sense or "feel" for the problem.

What a bonanza it would be for the student if we, the teachers, were able to enhance a solid analytical presentation with the teaching of experience, judgment, and common sense! Regretfully, although precious, the latter are facets that come with time and maturity, and are evasive and difficult to teach in a classroom. But try we must. It is in this context that this textbook was developed.

A concerted effort was made by the author to (1) offer a comprehensive and thorough analytical presentation, and (2) reflect and relate from his experience as a consultant on hundreds of projects acquired during his professional career. Furthermore, in order to enhance the concepts and procedures, photographs are included in the text in the belief that "a picture is worth a thousand words." Also, in recognition of the benefit one derives from illustrations, numerous example problems are included in order to explain the various procedures in problem solving. In the same context, numerous problems are provided at the end of each chapter for student assignment.

The first three chapters provide a condensed version of basic elements of soil mechanics, extracted predominantly from *Soil Mechanics,* by J. N. Cernica, published by John Wiley & Sons, Inc. The remaining chapters deal with the design of various types of foundation components (shallow and deep foundations), retaining structures (flexible and rigid), and site improvement. This is the material that is typically covered in a foundation design course during one semester. The material content is deemed as ample for demanding coverage, but not so overwhelming that a great percentage would normally be left uncovered.

In view of virtually total emphasis by the ACI code on the conventional English (fps) units, the emphasis herein is primarily on the English units in connection with the design of the concrete members. The equivalent SI units are provided where it was deemed advisable.

Although the objective is always to select a safe and economical design, one recognizes that frequently there is no unique design or methodology in Foundation Engineering. Hence, a concerted effort was made to tie in a thorough theoretical presentation with the practical aspects in Foundation Design.

At the end of each chapter, a somewhat limited but perhaps useful and relevant bibliography is included for those interested in further reading. Many of these publications were also used by the author as reference in formulating his observations and text content; these are cited by number in parentheses.

Although every attempt was made to eliminate errors, it is perhaps inevitable that errors were made and not detected. In this regard, if the reader should detect such errors, the author will be greatly indebted to the reader if these are brought to his attention so that they may be subsequently corrected. Similarly, any suggestions in the type and mode of coverage of material would be greatly appreciated.

The author wishes to thank the following professors for their reviews, observations and input: Dr. M. S. Aggour, University of Maryland; Dr. S. H. Armaleh, University of Arizona; Dr. C. Aryani, California State University at Sacramento; Dr. E. C. Drumm, University of Tennessee; Dr. C. L. Ho, Washington State University; Dr. R. D. Hryciw, University of Michigan; Dr. C. H. Juang, Clemson University; Dr. J. S. Lin, University of Pittsburgh; Dr. K. Tawfig, Florida State University.

Gratitude is also expressed to the following for their effort in connection with the preparation of the text: Jude Cernica, Tricia (Cernica) Lallo, Joe Stafford, Kent Knauf, Dr. Richard A. Muntean, and Dr. Shakyr Husain, Youngstown State University. Also, many thanks to Mary Ann Cantelmi and Shari Simko for their typing of the manuscript. Finally, I wish to express my sincere appreciation to many contributors of photographs, sketches and other input used in the text.

John N. Cernica
September 1, 1994

Contents

Symbols and Abbreviations

A	area; activity of clay	D	diameter; depth; dimension
A_v	area of voids		
Å	Angstrom units	D_r	relative density
a	distance; area	D_{10}	Hazen's effective size diameter at which 10 percent of the soil is finer
a_v	coefficient of compressibility		
B	width of base	D_{50}	diameter at which 50 percent of the soil is finer
b	width of footing or base; distance	d	depth; distance; diameter; dial reading
C_a	coefficient of secondary compression	E	Young's modulus
C_c	coefficient of compression	e	void ratio; eccentricity
C_r	slope of rebound curve for overconsolidated clays	e_{max}	void ratio of soil in loosest state
C_u	Hazen's uniformity coefficient	e_{min}	void ratio of soil in densest state
C_v	coefficient of consolidation	$e_o,\ e_i$	initial void ratio of mass
C_z	coefficient of curvature	F	friction force; factor of safety
c	cohesion; shape factor		
$c_d,\ \bar{c}$	developed cohesion	f	coefficient of friction; friction factor
c_e	effective or true cohesions	G	specific gravity

g	acceleration of gravity
H	height; hydrostatic head; thickness
h	hydraulic head; total head; height
h_c	height of capillary rise
I	indices; moment of inertia
I_p	plasticity index
I_f	flow index
I_t	toughness index
IV	influence values
i	hydraulic gradient; angle of inclination of slope with the horizontal
K	coefficient of lateral pressure
K_a	coefficient of active pressure
K_p	coefficient of passive pressure
K_o	coefficient of earth pressure at rest
$K_{F\text{-line}}$	line through p versus q
k	permeability coefficient
k_e	effective permeability
L	distance; length
L.L.	liquid limit
l	length
M	mass; moment
m	distance ratio
N	normal force; pressure index; various coefficients; standard penetration number
N_ϕ, N_q, N_γ	bearing capacity coefficients
n	number of equipotential drops; porosity; distance ratio; number
n_d	number of equipotential drops
n_f	number of flow paths
OCR	overconsolidation ratio
P	force; resultant pressure
P_a	resultant active pressure
P_p	resultant passive pressure
P.I.	plasticity index
p	pressure
Q	rate of flow; concentrated force; pile capacity; bearing capacity
Q_p	pile end-bearing resistance
Q_s	pile frictional resistance
Q_u	ultimate load
q	rate of discharge per unit area; stress or pressure
q_a	allowable bearing capacity
q_u	ultimate bearing capacity
R	radius; Reynold's number; resultant force; resistance
r	radius
S	degree of saturation; distance; shear strength resultant
s	shear strength per unit distance; distance; settlement
T	time factor; tensile force
T_a	surface tension
t	time
U	consolidation ratio; resultant neutral (pore-water pressure) force; unconfined compression
u	pore-water pressure; velocity
V	volume
V_s	volume of solids
V_v	volume of voids
V_w	volume of water
v	velocity
v_s	seepage velocity
W	weight
W_s	weight of solids
W_w	weight of water
w	water content
w_l	liquid limit
w_p	plastic limit
w_s	shrinkage limit
X, Y, Z	Cartesian coordinates
x, y, z	distances
$\bar{x}, \bar{y}$	centroidal distances
x_t	transformed x distance
Z	coordinate
z	depth
α	inclination of vectors; inclination of slopes; angle

α_m	maximum angle of stress obliquity		ρ	mass density; settlement
β	angle		σ	normal stress; total stress
γ	unit weight		$\bar{\sigma}$	effective normal stress
γ_b	buoyant unit weight		$\sigma_1, \sigma_2, \sigma_3$	principal stresses
γ_d	dry unit weight of soil		σ_t	combined or total pore-water stress and intergranular normal stress
γ_s	unit weight of solids			
γ_t, γ	total unit weight of mass			
γ_w	unit weight of water		Σ	sum
δ	angle; deflection; settlement		τ	shear stress
Δ	changes; increments		ϕ	friction angle
ε	strain		ϕ_d	developed friction angle
$\varepsilon_x, \varepsilon_y, \varepsilon_z$	strain in x, y, z directions, respectively		ϕ_e	effective friction angle
			$\bar{\phi}$	friction angle based on effective stress
θ	angle		ψ	angle
μ	Poisson's ratio; coefficient of viscosity		ω	angle

CHAPTER 1

Geotechnical Problems in Civil Engineering

1.1 INTRODUCTION

Geotechnical Engineering is a subdiscipline within Civil Engineering. It embraces the topics of soil mechanics and foundation design. It develops the groundwork for evaluating the interaction between the geological environment and man-made works including earth-supported structures, retaining walls, dams, and special foundations. It provides guidelines for evaluating the stability of slopes, for sampling and testing of soils, and methodologies for improving building sites. In short, Geotechnical Engineering covers all forms of soil-related problems.

It is common to separate Geotechnical Engineering into two parts: soil mechanics and foundation design. *Soil mechanics* encompasses topics associated with the physical index properties of the soil, water flow through soils, stress and deformation phenomena in soils, strength parameters, general bearing capacity, field exploration techniques, laboratory testing, and frequently a brief reflection on geology. *Foundation design* focuses more on the design of various types of foundations, including a variety of footings, piles, and drilled piers, as well as structural members such as retaining walls, sheet piles, and bracing, whose primary function is to provide lateral earth support. In this chapter, we shall briefly introduce some of the more common problems encountered in the soil mechanics and foundation design fields.

1.2 FOUNDATIONS

The component which interfaces the superstructure to the adjacent zone of soil or rock is referred to as the *foundation*. Its function is to transfer the loads above the foundation (*superstructure loads*) to the underlying soil formation without over-stressing the soil. Thus, a safe foundation design provides for a suitable factor of safety against shear failure of the soil and against excessive settlements.

Foundations are commonly divided into two categories: *shallow foundations* and *deep foundations*. It is common practice to use ordinary spread footings or wall footings as the foundation components when the soil formation has adequate strength for a safe bearing support. Such foundations are referred to as shallow foundations. Shallow foundations are discussed in Chapters 6 and 7. On the other hand, if the soil formation depicts either a low shear strength or is highly compressible, or both, shallow foundations may not be a viable option. Under such circumstances, if the weak strata cannot be improved within the context of a reasonable cost–benefit ratio, then the loads may have to be transmitted to a greater depth, perhaps to a stiffer stratum or to rock, by means of piles or drilled piers. The latter are generally referred to as deep foundations, and are discussed in Chapter 11.

1.3 EARTH-RETAINING STRUCTURES

One does not expect a vertical, steep, unsupported earth mass to remain standing very long; consequences from such a condition may be a shear or slip failure. Hence, under such circumstances, we may provide lateral supports via a variety of types and forms of retaining structures.

Typically, **rigid** retaining walls are constructed from concrete. Concrete retaining walls may be suitable for a large grade separation, bridge abutments, or concrete dams. They are usually suited for dry-land construction, where the bearing capacity of the soil is reasonably good, and where construction and water problems are not particularly adverse. It may be difficult, expensive, and perhaps even impossible to construct in areas where soils are extremely soft, and where the water table is very high and excavation is difficult. Concrete retaining walls are discussed in Chapter 9.

Sheet piles are specially formed steel plates used to construct specific shoring or retaining-wall installations (e.g., shoring of excavations, braced and cellular cofferdams, erosion precaution, wharf construction). Sheet piles are frequently used in areas of extremely soft formations or high water table, or for difficult excavation sites. Typical examples of such sites may be the edge of a river, lake, ocean, or other waterfront facilities. Sheet piles are installed by driving individual piles to a designated depth, and depending on the design, they may be cantilevered, braced, or tied. Examples of sheet pile installation, procedure for installment, and design of sheet pile are given in Chapter 10.

A reinforced-earth retaining wall is sometimes used in lieu of concrete walls, particularly for relatively small heights. Although economy may be a factor, the appeal of such walls is usually in the appearance of the finished "face" of the wall. These walls are discussed in Section 9.9.

Braced cuts in excavations are a form of retaining structures, consisting of generally temporary support during construction (e.g., tunnels, pits, cofferdams). These are discussed in Section 10.8.

1.4 SLOPES AND DAMS

A slope may be any laterally unsupported earth mass, natural or man-made, whose surface forms an angle with a horizontal. Hills and mountains, river banks and coastal formations, earth dams, highway cuts, trenches, and the like are examples of slopes.

Every slope experiences gravitational forces; it may also possibly be subjected to earthquakes, glacial forces, or water pressures. In turn, these phenomena may be direct influences on the stability of the slope. Slope masses that have experienced a sliding failure are common.

Generally, the "sliding planes" for slopes are somewhat circular rather than straight lines. The resisting forces to such sliding are a function of the normal forces on the slip plane, as well as soil properties such as cohesion, c, and angle of internal friction, ϕ. Thus, evaluating the stability of such masses entails an evaluation of the soil properties as well as the causes for impending motion; such causes may include not only the weight of the sliding mass but also external forces induced by perhaps earthquakes, water level fluctuation, or blasting.

Man-made lakes are typically formed by damming the water from streams or watersheds. Usually, a large percentage of the dam is constructed of soil; the spillway is typically of concrete. Hence, somewhat diverse design considerations, and rather restrictive material requirements and construction precautions, are encompassed in the construction of dams.

Though the dam's stability is an indispensable requirement, its ability to hold water is also unquestionably important. Quite obviously, the dam would be improperly designed and constructed if water loss via seepage were excessive.

Generally, the seepage loss through the concrete dam is negligible; however, the seepage loss through the soil formation under the concrete dam and the seepage through and under an earth dam are of concern.

Erosion is a problem associated with slopes and dams. Typically, the deposits associated with erosion are rather soft, compressible, and generally display poor strength (i.e., poor bearing capacity).

1.5 ALTERING/IMPROVING SOIL PROPERTIES

As the more "suitable" building sites are utilized, the problems associated with the poor sites appear to become more prevalent. Thus, the engineer may frequently be faced with the choice of one of the following: (1) adapt the design details to be compatible with the soil conditions (e.g., use piles, increase footing dimensions to compensate for low bearing capacity), (2) abandon the site in favor of one with more favorable soil characteristics, or (3) alter or improve the soil properties toward a designated goal (e.g., increase strength, reduce permeability, reduce compressibility).

The process of improving via altering the soil properties of a given site is broadly referred to as *soil stabilization*. Encompassed in the term stabilization are a number of techniques:

- Densification of soil via compaction, precompression, drainage, vibrations, or a combination of these
- Mixing and/or impregnation of the soil formation with chemicals or grouting, or using geofabrics to develop a more stable base for compaction
- Replacement of undesirable soil with a suitable one under controlled conditions

Not all methods and techniques mentioned are practical for all sites and for all types of soil. For example, the impregnation of a silt or clay strata with chemicals or cement grout is normally appreciably more difficult than injection of such materials into a granular layer; permeability of the stratum to be impregnated is of obvious importance. The construction procedures and equipment solutions are usually adopted in the context of site conditions (e.g., type of soil, water table) and feasibility of approach and economy. Site improvement is discussed in Chapter 5.

CHAPTER 2

Geotechnical Properties of Soils

2.1 INTRODUCTION

The typical composition of soil is a combination of solid particles, liquid, and gas. This combination is generally referred to as the *soil mass,* and it may range from a variety of soft, highly compressible silts and clays or organic matter, through firmer formations of sand, gravel, and rock.

The *solid* phase may encompass a wide range of shapes and may vary in size from large pieces of hard, dense rock or boulders to very tiny particles invisible to the naked eye. The *liquid* phase consists of water containing various amounts and types of dissolved electrolytes. The *gas* phase consists typically of air, although organic gases may exist in areas of high biological deposits. All these materials may occur over a wide range of compositions, densities, moisture, and air content. Indeed, at any given building site, such variations may exist over intervals of as small as a few inches. Hence, it is a task for the geotechnical engineer to properly classify the soil and to evaluate its properties for use in a particular project.

The physical and index properties of soils ensue directly from the interaction of the solids, liquids, and gases with each other and other forces or stresses induced by ongoing environmental and physical changes in the planet Earth. To understand and properly assess the characteristics of a given soil deposit, one needs to understand the constituency of these materials and their interactions in the context of the foundation design—indeed, not an easy task.

2.2 SOIL FORMATION AND DEPOSITS

The layers of the earth are generally divided into four major parts: the *outer crust*, the *mantle*, the *outer core*, and the *inner core*. As engineers, we are concerned almost exclusively with the outer portion of the earth, generally referred to as *regolith*. This is the formation that supports virtually all engineering structures.

Regolith was formed by the weathering or disintegration of the masses of rocks and minerals, and by chemical weathering or decomposition of various minerals in the parent rock. *Gradation* is a process whereby the physical topography (mountains, valleys, etc.) may be altered by the action of water, air, ice, or other weathering factors. *Diastrophism* designates a process whereby portions of the earth move relative to the other portions. *Vulcanism* refers to the action of the molten rock, both on the Earth's surface and within the Earth.

The rocks and minerals of the Earth's surface were the starting materials from which soils originated. Exposure to atmospheric conditions and volcanic and tectonic action has subsequently transformed these rocks and minerals into a more or less unconsolidated blanket over the Earth's surface. With time, this separated into an orderly sequence of layers or horizons; such a soil formation is known as regolith. These differentiated laminated layers of the regolith are the soils that are of particular interest to geotechnical engineers since portions of the regolith are common support media for structures.

Various processes are continually acting on the Earth's surface. *Disintegration* is related to freezing and thawing, the action of water, glaciation, and so on. *Decomposition* is related to oxidation or hydration. The combination of the mechanical and chemical processes is called *weathering*. Thus, soil is a product of such transformation.

The Geologic Cycle

Geologists tell us that there is an ongoing, rather inconspicuous, and very slow process whereby rocks are decomposed into soil, while some of the soil goes through various stages of conversion back into rock. This phenomenon is described generally as the *geological cycle*. *Igneous rock* is the oldest formation (the parent material), formed as the molten magma cooled, perhaps 4.5 billion years ago. These rocks became subject to attack by various environmental agents, with some of the rock subsequently decaying and disintegrating into a state of *residual soils*. Some of these deposits were ultimately transported and dropped in a form generally classified as *sedimentary deposits*. A portion of such deposits may cement and consolidate into *sedimentary rock*.

Subjected to various environmental impositions of pressures and heat, some of the sedimentary deposits may be transformed into *metamorphic rock*. Other parts of the sedimentary deposits may be subjected again to weathering and disintegration, transportation and redeposition, forming a new generation of secondary rocks and fragmented material and soils. Indeed, excessive pressures and heat may result in melting of virtually all rocks, resulting in new igneous rock. Hence, a new geological cycle begins again, as an ongoing process.

Clays (clay minerals) are part of the soil with the smallest grain size, less than 2 μm and are electrochemically very active. Although the majority of clay minerals are insoluble in acid, they generally have appreciable affinity for water, they are elastic when wet, water retentive, and coherent when dry. Frequently, clay deposits become major challenges to the foundation engineer; they may be subject to large volume changes (swelling, shrinkage, consolidation), and may be greatly affected by environmental factors such as moisture, loads, and chemicals. *Kaolinites, montmorillonites,* and *illites* constitute the three main groups of clay minerals.

2.3 SOIL DENSITIES AND VOLUME RELATIONSHIPS

Soils consist of a combination of solids and voids. In geotechnical engineering the term *voids* is assumed to encompass both the volume of water and the volume of air. Hence, a total soil mass would be assumed to be composed of solid soil particles and voids (air and water). For convenience, the mass is separated into these basic components as shown in Fig. 2.1. The volumetric relationship takes into account the three quantities: solids, water, and air. On the other hand, the weight of the air is negligible relative to that of the water and solids and, therefore, is neglected in the overall weight consideration. The following relationships are used to identify the index properties of the soils.

Void Ratio and Porosity

The *void ratio* e of the mass is defined as the ratio of the volume of voids V_v to the volume of solids V_s given by Eq. 2-1:

$$e = \frac{V_v}{V_s} \qquad (2\text{-}1)$$

The void ratio is expressed as a number and falls in the range of

$$0 < e < \infty$$

The *porosity* n of the soil mass is defined as the ratio of the volume of the voids V_v to the total volume of the mass V. It is given by Eq. 2-2:

$$n = \frac{V_v}{V} \times 100 \qquad (2\text{-}2)$$

Porosity is expressed as a percentage and falls in the range of

$$0 < n < 100$$

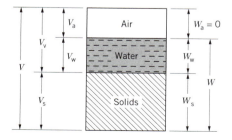

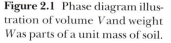

Figure 2.1 Phase diagram illustration of volume V and weight W as parts of a unit mass of soil.

The relationship between porosity and void ratio is given by Eq. 2-3:

$$n = \frac{V_v}{V} = \frac{V_v}{V_s + V_v} = \frac{V_v / V_s}{(V_s + V_v)/V_s} = \frac{e}{1 + e} \qquad (2\text{-}3)$$

where V_v = volume of voids
         V_s = volume of solids
         V = total volume of soil mass

It is common to assume that the volume of the solids V_s in a given mass remains a constant, regardless of any imposed conditions. The value of V_v, however, is altered by a change in the volumes of either air or water, or both. Hence, the void ratio varies in direct proportion to the change in V_v. On the other hand, in the expression for n (Eq. 2-3) one notes that both the numerator and the denominator are a function of the "variable" V_v. Thus, of the two, the void ratio is the more explicit and, therefore, the more widely used expression for the volumetric relationship between voids and solids.

The porosity for natural sands depends to a great degree on the shape of the particle and the uniformity of the particle's size, as well as on circumstances related to sedimentation and deposition. The porosity of most sand masses falls in the range of 25 to 50%. As mentioned above, the porosity cannot exceed 100%.

Although the void ratio can theoretically range from zero to infinity, the common range is between 0.5 and 0.9 for sand and gravel soils and between 0.7 and 1.5 for clays. It may, however, reach higher values, perhaps exceeding 3 or 4 for some colloidal-type clays.

Water Content and Degree of Saturation

The *water* or *moisture content w* is defined as the ratio of the weight of water to the weight of the solid particles. This is expressed by Eq. 2-4:

$$w = \frac{W_w}{W_s} \times 100 \qquad (2\text{-}4)$$

where W_w = weight of water
         W_s = weight of solids

The water content is expressed as a percentage and falls in the range of $0 < w < \infty$.

The *degree of saturation S*, expressed in percent, is defined as the ratio of the volume of water to the volume of voids. It may be written as

$$S = \frac{V_w}{V_v} \times 100 \qquad (2\text{-}5)$$

where V_w = volume of water
         V_v = volume of voids

One may view the degree of saturation as the volume of void spaces occupied by water relative to that which could be occupied if all the pores were full of water. From Eq. 2-5 one notes that the degree of saturation varies from $S = 0$ for a completely dry soil to $S = 100\%$ for a totally saturated state. From a practical point of view, however, the two extremes are seldom approached and never actually reached for soils in their natural states. For example, even for submerged soils, a certain percentage of air exists within a soil mass, and therefore the volume of water may approach but never quite equal the volume of the voids; that is, $S <$ 100% even for submerged conditions.

The water content can exceed 100%, as indicated by Eq. 2-4. In sands it generally varies between 10 and 30%, whereas in clay it may vary from perhaps 10 to over 300%—a value associated with very fine-grain, loosely deposited clays. The degree of saturation and the water content in a soil mass may have a significant effect on some of the characteristics and behavior of that soil. This is particularly true for a fine-grain soil. For example, a high water content may greatly reduce the shear strength of a clay stratum, or its bearing capacity, or both, or the amount and rate of consolidation may be appreciably influenced by the degree of saturation.

Specific Gravity

The general definition for *specific gravity* is the ratio between the unit weight of a substance and the unit weight of pure water at 4° C. This may be represented by Eq. 2-6.

$$G_{subs} = \frac{\gamma_{subs}}{\gamma_w} \tag{2-6}$$

where G_{subs} = specific gravity of a substance
γ_{subs} = unit weight of substance
γ_w = unit weight of water at 4° C (1 g/cm³ = 9.807 kN/m³)

The specific gravities of different minerals vary rather widely, with that of the majority of soil particles ranging between 2.6 and 2.8. Soils with high organic content will have lower values. The value commonly used for the specific gravity of soil particles is 2.7, and unless specified otherwise, this value would be assumed as a reasonable average in our example problems and discussions. Table 2.1 gives the specific gravities for a selected group of minerals.

The unit weight of water varies, but rather insignificantly, with temperature. For example, it varies from 1.0 g/cm³ = 9.807 kN/m³ at 4° C to approximate 0.996 g/cm³ = 9.768 kN/m³ at 25° C (77° F). For our purpose of calculations, and for the degree of accuracy generally expected in soil mechanics, a unit weight of 1 g/

Table 2.1 Specific Gravity of Some Selected Minerals

Mineral	Specific Gravity
Gypsum volcanic ash	2.32
Orthoclase	2.56
Kaolinite	2.61
Quartz	2.67
Calcite	2.72
Dolomite	2.87
Magnetite	5.17

$cm^3 = 9.807$ kN/m³ is a value deemed acceptable. Similarly, the corresponding unit weight of water in the *foot-pound-second* (fps) system is 62.42 lb/ft³.

Example 2.1

Given The following are useful relationships:

(a) Unit weight of mass $\gamma = \dfrac{G + Se}{1 + e} \gamma_w$

(b) Unit weight of mass $\gamma = \dfrac{1 + w}{1 + e} G\gamma_w = \dfrac{1 + w}{1 + e} \gamma_s$

(c) Dry unit weight $\gamma_d = \dfrac{G}{1 + e} \gamma_w \qquad \gamma_d = \dfrac{G}{1 + wG/S} \gamma_w$

(d) Submerged or *buoyant* unit weight $\gamma_b = \dfrac{G - 1}{1 + e} \gamma_w$

(e) $V_v = \dfrac{e}{1 + e} V$ (f) $V_s = \dfrac{1}{1 + e} V$

(g) $V_w = \dfrac{Se}{1 + e} V$ (h) $W_w = \dfrac{Se}{1 + e} V\gamma_w$

(i) $W_s = \dfrac{1}{1 + e} VG_m\gamma_w$ (j) $W = \dfrac{G + Se}{1 + e} V\gamma_w$

where γ = unit weight of mass
γ_b = submerged unit weight
γ_d = dry unit weight of soil
γ_s = unit weight of solids
W = total weight of soil mass
V = total volume of soil mass
G = specific gravity of *solids*

The other terms have already been defined.

Find Derive all of these expressions from basic definitions.

Procedure (a) $\gamma = \dfrac{W}{V} = \dfrac{W_s + W_w}{V_s + V_v} = \dfrac{(W_s + W_w)/V_s}{(V_s + V_v)/V_s} = \dfrac{G + V_w/V_s}{1 + e} \gamma_w$; but $V_w = SV_v$. Thus,

$\gamma = \dfrac{G + SV_v/V_s}{1 + e} \gamma_w$

Answer

$$\gamma = \frac{G + Se}{1 + e} \gamma_w$$

(b) $\gamma = \dfrac{W_s + W_w}{V_s + V_v} = \dfrac{(W_s + W_w)/W_s}{(V_s + V_v)/W_s} = \dfrac{1 + w}{(V_s + V_v)/\gamma_s V_s}$

$\gamma = \dfrac{1 + w}{1 + e}\gamma_s;\ \gamma_s = G\gamma_w$

Answer

$$\gamma = \dfrac{1 + w}{1 + e}\,G\gamma_w$$

(c) From (a) above, $S = 0$,

$$\gamma_d = \dfrac{G + 0}{1 + e}\,\gamma_w$$

Answer

$$\gamma_d = \dfrac{G}{1 + e}\,\gamma_w$$

(c′) $\gamma_d = \dfrac{W}{V} = \dfrac{W_s + W_w}{V_s + V_v}$

If dry, $W_w = 0$. Hence,

$$\gamma_d = \dfrac{W_s}{V_s + V_v} = \dfrac{W_s/V_s}{(V_s + V_v)/V_s} = \dfrac{G\gamma_w}{1 + V_v/V_s}$$

But $V_v = V_w/S;\ V_w = W_w/\gamma_w;\ V_s = W_s/G\gamma_w$ or

$$\gamma_d = \dfrac{G\gamma_w}{1 + V_w/V_s S} = \dfrac{G\gamma_w}{1 + W_s G/W_s S}$$

Answer

$$\gamma_d = \dfrac{G\gamma_w}{1 + wG/S}$$

(d) Using (a) above and $S = 1$, and an upward lift $\approx \gamma_w$,

$$\gamma_b = \gamma - \gamma_w = \dfrac{G + Se}{1 + e}\gamma_w - \gamma_w = \left(\dfrac{G + e}{1 + e} - 1\right)\gamma_w$$

$$\gamma_b = \left(\dfrac{G + e - 1 - e}{1 + e}\right)\gamma_w$$

Answer

$$\gamma_b = \frac{G-1}{1+e}\,\gamma_w$$

(e) $V_v = eV_s = e(V - V_v) = eV - eV_v$ and $V_v + eV_v = eV$
Thus,

Answer

$$V_v = \frac{eV}{1+e}$$

(f) $V_s = V - V_v = V - eV_s$ and $V_s(1 + e) = V$
Thus,

Answer

$$V_s = \frac{V}{1+e}$$

(g) $V_w = SV_v$; but $V_v = eV/(1 + e)$ from (e) above.
Thus,

$$V_w = S\left(\frac{eV}{1+e}\right)$$

Answer

$$V_w = \frac{SeV}{1+e}$$

(h) $W_w = W - W_s = V\gamma - V_s G\gamma_w$. From (a),

$$\gamma = \frac{G + Se}{1+e}\,\gamma_w$$

and from (f),

$$V_s = \frac{1}{1+e}\,V$$

Thus,

$$W_w = V\left(\frac{G + Se}{1+e}\right)\gamma_w - \left(\frac{1}{1+e}\,V\right)G\gamma_w$$

Answer

$$W_{\mathrm{w}} = \frac{Se}{1+e} V\gamma_{\mathrm{w}}$$

(i) $W_{\mathrm{s}} = W - W_{\mathrm{w}} = V\gamma_{\mathrm{m}} - V_{\mathrm{w}}\gamma_{\mathrm{w}} = V\left(\frac{G_{\mathrm{m}} + Se}{1+e}\right)\gamma_{\mathrm{w}} - \left(\frac{Se}{1+e}\right)V\gamma_{\mathrm{w}}$

Then,

Answer

$$W_{\mathrm{s}} = \frac{1}{1+e} G_{\mathrm{m}} V\gamma_{\mathrm{w}}$$

(j) $W = W_{\mathrm{w}} + W_{\mathrm{s}} = \frac{Se}{1+e} V\gamma_{\mathrm{w}} + \frac{1}{1+e} G_{\mathrm{m}} V\gamma_{\mathrm{w}}$

Thus,

Answer

$$W = \frac{Se + G_{\mathrm{m}}}{1+e} V\gamma_{\mathrm{w}}$$

Example 2.2

Given   A moist soil sample weighs 346 g. After drying at 105° C its weight is 284 g. The specific gravity of the mass and of the solids is 1.86 and 2.70, respectively.

Determine   (a) The water content. (b) The void ratio. (c) The degree of saturation. (d) The porosity.

Procedure   The air is assumed to be weightless; $\gamma_{\mathrm{w}} = 1$ g/cm³.

(a) $w = \dfrac{W_{\mathrm{w}}}{W_{\mathrm{s}}} = \left(\dfrac{346 - 284}{284}\right) \times 100$

Answer

$$w = 21.83\%$$

(b) $e = V_{\mathrm{v}}/V_{\mathrm{s}}$ and

$$V_{\mathrm{s}} = \frac{W_{\mathrm{s}}}{G\gamma_{\mathrm{w}}} = \frac{284}{(2.7)(1)} = 105.18 \text{ cm}^3$$

The volume of mass V is

$$V = \frac{W_m}{G_m \gamma_w} = \frac{346}{(1.86)(1)} = 186.02 \text{ cm}^3$$

and

$$V_v = V - V_s = 186.02 - 105.18 = 80.84 \text{ cm}^3$$

Thus,

$$e = \frac{V_v}{V_s} = \frac{80.84}{105.18}$$

Answer

$$e = 0.77$$

(c) $S = \dfrac{V_w}{V_v} = \left(\dfrac{346 - 284}{80.84}\right) \times 100$

Answer

$$S = 76.69\%$$

(d) $n = \dfrac{V_v}{V} = \left(\dfrac{80.84}{186.02}\right) \times 100$

Answer

$$n = 43.46\%$$

Example 2.3

Given Two sites are being considered as "borrow" sites. The in-place (in-situ) unit weight of the soil of the first site is 101 lb/ft³; the water content was found to be 10%. On the second site the unit weight of the soil was found to be 96 lb/ft³ and its water content 14%. The construction site required 35,000 yd³ of soil in a compacted state, at a unit weight of 126 lb/ft³, at a water content of 14%. The soil from the one site required some additional water. The cost for the "borrow" material was based on volumes of soil removed from the respective sites. That is, more volume would have to be removed from the site that had 96 lb/ft³ weight than from the other one. The unit price from each site was $3.00 for the material and $4.00 for transportation, for the respective cubic yards. In addition, for the material that required the additional water, $.50/yd³ was estimated to be the additional cost.

Find The cost for the material from each site. Assume $G = 2.65$ for each site.

Solution
$$\gamma = \left(\frac{1 + w}{1 + e}\right) G\gamma_w$$

Thus,

$$\text{site } \textcircled{1} \begin{cases} \gamma_1 = 101 = \left(\dfrac{1 + 0.10}{1 + e}\right)(2.65)(62.4) \\ e_1 = 0.802 \Rightarrow V_1 = 1.802\ V_s \end{cases}$$

$$\text{site } \textcircled{2} \begin{cases} \gamma_2 = 96 = \left(\dfrac{1 + 0.14}{1 + e}\right)(2.65)(62.4) \\ e_2 = 0.963 \Rightarrow V_2 = 1.963\ V_s \end{cases}$$

$$\begin{matrix} \text{construction} \\ \text{site} \end{matrix} \begin{cases} 126 = \left(\dfrac{1 + 0.14}{1 + e_c}\right)(2.65)(62.4) \\ e_c = 0.495 \Rightarrow V_c = 35{,}000 = 1.495\ V_s \end{cases}$$

$$V_s = 23.411 \text{ yd}^3 = \text{constant}$$

Hence,

$$V_1 = 1.8203(23.411) = 42{,}140 \text{ yd}^3$$
$$V_2 = 1.963(23.411) = 45{,}952 \text{ yd}^3$$

Costs

Answer

$$\text{site } \textcircled{1} = (42{,}140)(3 + 4 + 0.5) = \$316{,}050$$
$$\textcircled{2} = (45{,}952)(3 + 4) \qquad = \$321{,}664$$

Relative Density of Granular Soils

The state of compactness of a natural granular soil is commonly expressed by its *relative density* D_r. It is defined as

$$D_r = \left(\frac{e_{max} - e}{e_{max} - e_{min}}\right) \times 100 \tag{2-7}$$

or

$$D_r = \frac{\gamma_{d\,max}}{\gamma_d} \times \frac{\gamma_d - \gamma_{d\,min}}{\gamma_{d\,max} - \gamma_{d\,min}} \tag{2-7a}$$

where
e_{max} = void ratio of soil in loosest state
e_{min} = void ratio of soil in densest state
e = void ratio of soil deposit (in-situ state)
$\gamma_{d\,max}$ = dry unit weight of soil in densest state
$\gamma_{d\,min}$ = dry unit weight of soil in loosest state
γ_d = dry unit weight of soil deposit (in-situ state)

Table 2.2 Commonly Used
Designations Associated with
Relative Density for
Granular Soils

Designation	D_r (%)
Very loose	0–15
Loose	15–35
Medium dense	35–70
Dense	70–85
Very dense	85–100

In determining the void ratios, one is confronted with the problem of measuring solid volumes. Since it is much easier to measure unit weights, the second expression (Eq. 2-7a) is more appealing (see Example 2.4). The procedure for determining the unit weights is detailed in various ASTM standards (ASTM D-2049-69 for $\gamma_{d\,max}$ and $\gamma_{d\,min}$; ASTM D-2167-66, 1977; D-1556-66, 1974; D-2922-70, etc., for γ_d). Briefly, $\gamma_{d\,min}$ is determined by pouring, from a fixed height, dry sand into a mold in the loosest form. $\gamma_{d\,max}$ is determined by vibrating a sample subjected to a surcharge weight. The largest $\gamma_{d\,max}$ obtained from densifying either a dry or a saturated sample is used (dry or wet method). γ_d may be obtained by any of several ASTM approved methods.

Relative density is commonly used as a measure of density of compacted fills (e.g., part of the specification requirements), or as an indication of the state of compactness of in-situ soils. Indirectly it also reflects on the stability of a stratum. For example, a loose (small D_r) granular soil is rather unstable, especially if subjected to shock or vibrating loads; vibratory loads would ''compress'' it into perhaps a more dense and more stable formation (see Example 2.5 for perfect spheres).

The state of compactness and the relative density are related in a rather empirical way. For example, for very loose sand, D_r is very small; for very dense sand, D_r is very large. A commonly used range of values for D_r and the associated descriptions for the state of compactness are given in Table 2.2. During subsurface exploration the Standard Penetration Test (SPT) (described in Sec. 3.5) is commonly used to characterize the density of a natural soil (see Fig. 3.5).

Example 2.4 _____

Given A test of the density of soil in place was performed (explained in more detail in Section 5.2) by digging a small hole in the soil, weighing the extracted soil, and measuring the volume of the hole. The soil (moist) weighed 895 g; the volume of the hole was 426 cm³. After drying, the sample weighed 779 g. Of the dried soil, 400 g was poured into a vessel in a very loose state. Its volume was subsequently determined to be 276 cm³. That same 400 g was then vibrated and tamped to a volume of 212 cm³. $G = 2.71$; $\gamma_w = 1$ g/cm³.

Find D_r. (a) Via Eq. 2-7 and (b) via Eq. 2-7a.

Procedure

(a) $V_s = \dfrac{W_s}{G\gamma_w} = \dfrac{779}{(2.71)(1)} = 287.5 \text{ cm}^3$

$V_v = 426 - 287.5 = 138.5 \text{ cm}^3$

$e = \dfrac{V_v}{V_s} = \dfrac{138.5}{287.5} = 0.48$

$e_{max} = \dfrac{276 - 400/2.71}{400/2.71} = 0.88$

$e_{min} = \dfrac{212 - 400/2.71}{400/2.71} = 0.44$

Then

$$D_r = \dfrac{e_{max} - e}{e_{max} - e_{min}} = \left(\dfrac{0.88 - 0.48}{0.88 - 0.44}\right) \times 100$$

Answer

$$D_r = 91\%$$

(b) $\gamma_{d\,max} = \dfrac{400}{212}\gamma_w = 1.89$

$\gamma_{d\,min} = \dfrac{400}{276}\gamma_w = 1.45$

$\gamma_d = \dfrac{779}{426}\gamma_w = 1.83$

where units are relative. Then

$$D_r = \dfrac{1.89}{1.83} \times \left(\dfrac{1.83 - 1.45}{1.89 - 1.45}\right) \times 100$$

Answer

$$D_r = 90\%$$

Note that part (b) was determined without knowledge of the value of *G*.

Example 2.5

Given   A soil sample consists of sand grains uniform in size and spherical in shape.

Find   A general expression for the relative density, assuming this sand to be perfectly cohesionless.

Procedure   Ideally, the range of void ratios e_{min}/e_{max} for a sand of uniform size and round particles approaches that for perfect spheres. Hence, we shall assume the sand to consist of perfect spheres of equal size.

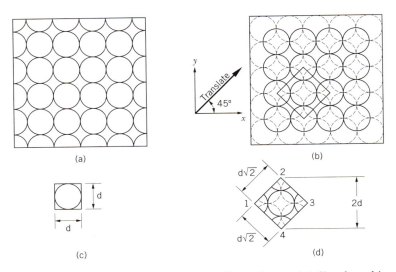

Figure 2.2 Sand sample, assuming perfect spheres. (a) Simple cubic packing. (b) Face-centered cubic. (c) Cubic element from simple cubic. (d) Cubic element from face-centered cubic.

Figure 2.2*a* is a plan view showing the loosest arrangement of identical spheres. This is known as *simple packing*, with each sphere making contact with its adjacent sphere (four from sides, one from top, and one from bottom). Each of the spheres may be assumed to fit within a cube of side dimension d, as shown in Fig. 2.2*c*. Thus, the volume of the cube V_C is d^3. The volume of the sphere V_S is $\pi d^3/6$. Thus, the maximum void ratio (loosest state) is

$$e_{\text{max}} = \frac{V_v}{V_S} = \frac{V_C - V_S}{V_S} = \frac{d^3 - \pi d^3/6}{\pi d^3/6} = 0.91$$

Now assume that the total column is permitted to translate horizontally along a 45° line with respect to the x axis such that each sphere falls within the pocket formed by four adjacent spheres in a layer below. This forms a *face-centered cubic*, as shown in Fig. 2.2*b*. Now let us extract a typical cube as shown in Fig. 2.2*d*. This is a top view, which shows a hemisphere on the top plane and four octants. There is a direct duplication of this arrangement at the bottom and four vertical sides of the cube. Hence, we note a total of six hemispheres and eight octants, or a total of four complete spheres. The volume of the cube, therefore, is equal to $(d\sqrt{2})^3 = 2d^3\sqrt{2}$, and that for the spheres equals $\frac{4}{6}\pi d^3 = \frac{2}{3}\pi d^3$. Hence the minimum void ratio (densest state) is

$$e_{\text{min}} = \frac{2d^3\sqrt{2} - \frac{2}{3}\pi d^3}{\frac{2}{3}\pi d^3} = 0.35$$

Substituting the corresponding values in our basic definition for relative density, we have

Answer

$$D_\mathrm{r} = \frac{0.91 - e}{0.56}$$

2.4 ATTERBERG LIMITS

In the previous section we developed some rather broad guidelines for describing the physical state of a granular stratum in terms of its relative density (Table 2.2). This is not an adequate description for the state of a very-fine-grain soil. For clays the usual classification is derived from their engineering properties under varying conditions of moisture. *Consistency* is a term that is frequently used to describe the degree of firmness (e.g., soft, medium, firm, or hard). The *Atterberg limits* are an empirically developed but widely used procedure for establishing and describing the consistency of soil. The limits were named after A. Atterberg (1), who first introduced the concept, although subsequent modifications were made by Terzaghi (23) and later by Casagrande (4, 5), who improved on the test procedures and amplified the relationship these limits hold to various soil types.

The consistency of cohesive soils is greatly affected by the water content of the soil. A gradual increase of the water content, for example, may transform a dry clay from perhaps a *solid* state to a *semisolid* state, to a *plastic* state, and, after further moisture increase, into a *liquid* state. The water contents at the corresponding junction points of these states are known as the *shrinkage limit,* the *plastic limit,* and the *liquid limit,* respectively. This is shown schematically in Fig. 2.3. The detailed test procedures for determining these limits may be found in most laboratory manuals dealing with soil tests or in ASTM, AASHTO standards, and the like.

The *liquid limit* w_1 (LL) is the water content at the point of transition of the clay sample from a liquid state to the plastic state, whereby it acquires a certain shearing strength (ASTM D-423). Briefly, the liquid limit of a clay is the water content, in percent, at which a grooved sample in a standard apparatus, cut by a standard tool, closes along the groove for approximately 10 mm when subjected to 25 drops in a liquid-limit apparatus. By trying a number of moisture contents, a series of points could be plotted on a semilogarithmic scale as shown in Fig. 2.4. For the sample shown, the liquid limit is 47.5.

The *plastic limit* w_p (PL) is the smallest water content at which the soil begins to crumble when rolled out into thin threads, approximately 3 mm in diameter (ASTM D-424). Briefly, the samples are rolled slowly at decreasing water content until the

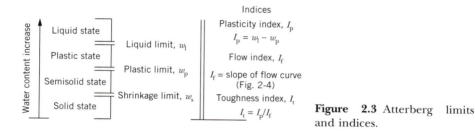

Figure 2.3 Atterberg limits and indices.

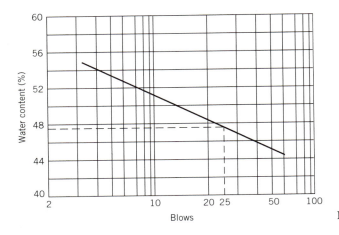

Figure 2.4 Liquid limit chart.

water content is reached at which a thread of approximately 3-mm diameter begins to crumble. This may be done by hand on a glass or some other smooth surface.

The *shrinkage limit* w_s (SL) is the smallest water content below which a soil sample will not reduce its volume any further, that is, it will not shrink any further with further drying (ASTM D-427). Briefly, the test is conducted by measuring the volume at the various water contents in the process of drying.

As mentioned before, the Atterberg limits are of a somewhat empirical nature, useful mostly for soil identification and classification. For example, the liquid limit appears to be directly proportional to the compressibility of the clay. Referring to the plasticity chart of Fig. 2.7, one notes that the inorganic clays lie above the *A* line, the organic ones below. Thus, by plotting the plasticity index and the liquid limit of the clay sample, we are able to approximate the soil classification as organic or inorganic. It is possible to have an inorganic clay and an organic clay fall within a common region. However, the strength of the dry inorganic soil is significantly greater than that of the organic. Furthermore, organic soils have a dark-gray or black color, and possess certain odors of organic matter.

Clay absorbs water to its surface. As a guide, therefore, the water absorbed by a soil provides some estimate of the amount of clay present in that soil. Skempton (20) proposed a relationship between the plasticity index and the percentage (by weight) of clay sizes finer than 2 μm. He called it the *activity of clay A*, which is expressed by

$$A = \frac{I_p}{\% \text{ finer than 2 } \mu\text{m}}$$

Information regarding the activity of clay may provide one with an inclination as to the type of clay present, and subsequently the behavioral nature of the soil. For example, the relative level of activity expected is low for kaolinite, medium for illite, and high for montmorillonite (e.g., certain results reported by Skempton indicate that the level of activity for kaolinite was less than 0.5; for illite, it was approximately 1; and for montmorillonite, greater than 7).

The following is an approximate relationship between the activity of clay (*A*) and its behavioral tendency to volume changes.

Activity, A	Classification
$A < 0.7$	inactive clay
$0.7 < A < 1.2$	normal clay
$A > 1.2$	active clay

Generally, the relative level of activity expected is low for kaolinite, and thus it is stable; medium for illite, and thus, normally stable; and high for montmorillonite and thus subject to large volume changes.

Because of the high degree of disturbance during the extraction, handling, and testing phases, a typical soil specimen may display properties that are appreciably different from those of an in-situ soil. This observation may be particularly applicable to the Atterberg limits. That is, except for the shrinkage limit, appreciable remolding and structural disturbances of the clay specimen are inherent in the test procedures, not to mention the disturbance from sampling and handling. Further evidence of the effects of disturbance is manifested clearly in the results from unconfined compression tests (ASTM D-2166-72). The strength of an undisturbed clay is several times that of the remolded clay. The ratio of the compressive strength of an undisturbed sample to that of a remolded sample is referred to as the degree of **sensitivity.** Hence, the results of tests from disturbed samples, including the Atterberg limits tests, should be viewed within the context of such limitations and should be regarded as only a part or a supplement to a more detailed evaluation program.

2.5 SOIL CLASSIFICATION

Up to now we have categorized different soils mostly by such general terms as coarse grained or fine grained, cohesive or cohesionless. We associated sands or gravels with the coarse-grained particles and regarded them as cohesionless; silts and clays were viewed as fine grained and cohesive. Such classifications are too general; they provide neither a reasonable delineation of these categories nor an acceptable description of a *mixture* of different size grains. More systematic and uniform identification means are needed for grouping soils for engineering purposes and to establish a more unified and rational basis for communicating their properties from one user to another.

The classification systems are empirical in nature. Most were developed to serve a specific need related to a particular type of engineering work. For example, the AASHTO (American Association of State Highway and Transportation Officials, formerly the Bureau of Public Roads) system provided a systematic grouping of various soils in accordance with their suitability for use in highway subgrades and embankment construction. The Unified Classification System evolved mostly in connection with Casagrande's work (5) on military airfields. The Corps of Engineers has developed a classification for soils that display similar frost behavior.

The early classification systems were generally based on grain size. Although such classifications are widely used and may prove useful in many instances, they are generally inadequate. For example, it is not good practice to predict the permeability of two soils of like grain sizes; the permeability is affected to a significant degree by the grain shape. Similarly, it may be folly to compare the compressibility of two clays of identical particle sizes, using size as the relevant characteristic, and ignore the mineral content, environmental factors, and behavioral nature of the clays. Hence, many proposals have been made to expand the classification system to include properties beyond those based on grain size.

Of the number of classification systems proposed over the past few decades, the Unified Classification System and the AASHTO system appear to be the most widely used by current-day practitioners. However, most adopt the grain-size characteristics as a basis for separating the ingredients into gravel, sand, silt, or clay. Also, the Atterberg limits are normally used as an additional criterion for identifying consistency and plasticity characteristics of the fine-grained particles.

Classification Based on Grain Size

Figure 2.5 shows the delineation between various different grain-size fractions (e.g., gravel, sand, silt, or clay) for some of a number of classification systems. The textural composition of coarse-grained soils is usually determined by screening the soil through a series of sieves of various sizes and weighing the material retained by each sieve. This is usually referred to as *sieve* or *mechanical analysis*. Material finer than the openings of a No. 200 mesh sieve (see Table 2.3) is generally analyzed by a method of *sedimentation*. The most common test, the *hydrometer test,* is based on the principle that grains of different sizes fall through a liquid at different velocities. The essence of this concept is that a sphere falling through a liquid will reach a terminal velocity expressed by Stoke's law: $v = (\gamma_s - \gamma_w)D/18\mu$, where γ_s and γ_w are unit weights of sphere and liquid, respectively, μ is the viscosity of the liquid, and D is the diameter of the sphere. The details for performing these tests may be found in laboratory manuals dealing with soil testing or ASTM D-442-63.

The results of a grain-size analysis are commonly represented in the form of a graph as shown in Fig. 2.6. The aggregate weight, as a percentage of the total weight, of all grains smaller than any given diameter is plotted on the ordinate using an arithmetic scale; the size of a soil particle, in millimeters, is plotted on the abscissa, which uses a logarithmic scale. Reference is made to Fig. 2.5, showing the typical results of a mechanical analysis test plotted in this case using the MIT system. One notes the range of particle sizes for the three basic designations of

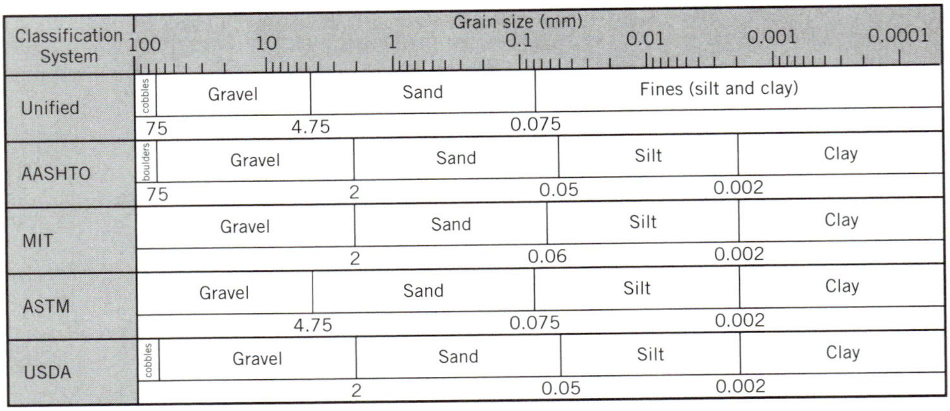

Classification System	Grain size (mm)						
	100 cobbles	10	1	0.1	0.01	0.001	0.0001
Unified	boulders	Gravel		Sand	Fines (silt and clay)		
		75	4.75		0.075		
AASHTO		Gravel		Sand	Silt		Clay
		75		2	0.05	0.002	
MIT		Gravel		Sand	Silt		Clay
				2	0.06	0.002	
ASTM		Gravel		Sand	Silt		Clay
			4.75		0.075	0.002	
USDA	cobbles	Gravel		Sand	Silt		Clay
				2	0.05	0.002	

Figure 2.5 Soil classification based on grain size.

Table 2.3 U.S. Standard Sieve Sizes

Sieve No.	4	6	8	10	16	20	30	40	50	60	100	140	200
Opening in.	0.187	0.132	0.0937	0.0787	0.0469	0.0331	0.0232	0.0165	0.0117	0.0098	0.0059	0.0041	0.0029
Opening mm	4.76	3.36	2.38	2.00	1.19	0.840	0.590	0.420	0.297	0.250	0.149	0.105	0.075

sand, silt, and clay, with each further subdivided into categories of coarse, medium, or fine grained. This scale of classification is typical of several arbitrarily chosen such scales.

Some relevant and useful information may be obtained from a grain-size curve, such as indicated in Fig. 2.6: (1) the total percentage of a given size, (2) the total percentage larger or finer than a given size, and (3) the uniformity or the range in grain-size distribution. One indication of the gradation is given by the *uniformity coefficient* C_u, attributed to Allen Hazen, and defined as

$$C_u = \frac{D_{60}}{D_{10}} \tag{2-8}$$

It expresses the ratio of the diameter of the particle size at 60% to the diameter of the particle size at 10% finer by weight on the grain-size distribution curve (see Example 2.6). A large coefficient corresponds to a large range in grain sizes, and the soil is regarded as well graded. A coefficient of 1 would represent soil sizes of

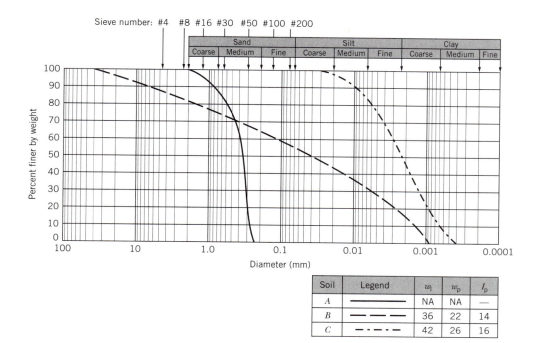

Soil	Legend	w_l	w_p	I_p
A	———————	NA	NA	—
B	— — — —	36	22	14
C	— · — · —	42	26	16

Figure 2.6 Grain-size distribution curves, MIT classification.

the same magnitude. Generally, soils whose coefficient is less than 4 are considered uniform.

Another is the *coefficient of curvature* (or *coefficient of concavity*) C_z, sometimes used as a measure of the shape of the grain-size distribution curve. It is defined as

$$C_z = \frac{D_{30}^2}{D_{10}D_{60}} \qquad (2\text{-}9)$$

For a C_z value of about 1, the soil is considered well graded. For a C_z much less or much larger than 1, the soil is viewed as poorly graded.

Unified Soil Classification System

Subsequent to the airfield classification system developed by Casagrande (4), the Bureau of Reclamation and the Army Corps of Engineers developed the Unified Classification System in 1952. The soils are divided into two main groups, coarse grained and fine grained, and defined by a set of two letters, a prefix and a suffix. The coarse-grained designation is assigned to soils for which over 50%, by weight, of the material is retained by the No. 200 sieve (0.075 mm). Within the coarse-grained group, the prefix G is assigned to the soil if more than 50% of the particles are retained by the No. 4 sieve (4.76 mm), and S if more than 50% pass through the No. 4 sieve. G or S is followed by a suffix that describes the gradation: W—well graded; P—poorly graded; M—containing silt; C—containing clay. For example, a well-graded gravel would be represented by GW; a poorly graded sand by SP. This is shown in Fig. 2.7.

The fine-grained designation represents the soils for which more than 50% pass the No. 200 (0.075 mm) sieve. These are divided into silts (M), clays (C), and organic silts or clays. The suffix following one of these designations is L—low plasticity, or H—high plasticity (L for a liquid limit $<$ 50%, H for a liquid limit $>$ 75%). The fine-grained soils are classified according to their plasticity index and liquid limit via the plasticity chart of Fig. 2.7. The A line separates the inorganic clays from the silts and organic soils. The general classification criteria are given in Fig. 2.7.

AASHTO Classification System

As mentioned previously, the AASHTO classification system is a widely used method for classifying soils for earthwork structures, particularly subgrades, bases, subbases, and embankments. Its present form represents the culmination of several revisions and alterations since its development by the U.S. Bureau of Public Roads in about 1929. It classifies soils into seven groups (A-1 through A-7) based on particle-size distribution, liquid limit, and plasticity index. This is shown in Fig. 2.8, including subgroups for A-1, A-2, and A-7. A brief description of the various groups is given in Fig. 2.9. The system separates soils into granular and silt-clay groups. A-1 through A-3 are granular, with 35% or less passing the No. 200 sieve (0.075 mm). If more than 35% passes through a No. 200 sieve, the material falls in the silt-clay group. These groups are evaluated via a formula referred to as a *group index* (GI):

$$GI = \underbrace{(F - 35)}_{1 \text{ to } 40}[0.2 + 0.005\underbrace{(LL - 40)}_{1 \text{ to } 20}] + 0.01\underbrace{(F - 15)}_{1 \text{ to } 40}\underbrace{(PI - 10)}_{1 \text{ to } 20} \qquad (2\text{-}10)$$

where F = percentage passing No. 200 sieve, expressed as a whole number. This percentage is based only on the material passing the No. 200 sieve.

   LL = liquid limit
   PI = plasticity index

The higher the value of GI, the less the suitability (e.g., a material of GI = 15 is less suitable than one whose GI = 1). Likewise, A-1 is more suitable than A-4, as indicated in Fig. 2.8.

Example 2.6

Given   The grain-size distribution shown in Fig. 2.6 and the corresponding values for w_l and w_p shown.

Find   The coefficient of uniformity for soil A.

Procedure   For soil A, D_{60} = 0.38 mm, D_{10} = 0.28 mm, and $C_u = D_{60}/D_{10}$ = 1.36. The soil is considered uniform in size.

Example 2.7

Given   The data shown in Fig. 2.6.

Find   Classification for the three soils based on the unified soil classification system.

Procedure   **For Soil A.**   From Fig. 2.6 we see that more than 50% (virtually 100%) is retained by the No. 200 sieve (0.075 mm). Thus the soil is coarse grained. Also, since more than 50% passes through the No. 4 sieve (4.75-mm size), the soil is sand, S. Furthermore, since less than 12% passes through the No. 200 sieve, the soil is poorly graded, P. Hence, the material is a poorly graded sand, SP.

   For Soil B.   From Fig. 2.6 we note that more than 50% (about 57%) passes the No. 200 sieve. Thus, the soil is fine grained. We determine the classification from the plasticity chart, Fig. 2.7. For w_l = 36 and I_p = 14, the material plots just above the A line. Thus, the material may be classified as SC.

   For Soil C.   Since more than 50% (about 100%) passes the No. 200 sieve, the soil is fine grained. For w_l = 42 and I_p = 16, the soil plots almost on the A line. Hence, an approximate classification is CL.

Example 2.8

Given   The data of Fig. 2.6

Find   The AASHTO classification for the three soils in Example 2.7.

Procedure   **For Soil A.**   The amount of soil passing the No. 10, 40, and 200 sieves is:

No. 10 (2-mm) sieve	100%
No. 40 (0.425-mm) sieve	70%
No. 200 (0.075-mm) sieve	0%

Hence the classification is A-3 (excellent to good category).

Major Divisions			Group Symbols	Typical Names
Coarse-Grained Soils More than 50% retained on No. 200 sieve*	Gravels 50% or more of coarse fraction retained on No. 4 sieve	Clean gravels −200 <5%	GW	Well-graded gravels and gravel–sand mixtures, little or no fines
			GP	Poorly graded gravels and gravel–sand mixtures, little or no fines
		Gravels with fines −200 >12%	GM	Silty gravels, gravel–sand–silt mixtures
			GC	Clayey gravels, gravel–sand–clay mixtures
	Sands More than 50% of coarse fraction passes No. 4 sieve	Clean sands −200 <5%	SW	Well-graded sands and gravelly sands, little or no fines
			SP	Poorly graded sands and gravelly sands, little or no fines
		Sands with fines −200 >12%	SM	Silty sands, sand–silt mixtures
			SC	Clayey sands, sand–clay mixtures
Fine-Grained Soils 50% or more passes No. 200 sieve*	Silts and Clays Liquid limit 50% or less		ML	Inorganic silts, very fine sands, rock flour, silty or clayey fine sands
			CL	Inorganic clays of low to medium plasticity, gravelly clays, sandy clays, silty clays, lean clays
			OL	Organic silts and organic silty clays of low plasticity
	Silts and Clays Liquid limit greater than 50%		MH	Inorganic silts, micaceous or diatomaceous fine sands or silts, elastic silts
			CH	Inorganic clays of high plasticity, fat clays
			OH	Organic clays of medium to high plasticity
Highly Organic Soils			PT	Peat, muck, and other highly organic soils

Figure 2.7 Unified Classification System, ASTM D-2487-69.

Classification Criteria

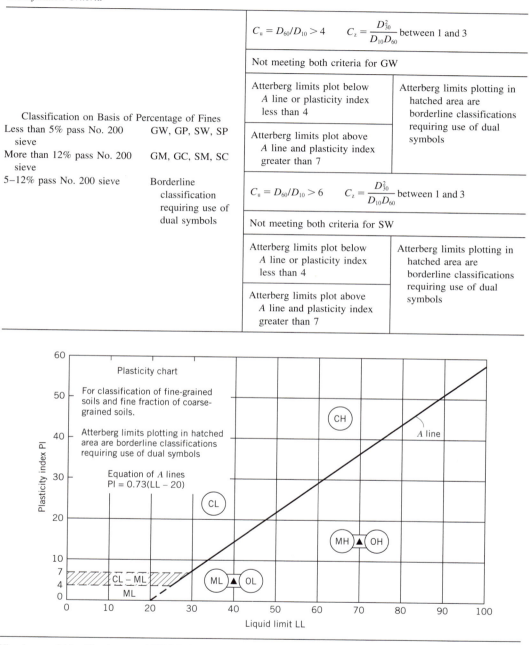

	$C_u = D_{60}/D_{10} > 4 \qquad C_z = \dfrac{D_{30}^2}{D_{10}D_{60}}$ between 1 and 3

Not meeting both criteria for GW

Atterberg limits plot below A line or plasticity index less than 4	Atterberg limits plotting in hatched area are borderline classifications requiring use of dual symbols
Atterberg limits plot above A line and plasticity index greater than 7	

Classification on Basis of Percentage of Fines

Less than 5% pass No. 200 sieve   GW, GP, SW, SP

More than 12% pass No. 200 sieve   GM, GC, SM, SC

5–12% pass No. 200 sieve   Borderline classification requiring use of dual symbols

$C_u = D_{60}/D_{10} > 6 \qquad C_z = \dfrac{D_{30}^2}{D_{10}D_{60}}$ between 1 and 3

Not meeting both criteria for SW

Atterberg limits plot below A line or plasticity index less than 4	Atterberg limits plotting in hatched area are borderline classifications requiring use of dual symbols
Atterberg limits plot above A line and plasticity index greater than 7	

Plasticity chart

For classification of fine-grained soils and fine fraction of coarse-grained soils.

Atterberg limits plotting in hatched area are borderline classifications requiring use of dual symbols

Equation of A lines
$PI = 0.73(LL - 20)$

Visual-manual identification, see ASTM D-2488.

* 0.075-mm sieve.

General Classification	Granular Materials (35% or less passing 0.075 mm)								Silt-Clay Materials (More than 35% passing 0.075 mm)			
	A-1		A-3	A-2				A-4	A-5	A-6	A-7-5 A-7-6
Group Classification	A-1-a	A-1-b		A-2-4	A-2-5	A-2-6	A-2-7				
Sieve analysis, percent passing:											
2.00 mm (No. 10)	50 max	—	—								
0.425 mm (No. 40)	30 max	50 max	51 min								
0.075 mm (No. 200)	15 max	25 max	10 max	35 max	35 max	35 max	35 max	36 min	36 min	36 min	36 min
Characteristics of fraction passing 0.425 mm (No. 40)											
Liquid limit		—	—	40 max	41 min	40 max	41 min	40 max	41 min	40 max	41 min
Plasticity index*	6 max		NP	10 max	10 max	11 min	11 min	10 max	10 max	11 min	11 min
Usual types of significant constituent materials	Stone fragments, gravel, and sand		Fine sand	Silty or clayey gravel and sand				Silty soils		Clayey soils	
General rating as subgrade	Excellent to good							Fair to poor			

* Plasticity index of A-7-5 subgroups is equal to or less than LL − 30. Plasticity index of A-7-6 subgroup is greater than LL − 30.

Figure 2.8 The AASHTO Classification of soils and soil-aggregate mixtures.

	Description of Classification Groups
Subgroup A-1a	Includes those materials consisting predominantly of stone fragments or gravel
Subgroup A-1b	Includes those materials consisting predominantly of coarse sand, either with or without a well-graded soil binder
Subgroup A-3	Fine beach sand or fine desert loess sand without silty or clay fines or with a very small amount of nonplastic silt
Subgroups A-2-4 and A-2-5	Include various granular materials containing 35% or less passing the 0.075-mm sieve and with a minus 0.425 mm in having the characteristics of the A-4 and A-5 groups
Subgroups A-2-6 and A-2-7	Include material similar to that described under subgrades A-2-4 and A-2-5, except that a fine portion contains plastic clay having the characteristics of the A-6 or A-7 group
Subgroup A-4	The typical materials of this group are the nonplastic or moderately plastic silty soils
Subgroup A-5	Similar to that described under group 2-4, except that it is usually of diatomaceous or micaceous character
Subgroup A-6	Usually a plastic clay having 75% or more passing the 0.075-mm sieve
Subgroup A-7-5	Includes materials with moderate plasticity indexes in relation to liquid limit
Subgroup A-7-6	Includes materials with high plasticity indexes in relation to liquid limit

Figure 2.9 Description of AASHTO groups.

For Soil B. The amount passing, for corresponding sieves is:

$$
\begin{array}{ll}
\text{No. 10 sieve} & 82\% \\
\text{No. 40 sieve} & 72\% \\
\text{No. 200 sieve} & 57\%
\end{array}
$$

For $w_l = 36$ and $I_p = 22$, the soil is A-6, clay.

For Soil C. The amount passing, for the respective sieves, is:

$$
\begin{array}{ll}
\text{No. 10 sieve} & 100\% \\
\text{No. 40 sieve} & 100\% \\
\text{No. 200 sieve} & 100\%
\end{array}
$$

For $w_l = 42$ and $I_p = 16$, the soil is A-7, clay.

The group index is

$$GI = (F - 35)[0.2 + 0.005(w_l - 40)] + 0.01(F - 15)(I_p - 10)$$
$$= (100 - 35)[0.2 + 0.005(42 - 40)] + 0.01(100 - 15)(16 - 10)$$
$$= 65(0.2 + 0.010) + 0.01(85)(6)$$
$$= 13.65 + 5.10$$

Answer

> **GI = 18.75—clay**

2.6 SOIL–WATER PHENOMENA

Darcy's Law

In 1856 the French hydraulic engineer Henri Darcy demonstrated experimentally that the rate of flow of water through a soil is proportional to the hydraulic gradient. Referring to Fig. 2.10 and assuming laminar flow, Darcy's law may be written as

$$Q = kiA$$

or

$$Q = k\left(\frac{\Delta h}{L}\right) A \qquad (2\text{-}11)$$

where Q = rate of flow
k = coefficient of permeability
i = gradient or head loss between two given points, $= (h_1 - h_2)/L$
A = total cross-sectional area of tube
Δh = difference in heads at the two ends of soil sample
L = length of sample

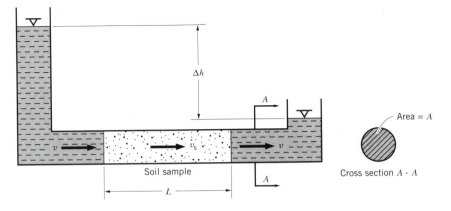

Figure 2.10 Uniform gravitational flow through soils.

As used in geotechnical engineering, *permeability* is that property of the soil that permits water to flow through it—through its voids (17). Thus, soils with large voids are more permeable than those whose voids are small. Furthermore, since most soils with large voids usually have large void ratios, we may deduce that, other factors not withstanding, permeability increases with increasing void ratios.

In Fig. 2.10 one notes that the "tube" velocity v of the water must necessarily be different from the velocity of the water through the soil voids, or *seepage velocity* v_s, since the tube area A is much larger than the cross-sectional area of the voids A_v. For continuity of flow, the quantity of flow Q must be the same throughout the system. Hence,

$$Q = Av = A_v v_s \qquad \text{(a)}$$

from which

$$v_s = \left(\frac{A}{A_v}\right) v = \left(\frac{AL}{A_v L}\right) v = \frac{V}{V_v} v \qquad \text{(b)}$$

or

$$v_s = \frac{1}{n} v \qquad \text{(2-12)}$$

or

$$v = n v_s \qquad \text{(2-12a)}$$

where V = total volume of mass
  V_v = volume of voids
  n = porosity ratio

It is rather obvious then that the velocity v is a *superficial* velocity; the use of v instead of v_s, however, is a convenient and widely used notation.

From Eq. 2-11, $Q/A = v = ki$. Hence, substituting in Eq. 2-12a,

$$v_s = \frac{ki}{n} \qquad \text{(2-13)}$$

The common units for k are those of velocity, usually centimeters per minute or centimeters per second.

From Eq. 2-13 we note that the coefficient of permeability k of a soil is the constant of proportionality between the tube velocity v and the hydraulic gradient i. This is not to imply a constant value for a given soil. Indeed, as we shall see in subsequent discussions, the value of k may vary quite greatly for a given soil with the direction of flow. Table 2.4 provides a rough approximation of the range of permeability values for different soils.

The values for permeability given in Table 2.4 should be used with caution. They are merely rough estimates for various types of soil. Nevertheless, for someone familiar with soil who is able to classify the soils (particularly granular soils) via visual examinations, these values may be useful for comparative purposes.

Table 2.4 Approximate Range of Values of
Coefficient of Permeability k

Particle Size	Coefficient of Permeability k ($\times 10^{-4}$ cm/s)
Sand	
Coarse	3000–5000
Medium	1000
Fine	50–150
Silt	
Coarse, sandy	1–20
Medium	0.1–1
Fine, clayey	0.01–0.1
Clay	
Coarse, silty	0.001–0.01
Medium	0.0001–0.001
Fine, colloidal	0.00001–0.0001

Capillarity

The forces that pull up water, if a fine-grain soil mass came in contact with water, above the free-water surface are known as *capillary forces*. The height of the water column thus drawn up or retained is called *capillary head*. The phenomenon that explains this rise is known as *capillarity*.

The basis of soil capillarity and related forces stems from the interaction of water and soil particles. The water rise above the free surface is attributed to a combination of *surface tension* (a molecular attraction of liquid) and the tendency of the water to "wet" the soil particles. That is, *cohesion* (molecular attraction of like particles) is responsible for the development of a state of tension of the surface water molecules, while *adhesion* (molecular attraction of unlike particles) results in the "wetting" of the soil particles.

Capillarity is at least in part responsible for some soil moisture above the water table. In other words, except for some entrapped moisture or some surface infiltration, soils above the groundwater table would eventually dry up were it not for capillary forces. Capillarity makes it possible for a dry fine-grained soil to draw up water to elevations well above the water table or to retain moisture above this table. Depending on a given set of circumstances, soil engineers may view such moisture as beneficial or detrimental. For example, capillary forces may develop an increase in intergranular pressure and, thereby, improve the stability and shear strength of some fine-grained soils. On the other hand, capillary moisture near the surface may cause pavement heave during frost through the formation and subsequent growth of ice crystals or ice lenses in colder regions.

Because of the complex nature of the soil voids, a theoretical prediction of capillary rise in soils is not only of dubious accuracy but likely to be even misleading. Perhaps the most reliable approach is by direct observation of this behavior, preferably in situ if possible. Allen Hazen suggested a formula for approximating the capillary rise. This is given by Eq. 2-14:

$$h_c = \frac{C}{eD_{10}} \qquad (2\text{-}14)$$

where D_{10} = Hazen's effective size (e.g., 10% size of grain-size analysis curve), in centimeters

   C = empirical constant, which usually varies from 0.1 to 0.5 cm^2

   e = void ratio

Another approximation, perhaps equally empirical, is the use of Eq. 2-15, where D is approximately $D_{10}/5$, given in millimeters. The numerator of Eq. 2-15 is adjusted as appropriate to account for the change in units:

$$h_c = \frac{0.0306}{\frac{1}{5}D_{10}(\text{mm})}; \quad \text{m} \qquad (2\text{-}15)$$

Seepage

The process of water flow through soil is commonly referred to as *seepage*. Figure 2.11 will be used as the basis of our discussion in connection with *seepage stresses* and *seepage forces*. In Fig. 2.11*b* we show a soil sample totally submerged. For convenience we shall assume total saturation, although, as we have noted in previous sections, total saturation is virtually impossible.

Figure 2.11*a* depicts the pressures associated with this arrangement. Assuming no water flow in the upward or downward directions, we note that a force from the hydrostatic head acts in all directions on a given particle of soil. Obviously, this force would tend to compress the particles, but would not tend to move them relative to each other. Thus, this force does not create any shear effects between soil particles. This type of pressure is commonly referred to as *neutral pressure* or *pore water pressure*. On the other hand, the weight of the particles above any given

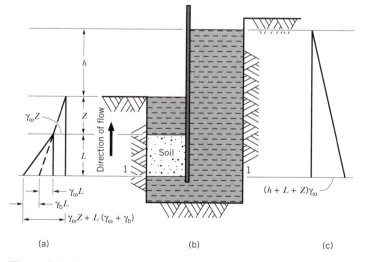

Figure 2.11 Water pressures on soil sample of cross section *A*.

level of soil is supported by the corresponding particles beneath. In the process, forces between particles are induced by this weight. It is rather evident that the lower particles will thus experience greater stress or pressure from such loads than is experienced by the higher level particles; they simply carry more overburden (see also Section 2.7). This is referred to as *intergranular pressure* or *effective pressure*. In Fig. 2.11*a*, $\gamma_w L$ is equal to the pore water pressure, and $\gamma_b L$ is the intergranular pressure.

Now let us reflect on Fig. 2.11*c*. This is a sketch of the hydrostatic pressures induced by the column of water. Since the pressure will act in all directions, the particles of soil at a given section, say 1–1 in Fig. 2.11*b* at the bottom of soil column, experienced downward as well as upward loads. This is indicated in Fig. 2.12 for an assumed cross section A. The horizontal pressures on the vertical faces of the sample cancel. Hence, by summing forces in the y direction, we get

$$F_{1-1} = A(h + L + Z)\gamma_w - A[\gamma_w Z + L(\gamma_w + \gamma_b)]$$

Simplifying,

$$F_{1-1} = (\gamma_w h - \gamma_b L)\,A \tag{2-16}$$

where $\gamma_w h A$ = seepage force

$$\gamma_b LA \quad = \text{ buoyant force} = \gamma_w L \left(\frac{G + Se}{1 + e} - 1 \right) A$$

F_{1-1} = net force at section 1–1

G, S, e are specific gravity of solids, degree of saturation, and void ratio, respectively.

The term $\gamma_w h$ in Eq. 2-16 represents the seepage pressure, while the product $\gamma_w h A$ represents the seepage force. Seepage forces may be visualized as being the result of the drag force by water against the soil particles and the associated reaction by the soil particles to the water. These forces are in the direction of flow.

Equation 2-11 represents the net force on the soil particles at section 1–1. One notes that without the seepage forces given by the expression $\gamma_w h A$, the net force at this section would be the buoyant force. Hence, the intergranular pressure would simply be $\gamma_b L$. Thus, it is readily apparent that the introduction of the seepage force alters the net force at this particular section. In turn, the seepage force alters the intergranular pressure in the soil mass. An important consequence of this effect will be illustrated in the following section.

Quick Condition

Reference is made to Eq. 2-16. As mentioned in the previous section, this equation represents the net force at section 1–1 of Fig. 2.11. By increasing the seepage

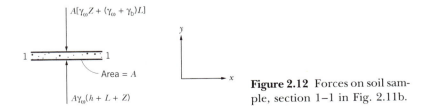

Figure 2.12 Forces on soil sample, section 1–1 in Fig. 2.11b.

pressures $\gamma_w h$, a point may be reached where the two terms in parentheses would be equal. That is, at this point the seepage forces would equal the buoyant forces. This may be viewed as a condition of impending upward movement. Obviously, any further increase in the seepage force would result in actual movement. Such a buoyant condition is generally referred to as a *quick* or a *boiling* condition (conditions that result in impending upward movement of soil and water).

At a point at which a quick condition exists, the net force would equal 0. Hence equating Eq. 2-16 to 0 and solving for the ratio of h to L for which boiling occurs, we have

$$0 = (\gamma_w h - \gamma_b L) A; \qquad A \neq 0$$

Thus,

$$0 = \gamma_w h - \gamma_b L = \gamma_w h - \gamma_w L \left(\frac{G + Se}{1 + e} - 1 \right)$$

Assuming total saturation ($S = 1$) and expanding, we get

$$h = L \left(\frac{G - 1}{1 + e} \right)$$

or

$$\frac{h}{L} = \frac{G - 1}{1 + e} \qquad\qquad (2\text{-}17)$$

By definition, the ratio of h to L is referred to as *gradient* whose magnitude is given by $(G - 1)/(1 + e)$. When its value equals 1, it is commonly known as the *critical gradient*—the condition for impending boiling. Hence, for some rather common values of $G = 2.7$ and $e = 0.7$ (approximately), h/L is about unity.

As mentioned above, the quick condition was based on a net or effective stress equal to 0. At this point the shear strength of the soil would theoretically appear to be lost. However, the zero effective stress does not necessarily mean boiling in cohesive soils since they display some shear strength even at zero effective stress. On the other hand, a fine-grained cohesionless soil (fine sand, for example) is most likely to be subject to boiling. Furthermore, the probability of boiling is greater for fine sand than for coarse sand or gravel strata. The coarse-grained soils display a greater porosity and permeability. Therefore, a larger supply of water would be needed to maintain a gradient of unity—the critical gradient. Hence, although it is theoretically possible to have a quick condition in coarse-grained soils, the volume of water necessary for the critical gradient makes such an occurrence unlikely.

Boiling frequency occurs in fine sands when the depth of excavation is a certain distance below the water table. That is, although the sides of the excavation are shored and properly supported laterally, the bottom of the excavation may flow upward when the critical gradient is reached. Sometimes a solution to such a problem may be found by driving sheet piling to depths such that L is large enough to reduce the gradient h/L to a value of 1 or less. Other common examples may include boiling due to artesian pressures and boiling near the downstream part of an earth dam.

Flow Nets

Seepage losses through the ground or through earth dams and levees and the related flow pattern and rate of energy loss, or dissipation of hydrostatic head, are frequently estimated by means of a graphical technique known as a *flow net* (7).

Figure 2.13 represents an example of a flow net. The path followed by a particle of water as it moves through a saturated soil mass is called a *flow line* or *stream line*. These are shown as solid lines. Assuming laminar flow (typical assumption for such analysis), the flow lines never cross each other. Each of the flow lines in Fig. 2.13 starts at a point where the hydrostatic head is equal to h and ends up on the free surface where the hydrostatic head is equal to 0. The viscous friction in the soil mass has dissipated a hydrostatic head of value h in each of these flow lines along its path of flow. Hence, along each flow line there must be a point where the total head or energy is the same for any other line. A line that connects these points of equal head is called an *equipotential line*. In Fig. 2.13 the dashed lines are identified as equipotential lines. It is readily apparent that an infinite number of flow lines and corresponding equipotential lines exists for any given condition.

The hydraulic gradient between two adjacent equipotential lines is the difference in head divided by the distance between these two lines, that is, $i = \Delta h / \Delta L$. The gradient is a maximum along a flow path perpendicular to the equipotential lines. Since ΔL is the shortest distance, i has a maximum value for any given Δh. Hence, the flow lines cross the equipotential lines at right angles, since in isotropic soil, flow occurs along paths of greatest or steepest gradient. Therefore, the flow lines and the equipotential lines in Fig. 2.13 form a family of mutually orthogonally-intersecting curves.

Although an infinite number of flow lines and equipotential lines could be drawn for any given condition, in drawing a flow net it is convenient to limit the number of flow lines and equipotential lines. The number should be greatly influenced by

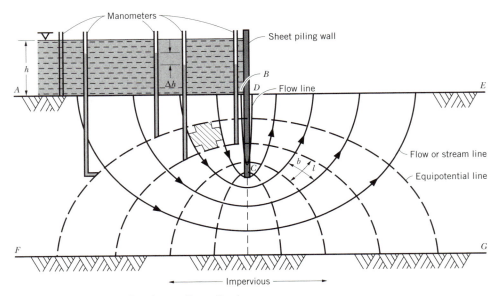

Figure 2.13 Flow net for sheet pile cofferdam.

one important consideration: *the geometric figures formed by the equipotential lines and flow lines in the construction of a flow net should approach a square shape as much as possible.* Obviously, not all of the blocks in Fig. 2.13 are squares. What is important, however, is to proportion the majority of these figures into approximate squares. If this is adhered to, then (1) the diagonals of the "squares" will have approximately equal lengths, and (2) the intersection of the equipotential and flow lines will be a 90° angle.

Figure 2.13 illustrates a flow net for a two-dimensional case. It assumes that all flow conditions in other parallel planes are similar. Although the flow of water through a soil mass usually poses a three-dimensional situation, the analysis for the three-dimensional case is rather complex and of limited practical value for the purpose of illustrating the fundamentals involved in flow net construction. Hence, our discussion will be limited to the two-dimensional case.

From Darcy's law the quantity of flow through any "square" is, $q = n_f \Delta q$. But,

$$\Delta q = kib = k\frac{\Delta h}{l} b \tag{a}$$

where b represents the distance between flow paths, and l is the distance between equipotential lines, as shown in Fig. 2.12. Since we regard these as square figures and therefore $l = b$, Eq. (a) may be written as $\Delta q = k\Delta h$. Also,

$$h = n_d \Delta h$$

Hence,

$$\Delta q = kh/n_d$$

In terms of q, we have

$$q = n_f \left(\frac{kh}{n_d}\right) \tag{b}$$

or

$$q = \frac{n_f}{n_d} kh \tag{2-18}$$

One notes that the value of n_f and therefore n_d may vary for any given situation, but the ratio n_f/n_d should remain a constant if the flow net is properly drawn.

In the case of anisotropic soils (say, $k_x > k_z$), the flow net is drawn for the transformed section. In Fig. 2.14 we see the same section to transformed and natural scales. The quantities of flow Δq_T and Δq_N through the two respective sections may be expressed by

$$\Delta q_T = k_e \frac{\Delta h}{l} b = k_e \Delta h \tag{a}$$

and

$$\Delta q_N = k_x \frac{\Delta h \, b}{l\sqrt{k_x/k_z}} = k_x \frac{\Delta h}{\sqrt{k_x/k_z}} \tag{b}$$

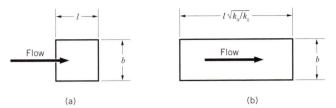

(a)   (b)

Figure 2.14 (a) Transformed scale. (b) Natural scale.

But $\Delta q_T = \Delta q_N$. Thus, we have

$$k_e \Delta h = k_x \frac{\Delta h}{\sqrt{k_x/k_z}} \tag{c}$$

Simplifying, we have

$$k_e = \sqrt{k_x k_z} \tag{2-19}$$

where k_x = coefficient of permeability in x direction
        k_z = coefficient of permeability in z direction
        k_e = effective coefficient of permeability

2.7 STRESSES IN SOILS

Stress is regarded as intensity of force, generally defined as load per unit area. Within the context of geotechnical engineering analysis it is convenient to view the in-situ soil stress, at a given depth, in terms of the components of total stress:

- Stresses induced by the weight of the soil above that level
- Fluid pressures
- Stresses introduced by externally applied loads (if any)

Effective Stress

In soils, stresses are separated into (1) *intergranular*—stress resulting from particle-to-particle contact, and (2) *pore water*—the stress induced by water-pressures. The former is commonly referred to as *effective stress;* the latter is frequently termed *neutral stress* (or *neutral pressure*). The sum of the effective and the neutral stresses is called the *total stress.*

Stresses Due to Surface Loads; Boussinesq's Equations

In 1885 Joseph Valentin Boussinesq advanced theoretical expressions for determining stresses at a point within an "ideal" mass due to surface point loads (2). They are based on the assumption that the mass is an (1) *elastic,* (2) *isotropic,* (3) *homogeneous,* and (4) *semi-infinite medium* that extends infinitely in all directions from a level surface. Boussinesq's equations provide a widely used basis for estimating the stresses within a soil mass caused by a concentrated load applied perpendicularly to the soil surface. In 1938 Westergaard (24) developed a solution for stresses within a soil mass by assuming the material to be reinforced by very rigid horizontal sheets that prevent any horizontal strain. Several load conditions based on Westergaard's development are mentioned in Appendix A. Within this chapter, Boussinesq's expressions will receive the primary focus.

Boussinesq's equations may be expressed in terms of either rectangular or polar coordinates. Referring to the elements in Fig. 2.15, the equations are as follows.

In rectangular coordinates:

$$\sigma_z = \frac{3Q}{2\pi}\frac{z^3}{R^5} \tag{2-20a}$$

$$\sigma_x = \frac{3Q}{2\pi}\left\{\frac{x^2 z}{R^5} + \frac{1-2\mu}{3}\left[\frac{1}{R(R+z)} - \frac{(2R+z)x^2}{R^3(R+z)^2} - \frac{z}{R^3}\right]\right\} \tag{2-20b}$$

$$\sigma_y = \frac{3Q}{2\pi}\left\{\frac{y^2 z}{R^5} + \frac{1-2\mu}{3}\left[\frac{1}{R(R+z)} - \frac{(2R+z)y^2}{R^3(R+z)^2} - \frac{z}{R^3}\right]\right\} \tag{2-20c}$$

$$\tau_{zx} = \frac{3Q}{2\pi}\frac{xz^2}{R^5} \tag{2-20d}$$

$$\tau_{xy} = \frac{3Q}{2\pi}\left[\frac{xyz}{R^5} - \frac{1-2\mu}{3}\frac{(2R+z)xy}{R^3(R+z)^2}\right] \tag{2-20e}$$

$$\tau_{yz} = -\frac{3Q}{2\pi}\frac{yz^2}{R^5} \tag{2-20f}$$

The expression for vertical stress, designated σ_z, is regarded as reasonably accurate and is widely used in problems associated with bearing capacity and settlement analysis. Hence, it is this expression that is primarily discussed in this chapter.

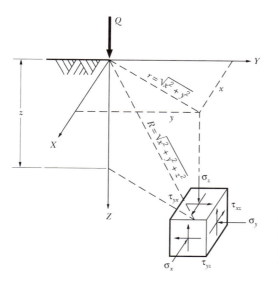

Figure 2.15 Stresses on elements due to concentrated load Q, in rectangular coordinate notation.

Table 2.5 Values of Boussinesq's Vertical Stress
Coefficient N_B

r/z	N_B	r/z	N_B	r/z	N_B
0.00	0.47746	2.50	0.00337	5.00	0.00014
0.25	0.41032	2.75	0.00223	5.25	0.00011
0.50	0.27332	3.00	0.00151	5.50	0.00009
0.75	0.15646	3.25	0.00105	5.75	0.00007
1.00	0.08440	3.50	0.00075	6.00	0.00006
1.25	0.04543	3.75	0.00054	6.25	0.00005
1.50	0.02508	4.00	0.00040	6.50	0.00004
1.75	0.01436	4.25	0.00030	6.75	0.00003
2.00	0.00854	4.50	0.00023	7.00	0.00003
2.25	0.00528	4.75	0.00018	7.25	0.00002

Equation 2-20a is more conveniently expressed in a slightly different form, as shown by Eq. 2-21.

$$\sigma_z = \frac{Q}{z^2} \frac{\left(\dfrac{3}{2\pi}\right)}{[(r/z)^2 + 1]^{5/2}} \qquad \text{or,} \qquad \sigma_z = \frac{Q}{z^2} N_B \qquad (2\text{-}21)$$

where

$$N_B = \frac{\dfrac{3}{2\pi}}{[(r/z)^2 + 1]^{5/2}}$$

Table 2.5 gives the values of Boussinesq's vertical stress coefficient for ratios of r/z.

Figure 2.16 shows a line load applied at the surface. For an element selected at an arbitrary fixed point in the soil mass, an expression for σ_z could be derived by integrating Boussinesq's expression for point load as given by Eq. 2-22. The line

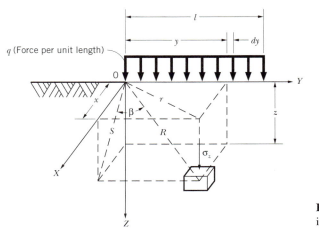

Figure 2.16 Vertical stress σ_z induced by line loads.

load is assumed to be of equal intensity q and applied at the surface. Thus,

$$\sigma_z = \frac{q/z}{2\pi(x^2/z^2 + 1)^2}\left[\frac{3(y^2/z^2)^{1/2}}{\sqrt{y^2/z^2 + 1 + x^2/z^2}} - \left(\frac{(y^2/z^2)^{1/2}}{\sqrt{y^2/z^2 + 1 + x^2/z^2}}\right)^3\right] \qquad \text{(a)}$$

If we let $(x/z) = m$ and $(y/z) = n$, Eq. (a) becomes

$$\sigma_z = \frac{q/z}{2\pi(m^2 + 1)^2}\left[\frac{3n}{\sqrt{n^2 + 1 + m^2}} - \left(\frac{n}{\sqrt{n^2 + 1 + m^2}}\right)^3\right] \qquad \text{(2-22)}$$

or

$$\sigma_z = \frac{q}{z}P_0 \qquad \text{(2-23)}$$

where

$$P_0 = \frac{1}{2\pi(m^2 + 1)^2}\left[\frac{3n}{\sqrt{n^2 + 1 + m^2}} - \left(\frac{n}{\sqrt{n^2 + 1 + m^2}}\right)^3\right]$$

Values for P_0 for various combinations of m and n are given in Table 2.6. In using Table 2.6, one notes that the values for m and n are *not* interchangeable. Furthermore, for values of m and n falling within the range of those given in the table, a straight-line interpolation may be assumed.

The application of this expression and the use of Table 2.6 are illustrated by Example 2.9. If the point at which the stress is desired lies between the two ends of the line, the effects of the load on that point are evaluated separately on each side of the point and subsequently added. On the other hand, if the point lies beyond the end of the line, the value of σ_z is that produced by the full length of the extended line minus the effect of the extension.

Example 2.9

Given $q = 100$ kN/m (22.5 kip/m = 6.85 kip/ft).

Find σ_z at points 0, 1, and 2 shown in Fig. 2.17.

Procedure   (a) Point 0. $m = x/z = 2/4$; $n = y/z = 3.2/4$. From Table 2.6, $P_0 = 0.15775$. Thus,

Answer

$$\sigma_{z0} = \frac{100 \text{ kN}}{4}(0.15775) = 3.944 \text{ kN/m}^2 \ (82.4 \text{ lb/ft}^2)$$

(b) Point 1. $\sigma_{z1} = \sigma_{zL}$ due to load on left $+ \sigma_{zR}$ due to load on right of point 1. From Fig. 2.18a and Table 2.6, $P_0 = 0.06534$. Thus,

$$\sigma_{zL} = \frac{100}{4}(0.06534) = 1.634$$

Table 2.6 Influence Values P_0 for Case of Line Load of Finite Length Uniformly Loaded (Boussinesq Solution)

m						n						
	0.1	0.2	0.3	0.4	0.5	0.6	0.7	0.8	0.9	1.0	1.2	1.4
0.0	0.04735	0.09244	0.13342	0.16917	0.19929	0.22398	0.24379	0.25947	0.27176	0.28135	0.29464	0.30277
0.1	0.04619	0.09020	0.13023	0.16520	0.19470	0.21892	0.23839	0.25382	0.26593	0.27539	0.28853	0.29659
0.2	0.04294	0.08391	0.12127	0.15403	0.18178	0.20466	0.22315	0.23787	0.24947	0.25857	0.27127	0.27911
0.3	0.03820	0.07472	0.10816	0.13764	0.16279	0.18367	0.20066	0.21429	0.22511	0.23365	0.24566	0.25315
0.4	0.03271	0.06406	0.09293	0.11855	0.14058	0.15905	0.17423	0.18651	0.19634	0.20418	0.21532	0.22235
0.5	0.02715	0.05325	0.07742	0.09904	0.11782	0.13373	0.14694	0.15775	0.16650	0.17354	0.18368	0.19018
0.6	0.02200	0.04322	0.06298	0.08081	0.09646	0.10986	0.12112	0.13045	0.13809	0.14430	0.15339	0.15931
0.7	0.01752	0.03447	0.05035	0.06481	0.07762	0.08872	0.09816	0.10608	0.11265	0.11805	0.12607	0.13140
0.8	0.01379	0.02717	0.03979	0.05136	0.06172	0.07080	0.07862	0.08525	0.09082	0.09546	0.10247	0.10722
0.9	0.01078	0.02128	0.03122	0.04041	0.04872	0.05608	0.06249	0.06800	0.07268	0.07663	0.08268	0.08687
1.0	0.00841	0.01661	0.02441	0.03169	0.03832	0.04425	0.04948	0.05402	0.05793	0.06126	0.06645	0.07012
1.2	0.00512	0.01013	0.01495	0.01949	0.02369	0.02752	0.03097	0.03403	0.03671	0.03905	0.04281	0.04558
1.4	0.00316	0.00626	0.00927	0.01213	0.01481	0.01730	0.01957	0.02162	0.02345	0.02508	0.02777	0.02983
1.6	0.00199	0.00396	0.00587	0.00770	0.00944	0.01107	0.01258	0.01396	0.01522	0.01635	0.01828	0.01979
1.8	0.00129	0.00256	0.00380	0.00500	0.00615	0.00724	0.00825	0.00920	0.01007	0.01086	0.01224	0.01336
2.0	0.00085	0.00170	0.00252	0.00333	0.00410	0.00484	0.00554	0.00619	0.00680	0.00736	0.00836	0.00918
2.5	0.00034	0.00067	0.00100	0.00133	0.00164	0.00194	0.00224	0.00252	0.00278	0.00303	0.00349	0.00389
3.0	0.00015	0.00030	0.00045	0.00060	0.00074	0.00088	0.00102	0.00115	0.00127	0.00140	0.00162	0.00183
4.0	0.00004	0.00008	0.00012	0.00016	0.00020	0.00024	0.00027	0.00031	0.00035	0.00038	0.00045	0.00051
5.0	0.00001	0.00003	0.00004	0.00006	0.00007	0.00008	0.00010	0.00011	0.00012	0.00013	0.00016	0.00018
6.0	0.00001	0.00001	0.00002	0.00002	0.00003	0.00003	0.00004	0.00005	0.00005	0.00006	0.00007	0.00008
8.0	0.00000	0.00000	0.00000	0.00001	0.00001	0.00001	0.00001	0.00001	0.00001	0.00001	0.00002	0.00002
10.0	0.00000	0.00000	0.00000	0.00000	0.00000	0.00000	0.00000	0.00000	0.00000	0.00000	0.00001	0.00001

					n					
1.6	1.8	2.0	2.5	3.0	4.0	5.0	6.0	8.0	10.0	∞
0.30784	0.31107	0.31318	0.31593	0.31707	0.31789	0.31813	0.31822	0.31828	0.31830	0.31831
0.30161	0.30482	0.30692	0.30966	0.31080	0.31162	0.31186	0.31195	0.31201	0.31203	0.31204
0.28402	0.28716	0.28923	0.29193	0.29307	0.29388	0.29412	0.29421	0.29427	0.29428	0.29430
0.25788	0.26092	0.26293	0.26558	0.26670	0.26750	0.26774	0.26783	0.26789	0.26790	0.26792
0.22683	0.22975	0.23169	0.23426	0.23535	0.23614	0.23638	0.23647	0.23653	0.23654	0.23656
0.19438	0.19714	0.19899	0.20147	0.20253	0.20331	0.20354	0.20363	0.20369	0.20371	0.20372
0.16320	0.16578	0.16753	0.16990	0.17093	0.17169	0.17192	0.17201	0.17207	0.17208	0.17210
0.13496	0.13735	0.13899	0.14124	0.14224	0.14297	0.14320	0.14329	0.14335	0.14336	0.14338
0.11044	0.11264	0.11416	0.11628	0.11723	0.11795	0.11818	0.11826	0.11832	0.11834	0.11835
0.08977	0.09177	0.09318	0.09517	0.09608	0.09677	0.09699	0.09708	0.09713	0.09715	0.09716
0.07270	0.07452	0.07580	0.07766	0.07852	0.07919	0.07941	0.07949	0.07955	0.07957	0.07958
0.04759	0.04905	0.05012	0.05171	0.05248	0.05310	0.05330	0.05338	0.05344	0.05345	0.05347
0.03137	0.03253	0.03340	0.03474	0.03542	0.03598	0.03617	0.03625	0.03630	0.03632	0.03633
0.02097	0.02188	0.02257	0.02368	0.02427	0.02478	0.02496	0.02504	0.02509	0.02510	0.02512
0.01425	0.01496	0.01551	0.01643	0.01694	0.01739	0.01756	0.01765	0.01768	0.01769	0.01771
0.00986	0.01041	0.01085	0.01160	0.01203	0.01244	0.01259	0.01266	0.01271	0.01272	0.01273
0.00424	0.00453	0.00477	0.00523	0.00551	0.00581	0.00593	0.00599	0.00603	0.00605	0.00606
0.00201	0.00217	0.00231	0.00258	0.00277	0.00298	0.00307	0.00312	0.00316	0.00317	0.00318
0.00057	0.00063	0.00068	0.00078	0.00086	0.00096	0.00102	0.00105	0.00108	0.00109	0.00110
0.00021	0.00023	0.00025	0.00029	0.00033	0.00038	0.00041	0.00043	0.00045	0.00046	0.00047
0.00009	0.00010	0.00011	0.00013	0.00014	0.00017	0.00019	0.00020	0.00022	0.00023	0.00023
0.00002	0.00002	0.00003	0.00003	0.00004	0.00005	0.00005	0.00006	0.00007	0.00007	0.00008
0.00001	0.00001	0.00001	0.00001	0.00001	0.00002	0.00002	0.00002	0.00003	0.00003	0.00004

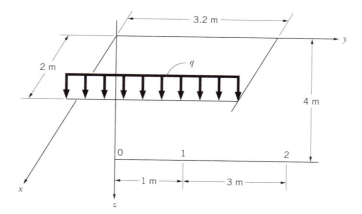

Figure 2.17 Line load.

From Fig. 2.18b and Table 2.6, $P_{0R} = 0.12578$. Thus,

Answer

$$\sigma_{zR} = \frac{100}{4}\,(0.12578) = 3.144$$

$$\sigma_{z1} = \sigma_{2L} + \sigma_{2R} = 4.778 \text{ kN/m}^2 \text{ (100 lb/ft}^2)$$

(c) Point 2. $\sigma_{z2} = \sigma_z$ due to actual plus extra load $-\,\sigma_z$ due to extra load. From Fig. 2.19a and Table 2.6, $P_0 = 0.1735$ and from Fig. 2.19b, $P_0 = 0.05325$. Thus,

Answer

$$\sigma_{z2} = \frac{100}{4}\,(0.1735 - 0.0532) = 3.01 \text{ kN/m}^2 \text{ (63 lb/ft}^2)$$

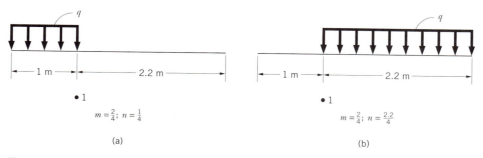

Figure 2.18 (a) Left load. (b) Right load.

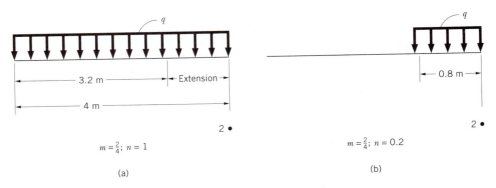

(a) $m = \frac{2}{4}$; $n = 1$

(b) $m = \frac{2}{4}$; $n = 0.2$

Figure 2.19 (a) Actual load plus extra load. (b) Extra load.

Pressure Caused by Uniformly Loaded Circular Area

Refering to Fig. 2.20, the vertical stress induced by a uniformly loaded circular area at any given depth z and horizontal distance a, as depicted in Fig. 2.20, can be determined with acceptable accuracy by extending Boussinesq's Eq. 2-20a. Although not attempted here, the analytical solution involves integrating Equation (a).

$$\sigma = \left(\frac{3qz^3}{2\pi}\right) \int_0^{2\pi} \int_0^r \frac{\rho \, d\beta \, d\rho}{(\rho^2 + z^2 + a^2 - 2a\rho \cos \beta)^{5/2}} \qquad (a)$$

However, charts and tables are available for relatively expedient procedures for evaluating σ_z, with the expression for stress taking the form typified by Eq. 2-24:

$$\sigma_z = qN_z(m, n) \qquad (2-24)$$

where N_z is a shape function of the dimensionless variables m and n. Foster and Ahlvin (10) developed the chart, depicted by Fig. 2.21 for $m = z/r$ and $n = a/r$. Table 2.7 provides a numerical relationship for the variables m and n. Example 2.10 illustrates the use of this chart.

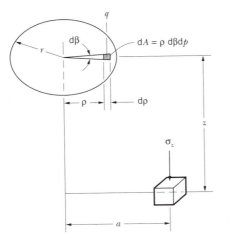

Figure 2.20 Vertical stress from loaded circular area.

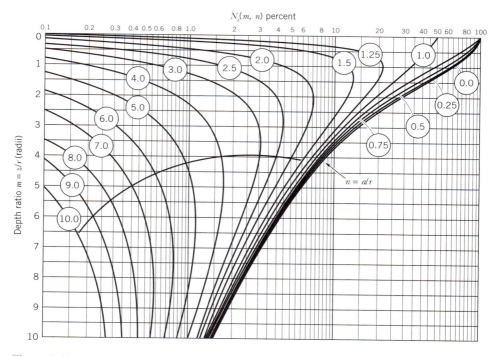

Figure 2.21 Vertical stress σ_z from uniformly loaded circular area. [By Foster and Ahlvin (10). Courtesy Highway Research Board.]

Example 2.10 _____

Given   A circular area, $r = 1.6$ m, induces a soil pressure at the surface of 100 kN/m².

Find   The pressure at a depth of (a) 2 m directly under the center of the circular area, and (b) 2 m below and 2 m away from the center of the circle.

Procedure   (a) From Fig. 2.21, for a depth ratio $z/r = 1.25$ radii and an offset distance $= 0$ from the centerline, $a/r = 0$,

$$\% \text{ of total} \approx 52\% \text{ of } 100$$

Thus,

Answer

$$\sigma_z = 52 \text{ kN/m}^2$$

(b) At a depth $= 2$ m ($m = z/r = 1.25$) and offset $= 2$ m ($n = a/r = 1.25$)

$$\sigma_z = 22\% \text{ of } 100$$

Table 2.7 Boussinesq's Coefficient, N_z, to be Used in Equation $\sigma_z = q\,N_z$ for Vertical Stress Induced by Circular-loaded Area, for Selected Values of $n > 0$

$m = z/r$	$n = a/r$											
	0.25	0.50	0.75	1.00	1.25	1.50	1.75	2.00	2.25	2.50	2.75	3.00
0.25	0.9824	0.9657	0.8825	0.4519	0.0723	0.0141	0.0042	0.0016	0.0008	0.0004	0.0002	0.0001
0.50	0.8961	0.8393	0.6913	0.4157	0.1683	0.0601	0.0235	0.0104	0.0052	0.0028	0.0016	0.0010
0.75	0.7625	0.6912	0.5569	0.3737	0.2063	0.1019	0.0499	0.0255	0.0138	0.0080	0.0048	0.0030
1.00	0.6254	0.5620	0.4584	0.3318	0.2134	0.1264	0.0724	0.0417	0.0247	0.0151	0.0095	0.0062
1.25	0.5068	0.4575	0.3816	0.2922	0.2062	0.1367	0.0874	0.0553	0.0352	0.0228	0.0151	0.0102
1.50	0.4112	0.3749	0.3202	0.2560	0.1925	0.1378	0.0954	0.0649	0.0441	0.0301	0.0208	0.0145
1.75	0.3362	0.3099	0.2706	0.2239	0.1765	0.1335	0.0981	0.0707	0.0505	0.0361	0.0259	0.0187
2.00	0.2777	0.2588	0.2302	0.1959	0.1601	0.1264	0.0971	0.0733	0.0547	0.0406	0.0301	0.0225
2.25	0.2320	0.2182	0.1972	0.1717	0.1444	0.1179	0.0939	0.0735	0.0568	0.0436	0.0334	0.0256
2.50	0.1960	0.1858	0.1702	0.1509	0.1299	0.1090	0.0894	0.0721	0.0575	0.0454	0.0357	0.0280
2.75	0.1673	0.1597	0.1479	0.1332	0.1169	0.1002	0.0842	0.0697	0.0569	0.0461	0.0371	0.0298
3.00	0.1441	0.1384	0.1294	0.1180	0.1052	0.0919	0.0788	0.0666	0.0556	0.0460	0.0378	0.0309
3.25	0.1253	0.1209	0.1139	0.1051	0.0949	0.0842	0.0734	0.0632	0.0537	0.0452	0.0379	0.0315
3.50	0.1098	0.1064	0.1009	0.0939	0.0858	0.0771	0.0682	0.0596	0.0515	0.0441	0.0375	0.0317
3.75	0.0969	0.0942	0.0899	0.0843	0.0778	0.0706	0.0633	0.0560	0.0491	0.0426	0.0367	0.0315
4.00	0.0861	0.0840	0.0805	0.0760	0.0707	0.0648	0.0587	0.0526	0.0466	0.0410	0.0358	0.0311
4.25	0.0770	0.0753	0.0725	0.0688	0.0644	0.0596	0.0545	0.0493	0.0441	0.0392	0.0346	0.0304
4.50	0.0692	0.0678	0.0655	0.0625	0.0589	0.0549	0.0506	0.0461	0.0417	0.0374	0.0334	0.0296
4.75	0.0626	0.0614	0.0595	0.0570	0.0540	0.0506	0.0470	0.0432	0.0394	0.0357	0.0321	0.0287
5.00	0.0568	0.0558	0.0543	0.0522	0.0497	0.0468	0.0437	0.0405	0.0372	0.0339	0.0307	0.0277
5.25	0.0518	0.0510	0.0497	0.0479	0.0458	0.0434	0.0407	0.0379	0.0351	0.0322	0.0294	0.0267
5.50	0.0474	0.0467	0.0456	0.0441	0.0423	0.0402	0.0380	0.0356	0.0331	0.0306	0.0281	0.0257
5.75	0.0430	0.0425	0.0415	0.0403	0.0392	0.0374	0.0354	0.0338	0.0314	0.0289	0.0268	0.0246
6.00	0.0401	0.0396	0.0388	0.0377	0.0364	0.0349	0.0332	0.0314	0.0295	0.0275	0.0256	0.0236
6.25	0.0370	0.0366	0.0360	0.0350	0.0339	0.0326	0.0311	0.0295	0.0278	0.0261	0.0244	0.0226
6.50	0.0343	0.0340	0.0334	0.0326	0.0316	0.0305	0.0292	0.0278	0.0263	0.0248	0.0232	0.0217
6.75	0.0319	0.0316	0.0311	0.0304	0.0295	0.0285	0.0274	0.0262	0.0249	0.0235	0.0221	0.0208
7.00	0.0297	0.0295	0.0290	0.0284	0.0277	0.0268	0.0258	0.0247	0.0236	0.0224	0.0211	0.0199
7.25	0.0278	0.0275	0.0272	0.0266	0.0260	0.0252	0.0243	0.0234	0.0223	0.0213	0.0201	0.0190
7.50	0.0260	0.0258	0.0255	0.0250	0.0244	0.0237	0.0229	0.0221	0.0212	0.0202	0.0192	0.0182
7.75	0.0244	0.0242	0.0239	0.0235	0.0230	0.0224	0.0217	0.0209	0.0201	0.0192	0.0183	0.0174
8.00	0.0229	0.0228	0.0225	0.0221	0.0217	0.0211	0.0205	0.0198	0.0191	0.0183	0.0175	0.0167
8.25	0.0216	0.0214	0.0212	0.0209	0.0205	0.0200	0.0194	0.0188	0.0182	0.0175	0.0167	0.0160
8.50	0.0204	0.0202	0.0200	0.0197	0.0194	0.0189	0.0184	0.0179	0.0173	0.0167	0.0160	0.0153
8.75	0.0192	0.0191	0.0189	0.0187	0.0183	0.0180	0.0175	0.0170	0.0165	0.0159	0.0153	0.0147
9.00	0.0182	0.0181	0.0179	0.0177	0.0174	0.0171	0.0167	0.0162	0.0157	0.0152	0.0147	0.0141
9.25	0.0172	0.0171	0.0170	0.0168	0.0165	0.0162	0.0158	0.0154	0.0150	0.0145	0.0140	0.0135
9.50	0.0164	0.0163	0.0161	0.0159	0.0157	0.0154	0.0151	0.0147	0.0143	0.0139	0.0135	0.0130
9.75	0.0155	0.0155	0.0153	0.0152	0.0150	0.0147	0.0144	0.0141	0.0137	0.0133	0.0129	0.0125
10.00	0.0148	0.0147	0.0146	0.0145	0.0143	0.0140	0.0138	0.0135	0.0131	0.0128	0.0124	0.0120

$$n = a/r$$

$m = z/r$	3.25	3.50	3.75	4.00	4.50	5.00	5.50	6.00	7.00	8.00	9.00	10.00
0.25	0.0001	0.0001	0.0000	0.0000	0.0000	0.0000	0.0000	0.0000	0.0000	0.0000	0.0000	0.0000
0.50	0.0007	0.0004	0.0003	0.0002	0.0001	0.0001	0.0000	0.0000	0.0000	0.0000	0.0000	0.0000
0.75	0.0020	0.0014	0.0010	0.0007	0.0004	0.0002	0.0001	0.0001	0.0000	0.0000	0.0000	0.0000
1.00	0.0042	0.0029	0.0021	0.0015	0.0008	0.0005	0.0003	0.0002	0.0001	0.0000	0.0000	0.0000
1.25	0.0071	0.0050	0.0036	0.0027	0.0015	0.0009	0.0006	0.0004	0.0002	0.0001	0.0000	0.0000
1.50	0.0103	0.0075	0.0055	0.0041	0.0024	0.0014	0.0009	0.0006	0.0003	0.0001	0.0001	0.0000
1.75	0.0137	0.0101	0.0076	0.0058	0.0034	0.0021	0.0014	0.0009	0.0004	0.0002	0.0001	0.0001
2.00	0.0169	0.0128	0.0097	0.0075	0.0046	0.0029	0.0019	0.0013	0.0006	0.0003	0.0002	0.0001
2.25	0.0197	0.0152	0.0118	0.0092	0.0058	0.0037	0.0025	0.0017	0.0008	0.0004	0.0002	0.0002
2.50	0.0220	0.0174	0.0137	0.0109	0.0070	0.0046	0.0031	0.0021	0.0011	0.0006	0.0003	0.0002
2.75	0.0239	0.0192	0.0154	0.0124	0.0082	0.0055	0.0038	0.0026	0.0014	0.0007	0.0003	0.0003
3.00	0.0252	0.0206	0.0168	0.0138	0.0093	0.0064	0.0044	0.0031	0.0016	0.0009	0.0004	0.0003
3.25	0.0262	0.0217	0.0180	0.0149	0.0103	0.0072	0.0051	0.0036	0.0020	0.0011	0.0005	0.0004
3.50	0.0267	0.0224	0.0188	0.0158	0.0112	0.0079	0.0057	0.0041	0.0023	0.0013	0.0007	0.0005
3.75	0.0269	0.0229	0.0195	0.0165	0.0119	0.0086	0.0063	0.0046	0.0026	0.0015	0.0008	0.0006
4.00	0.0268	0.0231	0.0199	0.0170	0.0125	0.0092	0.0068	0.0051	0.0029	0.0017	0.0009	0.0007
4.25	0.0266	0.0231	0.0201	0.0174	0.0130	0.0097	0.0073	0.0055	0.0032	0.0019	0.0011	0.0008
4.50	0.0261	0.0229	0.0201	0.0176	0.0134	0.0102	0.0077	0.0059	0.0035	0.0021	0.0012	0.0009
4.75	0.0255	0.0226	0.0200	0.0176	0.0136	0.0105	0.0081	0.0063	0.0038	0.0024	0.0014	0.0010
5.00	0.0249	0.0222	0.0198	0.0175	0.0138	0.0108	0.0084	0.0066	0.0041	0.0026	0.0015	0.0011
5.25	0.0241	0.0217	0.0195	0.0174	0.0139	0.0110	0.0087	0.0069	0.0043	0.0028	0.0017	0.0012
5.50	0.0234	0.0212	0.0191	0.0172	0.0139	0.0111	0.0089	0.0071	0.0045	0.0029	0.0018	0.0013
5.75	0.0225	0.0206	0.0185	0.0169	0.0138	0.0112	0.0090	0.0073	0.0047	0.0031	0.0019	0.0014
6.00	0.0218	0.0200	0.0182	0.0166	0.0137	0.0112	0.0092	0.0075	0.0049	0.0033	0.0021	0.0015
6.25	0.0210	0.0193	0.0178	0.0163	0.0136	0.0112	0.0092	0.0076	0.0051	0.0034	0.0022	0.0016
6.50	0.0202	0.0187	0.0172	0.0159	0.0134	0.0111	0.0093	0.0077	0.0052	0.0036	0.0024	0.0017
6.75	0.0194	0.0180	0.0167	0.0155	0.0131	0.0110	0.0092	0.0077	0.0054	0.0037	0.0025	0.0018
7.00	0.0186	0.0174	0.0162	0.0151	0.0129	0.0109	0.0092	0.0078	0.0055	0.0038	0.0026	0.0019
7.25	0.0179	0.0168	0.0157	0.0146	0.0126	0.0108	0.0092	0.0078	0.0055	0.0039	0.0027	0.0020
7.50	0.0172	0.0162	0.0152	0.0142	0.0123	0.0106	0.0091	0.0078	0.0056	0.0040	0.0028	0.0021
7.75	0.0165	0.0156	0.0147	0.0138	0.0120	0.0105	0.0090	0.0077	0.0057	0.0041	0.0029	0.0022
8.00	0.0159	0.0150	0.0142	0.0133	0.0118	0.0103	0.0089	0.0077	0.0057	0.0042	0.0030	0.0022
8.25	0.0152	0.0145	0.0137	0.0129	0.0115	0.0101	0.0088	0.0076	0.0057	0.0042	0.0031	0.0023
8.50	0.0146	0.0139	0.0132	0.0125	0.0112	0.0099	0.0087	0.0075	0.0057	0.0043	0.0032	0.0024
8.75	0.0141	0.0134	0.0128	0.0121	0.0109	0.0096	0.0085	0.0074	0.0057	0.0043	0.0032	0.0024
9.00	0.0135	0.0129	0.0123	0.0117	0.0106	0.0094	0.0084	0.0073	0.0057	0.0043	0.0033	0.0025
9.25	0.0130	0.0124	0.0119	0.0114	0.0103	0.0092	0.0082	0.0073	0.0057	0.0044	0.0033	0.0025
9.50	0.0125	0.0120	0.0115	0.0110	0.0100	0.0090	0.0081	0.0072	0.0056	0.0044	0.0034	0.0026
9.75	0.0120	0.0116	0.0111	0.0106	0.0097	0.0088	0.0079	0.0071	0.0056	0.0044	0.0034	0.0026
10.00	0.0116	0.0112	0.0107	0.0103	0.0094	0.0086	0.0077	0.0069	0.0055	0.0044	0.0034	0.0027

Thus,

$$\sigma_z = 22 \text{ kN/m}^2$$

Pressure Caused by Uniformly Loaded Rectangular Area

From Boussinesq's Equation 2-20a the vertical stress under a *corner* of a rectangular area uniformly loaded with a uniform load of intensity q (Fig. 2.22) can be expressed as

$$\sigma_z = \frac{3qz^3}{2\pi} \int_0^a \int_0^b \frac{dy\,dx}{(x^2 + y^2 + z^2)^{5/2}} \tag{a}$$

The integral is difficult and far too long to provide a practical benefit here. The integration was performed by Newmark (14) with the following results:

$$\sigma_z = \frac{q}{4\pi} \left[\frac{2mn\sqrt{m^2 + n^2 + 1}}{m^2 + n^2 + 1 + m^2 n^2} \frac{m^2 + n^2 + 2}{m^2 + n^2 + 1} + \sin^{-1} \frac{2mn\sqrt{m^2 + n^2 + 1}}{m^2 + n^2 + 1 + m^2 n^2} \right] \tag{2-25}$$

where $m = a/z$ and $n = b/z$. Equation 2-25 can also be expressed as

$$\sigma_z = q f_z(m, n) \tag{2-26}$$

where $f_z(m, n)$ is the shape function of the dimensionless ratios m and n. The influence values for various combinations of m and n can be found directly from Table 2.8.

When the point at which the stress is desired does not fall below a *corner* of the area, the area is "adjusted" into rectangles, as shown in Fig. 2.23, such that corners become located over the point in question. Subsequently, the effects are superimposed.

Figure 2.23 shows different locations of the point in question relative to the loaded area (solid boundaries) and the effective load combinations. From this figure, the load is calculated as follows:

(a) Load = load as determined directly from Table 2.7.

(b) Load = load from *AFIE + FBGI + GCHI + HDEI.*

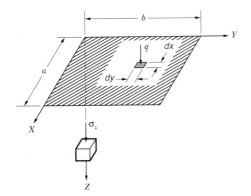

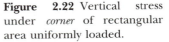

Figure 2.22 Vertical stress under *corner* of rectangular area uniformly loaded.

Table 2.8 Influence Values $f_z(m, n)$ for Case of Rectangular Area Uniformly Loaded (Boussinesq Solution)

m	0.1	0.2	0.3	0.4	0.5	0.6	0.7	0.8	0.9	1.0	1.2	1.4	1.6	1.8	2.0	2.5	3.0	4.0	5.0	6.0	8.0	10.0	∞
0.1	0.00470	0.00917	0.01324	0.01678	0.01978	0.02223	0.02420	0.02576	0.02698	0.02794	0.02926	0.03007	0.03058	0.03090	0.03111	0.03138	0.03150	0.03158	0.03160	0.03161	0.03162	0.03162	0.03162
0.2	0.00917	0.01790	0.02585	0.03280	0.03866	0.04348	0.04735	0.05042	0.05283	0.05471	0.05733	0.05894	0.05994	0.06058	0.06100	0.06155	0.06178	0.06194	0.06199	0.06201	0.06202	0.06202	0.06202
0.3	0.01324	0.02585	0.03735	0.04742	0.05593	0.06294	0.06859	0.07308	0.07661	0.07938	0.08323	0.08561	0.08709	0.08804	0.08867	0.08948	0.08982	0.09006	0.09014	0.09016	0.09018	0.09019	0.09019
0.4	0.01678	0.03280	0.04742	0.06024	0.07111	0.08009	0.08735	0.09314	0.09770	0.10129	0.10631	0.10941	0.11135	0.11260	0.11342	0.11450	0.11495	0.11527	0.11537	0.11541	0.11543	0.11544	0.11544
0.5	0.01978	0.03866	0.05593	0.07111	0.08403	0.09472	0.10340	0.11034	0.11584	0.12018	0.12626	0.13003	0.13241	0.13395	0.13496	0.13628	0.13684	0.13724	0.13736	0.13741	0.13744	0.13745	0.13745
0.6	0.02223	0.04348	0.06294	0.08009	0.09472	0.10688	0.11679	0.12474	0.13105	0.13605	0.14309	0.14749	0.15027	0.15207	0.15326	0.15483	0.15550	0.15598	0.15612	0.15617	0.15621	0.15622	0.15623
0.7	0.02420	0.04735	0.06859	0.08735	0.10340	0.11679	0.12772	0.13653	0.14356	0.14914	0.15703	0.16199	0.16515	0.16720	0.16856	0.17036	0.17113	0.17168	0.17185	0.17191	0.17195	0.17196	0.17197
0.8	0.02576	0.05042	0.07308	0.09314	0.11034	0.12474	0.13653	0.14607	0.15370	0.15978	0.16843	0.17389	0.17739	0.17967	0.18119	0.18321	0.18407	0.18469	0.18488	0.18496	0.18500	0.18502	0.18502
0.9	0.02698	0.05283	0.07661	0.09770	0.11584	0.13105	0.14356	0.15370	0.16185	0.16835	0.17766	0.18357	0.18737	0.18986	0.19152	0.19375	0.19470	0.19540	0.19561	0.19569	0.19574	0.19576	0.19577
1.0	0.02794	0.05471	0.07938	0.10129	0.12018	0.13605	0.14914	0.15978	0.16835	0.17522	0.18508	0.19139	0.19546	0.19814	0.19994	0.20236	0.20341	0.20417	0.20440	0.20449	0.20455	0.20457	0.20459
1.2	0.02926	0.05733	0.08323	0.10631	0.12626	0.14309	0.15703	0.16843	0.17766	0.18508	0.19584	0.20278	0.20731	0.21032	0.21235	0.21512	0.21633	0.21722	0.21749	0.21760	0.21767	0.21769	0.21770
1.4	0.03007	0.05894	0.08561	0.10941	0.13003	0.14749	0.16199	0.17389	0.18357	0.19139	0.20278	0.21020	0.21509	0.21836	0.22058	0.22364	0.22499	0.22600	0.22632	0.22644	0.22652	0.22654	0.22656
1.6	0.03058	0.05994	0.08709	0.11135	0.13241	0.15027	0.16515	0.17739	0.18737	0.19546	0.20731	0.21509	0.22025	0.22372	0.22610	0.22940	0.23088	0.23200	0.23235	0.23249	0.23258	0.23261	0.23263
1.8	0.03090	0.06058	0.08804	0.11260	0.13395	0.15207	0.16720	0.17967	0.18986	0.19814	0.21032	0.21836	0.22372	0.22736	0.22986	0.23336	0.23496	0.23617	0.23656	0.23671	0.23681	0.23684	0.23686
2.0	0.03111	0.06100	0.08867	0.11342	0.13496	0.15326	0.16856	0.18119	0.19152	0.19994	0.21235	0.22058	0.22610	0.22986	0.23247	0.23613	0.23782	0.23912	0.23954	0.23970	0.23981	0.23985	0.23987
2.5	0.03138	0.06155	0.08948	0.11450	0.13628	0.15483	0.17036	0.18321	0.19375	0.20236	0.21512	0.22364	0.22940	0.23336	0.23613	0.24010	0.24196	0.24344	0.24392	0.24412	0.24425	0.24429	0.24432
3.0	0.03150	0.06178	0.08982	0.11495	0.13684	0.15550	0.17113	0.18407	0.19470	0.20341	0.21633	0.22499	0.23088	0.23496	0.23782	0.24196	0.24394	0.24554	0.24608	0.24630	0.24646	0.24650	0.24654
4.0	0.03158	0.06194	0.09006	0.11527	0.13724	0.15598	0.17168	0.18469	0.19540	0.20417	0.21722	0.22600	0.23200	0.23617	0.23912	0.24344	0.24554	0.24729	0.24791	0.24817	0.24836	0.24841	0.24846
5.0	0.03160	0.06199	0.09014	0.11537	0.13736	0.15612	0.17185	0.18488	0.19561	0.20440	0.21749	0.22632	0.23235	0.23656	0.23954	0.24392	0.24608	0.24791	0.24857	0.24886	0.24907	0.24914	0.24919
6.0	0.03161	0.06201	0.09016	0.11541	0.13741	0.15617	0.17191	0.18496	0.19569	0.20449	0.21760	0.22644	0.23249	0.23671	0.23970	0.24412	0.24630	0.24817	0.24886	0.24916	0.24939	0.24946	0.24952
8.0	0.03162	0.06202	0.09018	0.11543	0.13744	0.15621	0.17195	0.18500	0.19574	0.20455	0.21767	0.22652	0.23258	0.23681	0.23981	0.24425	0.24646	0.24836	0.24907	0.24939	0.24964	0.24972	0.24980
10.0	0.03162	0.06202	0.09019	0.11544	0.13745	0.15622	0.17196	0.18502	0.19576	0.20457	0.21769	0.22654	0.23261	0.23684	0.23985	0.24429	0.24650	0.24841	0.24914	0.24946	0.24972	0.24981	0.24989
∞	0.03162	0.06202	0.09019	0.11544	0.13745	0.15623	0.17197	0.18502	0.19577	0.20458	0.21770	0.22656	0.23263	0.23686	0.23987	0.24432	0.24654	0.24846	0.24919	0.24952	0.24980	0.24989	0.25000

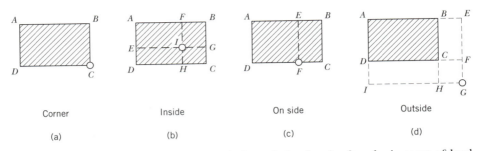

Corner	Inside	On side	Outside
(a)	(b)	(c)	(d)

Figure 2.23 Sketch showing the rectangle boundaries for the four basic cases of load superposition.

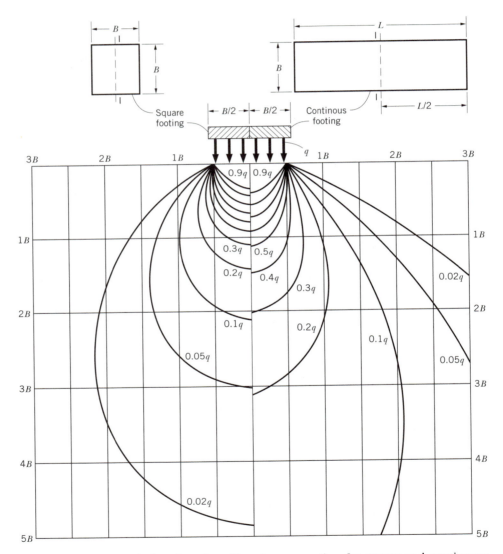

Figure 2.24 Pressure isobars based on Boussinesq equation for square and continuous footings, used to find pressures along line 1–1 as shown.

(c) Load = load from *AEFD* + *EBCF.*
(d) Load = load from *GIAE* − *GHBE* − *GIDF* + *GHCF.*

Figure 2.24 is a convenient and expedient chart for estimating pressure intensities at various depths, *z*, for square and long footings (e.g., the more common types). It only applies along lines 1-1 as shown. The isobars depict the intensity at various depths, *z*, where the depth is represented as a ratio of the dimension B. For example, at a depth *z* = 1B, the pressure intensity at a section through the center of a square footing (e.g., sec. 1-1) is approximately 0.35 of the surface pressure, and approximately 0.55 for a continuous footing.

Newmark's Influence Chart

The procedures outlined in the preceding sections for the determination of vertical stresses σ_z induced by uniformly loaded rectangular or circular areas are rather clumsy when applied to irregularly shaped areas. Newmark devised a graphical procedure for computing stresses induced by irregularly shaped loaded areas (14, 15).

Newmark's procedure evolves from the expression for the vertical stress under the center of a loaded circular area, given by Eq. (a).

$$\frac{\sigma_z}{q} = W_0 = \left[1 - \frac{1}{(r^2/z^2 + 1)^{3/2}} \right] \tag{a}$$

The values of *r/z* represent concentric circles of relative radii. Plotted for a selected scale for *z*, these circles are shown in Fig. 2.25, with the last circle not shown since *r/z* = ∞.

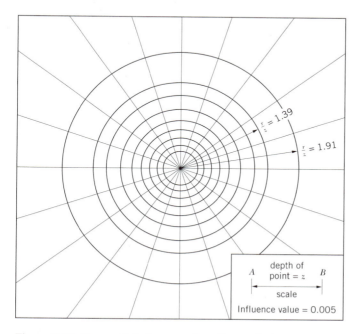

Figure 2.25 Newmark influence chart for vertical stress at any depth *z* = *AB.*

Table 2.9 Values of r/z for Selected Values of σ_z/q

σ_z/q	0	0.10	0.20	0.30	0.40	0.50	0.60	0.70	0.80	0.90	1.00
r/z	0	0.27	0.40	0.52	0.64	0.77	0.92	1.11	1.39	1.91	∞

Now divide the circles by evenly spaced rays emanating from the center, for convenience, say 20. Thus, we obtian a total of (10 circles) (20 rays) = 200 influence units. Therefore, the influence value IV is

$$\text{IV} = \frac{1}{(\text{no. circles})(\text{no. rays})} \tag{b}$$

In our case

$$\text{IV} = \frac{1}{10 \times 20} = 0.005$$

The stress at a depth z for a specific point is

$$\sigma_z = q \times \text{IV} \times (\text{number of influence units}) \tag{2-27}$$

To use this chart, one draws an outline of the loaded surface to a scale such that the distance AB from Fig. 2.25 equals the depth of the point in question. The point beneath the loaded area for which the vertical stress is sought is then located over the center of the chart. Hence, the area will encompass a number of influence units on the chart (in our case each unit has a value of 0.005). Thus, by counting the number of influence units and by using Eq. 2-27, one may proceed to determine the stress at the given point. Example 2.11 may serve to illustrate this procedure.

One may note that while the values of r/z indicated in Table 2.9 remain fixed for the selected values of σ_z/q, the scale for the influence chart was arbitrarily chosen and can, therefore, be altered as needed. Similarly, the number of rays or the number of radii may also vary as desired, thereby varying the influence values for these charts.

Example 2.11 _____

Given The T-shaped foundation is loaded with a uniform load of 100 kN/m² ($\approx$2 kips/ft²) (see Fig. 2.26).

Find The pressure at 6 m below point *G*.

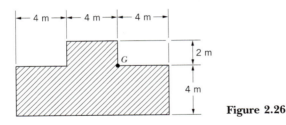

Figure 2.26

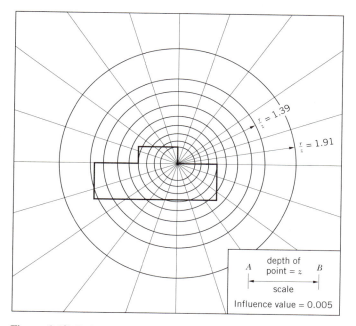

Figure 2.27 Point G is over the center of the chart.

Procedure   The scale is determined such that the distance AB in Fig. 2.27 represents 6 m. Hence, the T-shaped area is redrawn to scale with point G placed over the center of the chart. The number of "squares" encompassed by Fig. 2.27 is 66. Hence, the total pressure at G is $0.005 \times 66 \times 100$ kN/m².

Answer

Total pressure = 33 kN/m²

2.8 COMPRESSIBILITY OF SOIL

Settlement

Settlement is the direct result of the decrease in the soil volume. Consolidation is the rate of volume decrease with time. Hence, we view consolidation as a time-related process involving compression, stress transfer, and water drainage.

Cohesionless soil (e.g., sand, gravel) will generally compress during a relatively short time (8, 16). In fact, quite frequently most of the anticipated settlement of a foundation resting on cohesionless soil has taken place during the construction phase of the structure. Furthermore, the compression by induced vibrational effects can be accomplished with much greater ease with such soils than with cohesive soil.

Unlike cohesionless soils, a saturated cohesive soil mass (e.g., fine silts, clays) with low permeability will compress quite slowly since the expulsion of the pore

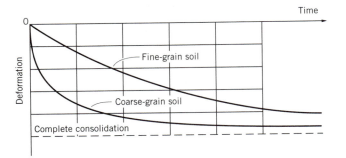

Figure 2.28 Time–consolidation relationship under constant load.

water from such soils occurs at a significantly slower rate than from cohesionless ones. In other words, because considerably longer time is required for water to exit from the voids of a cohesive material, the total compression of a saturated clay stratum takes place during a longer period of time than for cohesionless soils, as depicted by Fig. 2.28.

The total settlement of a stratum is generally regarded to be the result of a two-phase process:

- Immediate settlement
- Consolidation settlement

Consolidation

Immediate settlements are those that occur rather rapidly, perhaps within hours or days after the load is applied. The consolidation settlement is a time-dependent deformation that occurs in saturated or partially saturated silts or clays. These are soils that have low coefficients of permeability and are slow to dissipate the pore water.

It is common to further designate the consolidation as *primary consolidation* and *secondary consolidation* (or **creep**). Of the two, the primary consolidation is generally much larger. Secondary consolidation is speculated to be due to the plastic deformation of the soil as a result of some complex colloid-chemical processes whose roles in this regard are mostly hypothetical at this point. Also, the primary consolidation is easiest to predict, occurs at a faster rate than secondary consolidation, and is usually the more important of the two. In highly organic soils, however, secondary consolidation may be important, indeed.

Figure 2.29 depicts the development of compression with time. The initial segment of settlement is usually designated as *instantaneous* or *elastic* deformation, and is attributed to the elastic deformation of the soil without a change in the water content. It usually takes place during a relatively short time after the load has been applied. With time the water escapes from the pores, gradually dissipating the excess hydrostatic pore-water pressure. The resulting deformation is known as primary consolidation. It is largest of the three phases and the one singled out in importance for most inorganic clays.

Evidence exists that indicates that some additional consolidation in clays can occur even after the excess pore-water pressure appears to have dissipated. This is generally referred to as secondary consolidation.

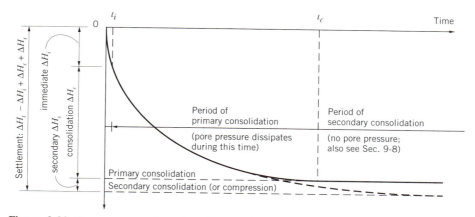

Figure 2.29 Components of total settlement ΔH_t, for a cohesive soil under constant load.

Load-Deformation Characteristics of Soils

The load-deformation relationship for clays is generally represented in terms of the *pressure-void ratio,* as depicted by Fig. 2.30, which represents the results of a typical laboratory one-dimensional consolidation test. For the sake of comparison, the pressure is plotted to natural scale in Fig. 2.30a and to logarithmic scale in Fig. 2.30b. The logarithmic plotting of the pressure is the more desirable of the two methods.

The steep straight line of the portion of the pressure-void ratio curve (*CDE* in Fig. 2.30b) assumes the form

$$y = mx + c \qquad \text{(a)}$$

or

$$e = e_0 - C_c \log_{10} \frac{P}{P_0} \qquad \text{(2-28)}$$

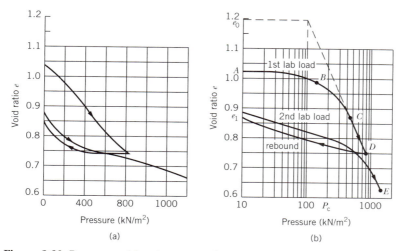

Figure 2.30 Pressure-void ratio curves of an undisturbed precompressed clay soil. (a) Arithmetic scale. (b) Logarithmic scale.

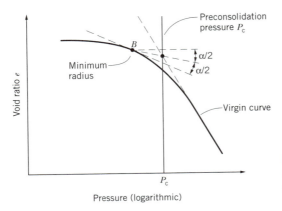

Figure 2.31 Casagrande procedure for determining preconsolidation pressure.

where C_c is termed the *compression index*. It is the slope of the straight-line portion of the curve on a semilogarithmic plot, with a minus sign indicating a negative slope. In Eq. 2-28, e_0 represents the void ratio at a corresponding effective pressure, say P_0. The compression index can be roughly approximated via an expression proposed by Terzaghi and Peck (22) related to the liquid limit of the clay:

$$C_c \approx 0.009(\text{LL} - 10) \tag{2-29}$$

Determination of the Preconsolidation Pressure

In Fig. 2.30*b*, pressure P_c represents the maximum probable pressure the soil experienced in situ. It is determined from the *e*–log*p* curve (laboratory test curve). Perhaps the most widely used procedure for determining P_c was proposed by Casagrande (3); it will be explained here. Assume that point *B* in Fig. 2.31 represents the point of maximum curvature (smallest radius) on an isolated segment of the curve shown in Fig. 2.30*b*. At point *B* draw a tangent to the curve and a horizontal line. Bisect the angle formed by the tangent and the horizontal line. The preconsolidated load P_c is given by the point where a line that bisects this angle meets the tangent to the virgin curve (e.g., *CDE* in Fig. 2.30*b*).

The *present effective pressure* P_0 induced by the existent overburden is computed as the weight of the column of soil above the point evaluated (i.e., $P_0 = \gamma h$); it is also known as the *geostatic pressure*. When P_0 equals P_c, the clay is said to be *normally consolidated*. When P_c is larger than P_0, the clay is known as *overconsolidated*. The ratio of P_c to P_0 is known as the *overconsolidated ratio* (OCR); that is, OCR $= P_c/P_0$.

Rate of Consolidation

In 1923 Karl Terzaghi advanced a rigorous mathematical solution of the process of consolidation of soils.* Equation 2-30 is *Terzaghi's consolidation equation:*

$$C_v \frac{\partial^2 u}{\partial z^2} = \frac{\partial u}{\partial t} \tag{2-30}$$

* See Appendix C.

where C_v, called the *coefficient of consolidation,* is

$$C_v = \frac{k}{\gamma_w}\left(\frac{1 + e}{a_v}\right)$$

a_v = *coefficient of compressibility;* $\quad a_v = \partial e / \partial u$
$\partial^2 u / \partial t^2$ = change in hydraulic pressure gradient, in N/cm^4
$\partial u / \partial t$ = rate of change of hydrostatic excess pore-water pressure, in $N/cm^2 \cdot s$

The solution to Eq. 2-30 for a constant initial pore pressure u_0 satisfying the following conditions

$$
\begin{aligned}
&\text{at} \quad z = 0, \quad\quad u = 0 \\
&\text{at} \quad z = 2H, \quad u = 0 \\
&\text{at} \quad t = 0, \quad\quad u = u_0
\end{aligned}
$$

is

$$u = \sum_{m=1}^{\infty} \frac{2u_0}{M}\left(\sin \frac{Mz}{H}\right) e^{-M^2 T} \tag{2-31}$$

where $\quad M = (\pi/2)(2m + 1)$, m = any integer, $1, 2, 3, \ldots$
$\quad\quad\quad H$ = one-half thickness of consolidating stratum and

$$T = \frac{C_v t}{H^2} = \text{time factor} \tag{2-32}$$

The *percentage consolidation U* at a given depth and time t is

$$U = \left(\frac{u_i - u}{u_i}\right) \times 100 = \left(1 - \frac{u}{u_i}\right) \times 100 \tag{2-33}$$

where $\quad u$ = hydrostatic excess pressure at time t and depth z
$\quad\quad\quad u_i$ = initial hydrostatic excess pressure at time $t = 0$ and depth z

The basic U–T relationship can be developed by combining Eqs. 2-31 and 2-33. Thus, we have

$$U = 1 - \sum_{m=0}^{\infty} \frac{2}{M}\left(\sin \frac{Mz}{H}\right) e^{-M^2 T} \tag{2-34}$$

One may obtain a numerical relationship between U and T by substituting successive values of m from 0 to ∞, as indicated in Fig. 2.32.

Equation 2-34 is represented graphically by Fig. 2.32 for a uniform (i.e., rectangular vs. say trapezoidal or sinusoidal shape distribution) initial excess pore pressure, in a doubly-drained layer. One notes that the consolidation proceeds most rapidly at the drained faces (i.e., at $z/H = 0$ and at $z/H = 2$), and slowest at the midpoint (at $z/H = 1$). For example, for a time factor T equal to 0.2, consolidation is less than 25% complete at the midpoint ($z/H = 1$), but is about 45% complete at a depth of 25% ($z/H = 0.5$).

Seldom are we interested in the amount of consolidation at a given depth of a clay stratum, as depicted in Fig. 2.32. Instead, we are usually interested in the average degree of consolidation U for the entire thickness of the clay layer. It is the consolidation of the entire layer that is relevant to predicting the total settlement at a given time.

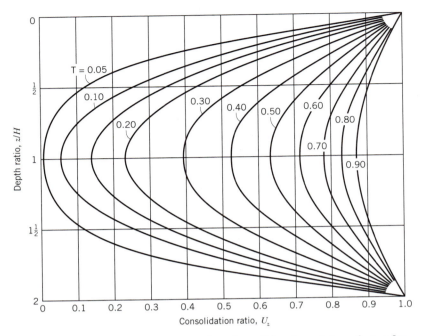

Figure 2.32 Consolidation ratio as a function of depth and time factor for uniform initial excess pore pressure in a doubly-drained layer.

For a constant initial hydrostatic excess pore pressure, the average degree of consolidation over the entire depth of the stratum is

$$U = 1 - \sum_{m=0}^{\infty} \frac{2}{M^2} e^{-M^2 T} \qquad (2\text{-}35)$$

Equation 2-35 is represented by Fig. 2.33.

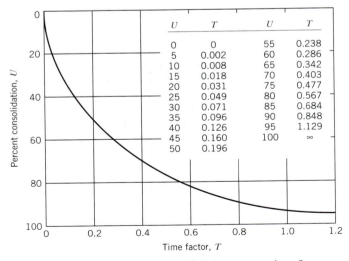

U	T	U	T
0	0	55	0.238
5	0.002	60	0.286
10	0.008	65	0.342
15	0.018	70	0.403
20	0.031	75	0.477
25	0.049	80	0.567
30	0.071	85	0.684
35	0.096	90	0.848
40	0.126	95	1.129
45	0.160	100	∞
50	0.196		

Figure 2.33 Average percent consolidation versus time factor.

It has been found that the following empirical relationships can be used, with acceptable accuracy, in place of Eq. 2-35:

$$T = \frac{\pi}{4} U^2 \qquad \text{when} \qquad U < 50 \tag{2-36a}$$

and

$$T = -0.933 \log_{10}(1 - U) - 0.0851 \qquad \text{when} \qquad U > 50 \tag{2-36b}$$

Determination of Coefficient of Consolidation C_v

Only two methods for determining the coefficient of consolidation will be discussed here, one by Casagrande and another by Taylor. Although empirical in nature, they are rather simple and expedient and give results that are comparable to some developed from theoretical expressions.

Figures 2.34 and 2.35 represent the same data from a laboratory test, plotted to two different scales. That is, both represent strain or percent consolidation of the same clay sample. These graphs will be used to determine the coefficient of consolidation by the two different methods as described below.

The Logarithm-of-Time Fitting Method   Figure 2.34 depicts the relationship between strain (percent consolidation) and log time for one stage of load.

The first part of the curve approximates a parabola. Hence, the 0% consolidation coordinate may be obtained in the following manner: Select a point of time t at the early stage of consolidation. Next determine the difference in the ordinate between that point and the one corresponding to $4t$. This is designated as δ in Fig. 2.34. Then lay off this δ value above the curve at point t to obtain the corrected 0 point.

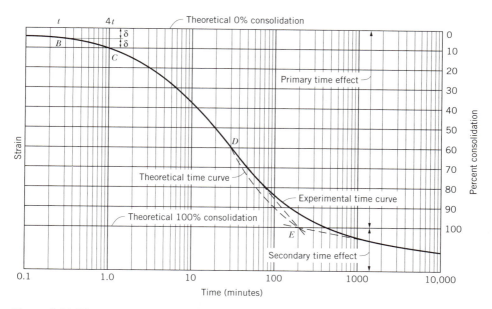

Figure 2.34 Time versus consolidation.

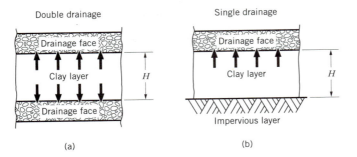

Figure 2.35 (a) Double drainage. (b) Single drainage.

Casagrande suggested that the point of 100% primary consolidation may be determined by the intersection of the tangent to the straight-line portion of the primary consolidation curve and the tangent of the straight-line portion of the secondary consolidation portion of the curve. With the 0 and 100% consolidation points established, one may divide the ordinate into ten equal parts to translate the dial readings into percent consolidation.

With the 0 and 100% primary compression points located, the coefficient of consolidation may be computed from Eq. 2-31 and at the 50% point by noting that the time factor at the 50% consolidation corresponds to the value of 0.196. Hence, corresponding with this time factor, Eq. 2-33 may be rearranged to express the coefficient of consolidation given by Eqs. 2-37a and 2-37b for the cases of drainage shown in Fig. 2.35.

For double drainage (Fig. 2.35a),

$$C_v = \frac{0.196H^2}{4t_{50}} \tag{2-37a}$$

and for single drainage (Fig. 2.35b),

$$C_v = \frac{0.196H^2}{t_{50}} \tag{2-37b}$$

Square-Root-of-Time Fitting Method   Proposed by D. Taylor, the 0% and the 100% consolidation can be determined by first plotting the dial readings on the ordinate axis to a natural scale and then the corresponding values on the abscissa as the square root of time, as indicated in Fig. 2.36.

Taylor observed that the first portion of the curve, perhaps past 50 to 60% of consolidation, is essentially a straight line, while at 90% consolidation the abscissa is 1.15 times the abscissa of the straight line intersecting the 90% consolidation. Hence, Taylor suggested the following method for determining the points of 0% and 100% consolidation, and for subsequently determining the coefficient of consolidation.

For the data plotted as shown in Fig. 2.36 draw a tangent to the straight-line portion of the curve. A second line, coinciding with the point of 0 time, is drawn such that the abscissa of this line are 1.15 times as great as those of the first line. The 90% consolidation is assumed to be the point of intersection of the time-

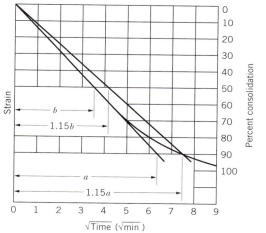

Figure 2.36 Time versus consolidation.

consolidation curve with the second straight line drawn. Hence, a scale for the percent consolidation could be drawn by dividing the distance between 0 and 90% into nine equal parts, as shown in Fig. 2.36.

The coefficient of consolidation for the curve may, therefore, be determined from Eq. 2-32. Noting that T_{90} from Fig. 2.33 has a value of 0.848, the coefficient of consolidation expressions are given by Eqs. 2-38a and 2-38b.

For *double drainage* (Fig. 2.35*a*),

$$C_v = \frac{0.848H^2}{4t_{90}}$$

(2-38a)

For *single drainage* (Fig. 2.35*b*),

$$C_v = \frac{0.848H^2}{t_{90}}$$

(2-38b)

Again, note that the time-consolidation relationships indicated in Figs. 2.34 and 2.36 represent the behavior of a specific soil sample subjected to only one load range. Hence, the coefficient of consolidation will be altered by different time-consolidation curves.

Example 2.12 _____

Given   Figures 2.34 and 2.36 represent time-consolidation relationships for the same clay sample, 3 cm thick, subjected to a given pressure range under double drainage.

Find   (a) The coefficient of consolidation C_v for the sample by the *two* methods described.

(b) The time necessary for 60% consolidation of the same sample if it were 2 m thick. Assume an average C_v.

Procedure   (a) From Eq. 2-37a,

$$C_v = \frac{0.848H^2}{4t_{90}} = \frac{0.848(3)^2}{4t_{90}}$$

From Fig. 2.36, at 90% consolidation, $\sqrt{t} = 7.6$ and $t = 57.76$ min. From Fig. 2.34, $t_{50} = 18$. Thus, based on the time-fitting method,

$$C_{v90} = \frac{0.848(3)^2}{4 \times 57.76 \times 60}$$

Answer

$$C_{v90} = 5.505 \times 10^{-4} \text{ cm}^2/\text{s}$$

Based on the logarithm-of-time method,

$$C_{v50} = \frac{0.196(3)^2}{4 \times 18 \times 60}$$

Answer

$$C_{v50} = 4.084 \times 10^{-4} \text{ cm}^2/\text{s}$$

(b) Average

$$C_v = \frac{5.505 + 4.084}{2} \times 10^{-4} = 4.795 \times 10^{-4} \text{ cm}^2/\text{s}$$

From Fig. 2.33 (tabulated values), $T_{60} = 0.286$. Thus,

$$t_{60} = \frac{T_{60}H^2}{4C_v} = \frac{0.286(200)^2}{4 \times 4.795 \times 10^{-4}}$$

$$t_{60} = 5.97 \times 10^6 \text{ s} = 9.94 \times 10^4 \text{ min}$$

Answer

$$t_{60} = 69 \text{ days}$$

Normally Consolidated Clay

Figure 2.37 represents a schematic of a homogeneous totally saturated clay mass. For the sake of illustration, the total volume is divided into totally saturated voids and solids. As mentioned in the previous discussion, the relative amount of deformation of the solids is negligible. Hence, it is the change in the void volume that is assumed to be the cause of the settlement. Thus, the total settlement S equals the total *change* of stratum thickness ΔH:

$$S = \Delta H = H_i - H_f = h_1 - h_2 = (h_1 - h_2)\frac{H_i}{H_i} = H_i\left(\frac{h_1 - h_2}{h_s + h_1}\right)$$

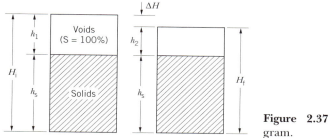

Figure 2.37 Soil phase diagram.

or

$$S = \Delta H = H_i \frac{(h_1 - h_2)\,/\,h_s}{(h_s + h_1)\,/\,h_s} = H_i \left(\frac{e_1 - e_f}{1 + e_1} \right) \tag{2-39}$$

where the quantities H_i, H_f, h_1, h_2, h_s are shown in Fig. 2.37. In terms of pressure, assuming a straight-line relationship (semilog scale) between e and P, we have from Eq. 2-28,

$$e_0 - e \cong e_1 - e_f = C_c \log_{10} \frac{P}{P_0} = C_c \log_{10} \frac{P_0 + \Delta P}{P_0}$$

Substituting in Eq. 2-39,

$$S = \Delta H = \left(\frac{H_i}{1 + e_1} \right) C_c \log_{10} \left(\frac{P_0 + \Delta P}{P_0} \right) \tag{2-40}$$

Either Eq. 2-39 or Eq. 2-40 is a suitable expression for settlement for a normally consolidated clay, although Eq. 2-40 is perhaps the more convenient of the two since e_f is not needed. P_0 and ΔP can be calculated via methods detailed in Section 2.7.

The expression for secondary settlement ΔH_s can be estimated from Eq. 2-41:

$$\Delta H_s = H C_\alpha \frac{t_1 + \Delta t}{t_1} \tag{2-41}$$

where C_α, called the *coefficient of secondary compression,* is given by Eq. 2-42:

$$C_\alpha = \frac{\varepsilon_1 - \varepsilon_2}{\log(t_2 / t_1)} \tag{2-42}$$

where ε_1 and ε_2 are *strains* at times t_1 and t_2, respectively. Some common ranges of C_α are given in Table 2.10.

Table 2.10 Some Common Values of C_α

Overconsolidated clays	0.0005–0.0015
Normally consolidated clays	0.005 –0.03
Organic soils, peats	0.04 –0.1

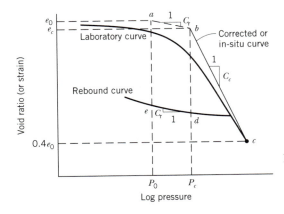

Figure 2.38 Determination of C_c for overconsolidated clays. [After Schmertmann (18).]

Overconsolidated Clay

When the soil is overconsolidated (i.e., $P_c > P_0$), the compression index C_c is different from that used in Eq. 2-40. An approach for determining C_c for such a case was developed by Schmertmann (18). Referring to Fig. 2.38 the procedure is as follows:

1. Plot the laboratory one-dimensional consolidation curve as shown.
2. Compute P_0 and e_0. Point a is the coordinate point for these values.
3. Determine P_c. One method was described previously (see Fig. 2.31).
4. Determine the slope C_r of the rebound-curve segment de. Use that slope to draw line ab. Thus, point b is determined.
5. From an ordinate value of $0.4e_0$, draw a horizontal line to intersect the extension of the laboratory curve. This determines point c.
6. Connect points b and c with a straight line. The slope of this line is C_c for overconsolidated clays, as suggested by Schmertmann.

Example 2.13

Given A footing is placed on the silty clay stratum, with properties shown in Fig. 2.39. Then a 4-ft fill layer is added, as shown.

Find (a) Estimate of ΔH.

(b) Time span for 6 in. of ΔH to occur.

Solution
$$\Delta H = S = \left(\frac{H}{1+e}\right) C_c \log \left(\frac{P + \Delta P}{P}\right)$$

To find e:

$$G_m = \frac{102}{62.4} = 1.63$$

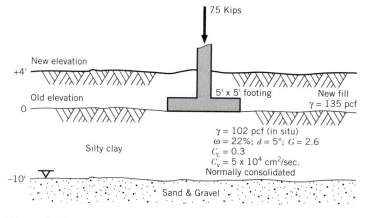

Figure 2.39

Assume

$$W_s = 100 \text{ g} \Rightarrow V_s = \frac{100}{2.6} = 38.5$$

$$W_w = 22 \text{ g}; \; V_w = 22 \text{ cm}^3$$

$$V = \frac{W}{G} = \frac{122}{1.63} = 74.8 \text{ cm}^3 \Rightarrow V_v = 36.3 \text{ cm}^3$$

$$e = \frac{36.3}{38.5} = 0.94$$

At midpoint of clay:

$$P = 102 \times 5 = 510 \text{ psf}$$
$$\Delta P = \Delta P @ \text{ new fill wt} + \Delta P @ \text{ column load}$$
$$@ \text{ new fill: } \Delta P = 135 \times 4 \text{ ft} = 540 \text{ psf}$$

$$@ \text{ col. ld.: } \Delta P = \left(\frac{75{,}000}{5 \times 5}\right)(0.35) \text{—via Boussinesq for } z = B \text{ under } \mathbb{C}_L$$

$$= 1050 \text{ psf}$$

Hence,

$$\Delta P = 540 + 1050 = 1590 \text{ psf}$$

Thus,

Answer

$$\Delta H = 5 = \left(\frac{120 \text{ in.}}{1 + 0.94}\right)(0.3)\left[\log\left(\frac{510 + 1590}{510}\right)\right] = 11.4 \text{ in.}$$

For 6 in.,

$$\% = \frac{6}{1.4} \cong 53\%$$

$$t_{53} = \frac{T_{53}(H)^2}{4\,C_v}; \qquad H = cm = 120\ \text{in.} \times 2.54\ \text{cm/in.} = 304.8$$

$$t_{53} = \frac{0.22(304.8)^2}{4 \times 5 \times 10^{-4}} = 1022 \times 10^4\ \text{sec} = 17 \times 10^4\ \text{min}$$

or

Answer

$$t_{53} = 118\ \text{days}$$

2.9 SHEAR STRENGTH OF SOIL

The shear strength of the soil may be attributed to three basic components:

1. Frictional resistance to sliding between solid particles
2. Cohesion and adhesion between soil particles
3. Interlocking and bridging of solid particles to resist deformation

It is neither easy nor practical to clearly delineate the effects of these components on the shear strength of the soil. This becomes more apparent when one relates these components to the many variables that directly or indirectly influence them, not to mention the lack of homogeneity and uniformity of the characteristics that typify most soil masses. For example, these components may be influenced by changes in the moisture content, pore pressures, structural disturbance, fluctuation in the groundwater table, underground water movement, stress history, time, and perhaps chemical action or environmental conditions.

Mohr's Theory of Failure

The shear strength of soil is generally regarded as the resistance to deformation by continuous shear displacement of soil particles along *surfaces of rupture*. That is, the shear strength of the soil is not regarded solely in terms of its ability to resist peak stresses, but it must be viewed in the context of deformation that may govern its performance. In that light, therefore, shear failure is necessarily viewed as the state of deformation when the functional performance of the soil mass is impaired.

There are a number of different theories as to the nature and extent of the state of stress and deformation at the time of failure. Failure of a soil mass, particularly cohesionless soil, which develops its strength primarily from solid frictional resistance between the interlocking of grains, appears to be best explained by *Mohr's rupture theory*. According to Mohr's theory, the shear stress in the plane of slip reaches at the limit a maximum value, which depends on the normal stress acting in the same planes and the properties of the material. This represents the combina-

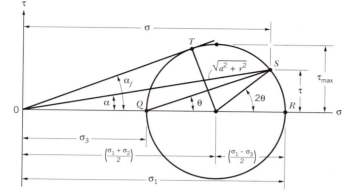

Figure 2.40 Some basic Mohr circle relationships.

tion of normal and shear stresses that results in a maximum angle of obliquity α_f (see Fig. 2.40).

Hence, based on Mohr's theory, the shear strength of a given soil is a function of the normal stress and the soil properties (e.g., strength parameters c and ϕ). For cohesionless soils, for example, the angle of obliquity α_f reaches a maximum or limiting value equal to the friction angle ϕ, as indicated in Fig. 2.41. Hence, for this special case, where the cohesion intercept is zero ($c = 0$), the shear strength could be expressed as

$$s = \tau_f = \sigma_f \tan \phi \tag{2-43}$$

where s represents a convenient and widely used symbol to denote shear strength in soils. One may note that for $\alpha_f > 0$, the value for s is always less than τ_{max}.

Figure 2.42a depicts a general s–σ relationship, while Fig. 2.42b illustrates a linear s–σ relationship. The linearity may not be quite correct (e.g., see Fig. 2.43), but is reasonably accurate and mathematically quite convenient.

Now let us assume that the function $s = f(\sigma)$ represents the shear strength of the soil at impending failure. This may be represented by the heavy line in Fig. 2.43 designated as *Mohr's envelope*. Based on Mohr's theory, any combination of

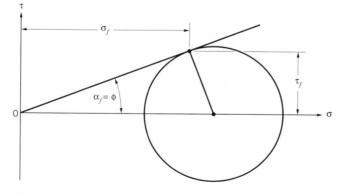

Figure 2.41 Relationship between shear strength, normal stress, and angle of obliquity α.

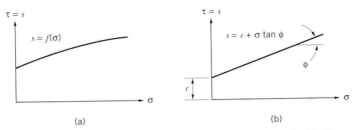

Figure 2.42 Shear strength according to (a) Mohr and (b) Coulomb.

stresses that falls within this envelope represents a stable condition. All five Mohr circles represented in Fig. 2.43 fall in this category. It is to be noted, however, that circles I–IV are tangent to the Mohr envelope, thereby depicting a condition of impending failure. The corresponding points of tangency represent the resultant's shear and normal stresses for the respective angles of obliquity. Also, the corresponding shear strengths for the various cases are the shear stresses at the points of tangency. Circle V, on the other hand, is well within Mohr's envelope. The maximum shear stress for this condition is below the shear strength of the soil.

The friction angle decreases slightly in most soils, and particularly granular soils, with increasing confining stresses. Hence, Mohr's envelope is slightly curved, a best fitted curve, as indicated in Fig. 2.42a. However, the variation from a straight line in most instances is relatively small. Thus, it is mathematically convenient to represent the Mohr envelope by a straight line, sloped at an angle ϕ. This is shown in Fig. 2.42b as first proposed by Coulomb in 1776 in connection with his investigations of retaining walls.

Figure 2.44c shows a Mohr circle and the corresponding strength envelope for an element stressed as shown in Fig. 2.44a. For this case the expression for the shear strength s may be given as

$$s = c + \sigma \tan \phi \tag{2-44}$$

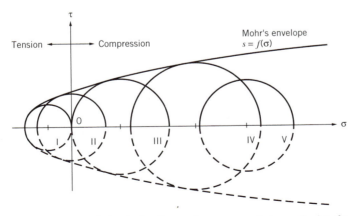

Figure 2.43 Mohr's circles for various cases of stress: I, simple tension; II, pure shear; III, simple compression; IV and V, biaxial compression.

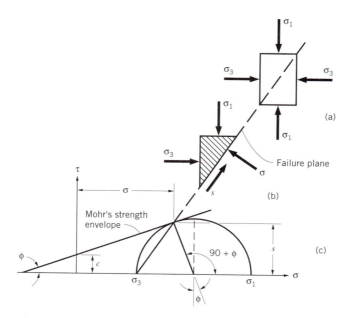

Figure 2.44 (a) Element; $\sigma_1 > \sigma_3$. (b) Normal and shear stresses on plane of failure. (c) Mohr circle for stress condition shown in (a).

where c = *cohesion* or cohesion intercept

ϕ = angle of internal friction

Also from Fig. 2.44c, the orientation of the failure plane with reference to the plane of major principal stresses in a granular soil may be readily established as $2\theta_{cr} = 90 + \phi$, or

$$\theta_{cr} = 45° + \frac{\phi}{2} \tag{2-45}$$

The normal and shear stresses on the plane of failure are shown in Fig. 2.44b.

Although it is difficult to measure it very accurately, pore-water pressure has long been recognized as an important factor in shear strength evaluation. Hvorslev's work (12) provided relevant data to support the use of effective stress parameters in describing the shear strength of soils. Sometimes referred to as the Mohr–Coulomb–Hvorslev equation, the relationship is

$$s = \bar{c} + \bar{\sigma} \tan \overline{\phi} \tag{2-46}$$

where $\bar{c}$ = effective cohesion value

$\bar{\sigma}$ = effective normal stress = $\sigma - u$

u = pore-water pressure

$\overline{\phi}$ = effective angle of internal friction

Equation 2-46 does not imply, however, any unique relationship between effective stress and shear strength. Many other factors enter into the picture. Void ratio at failure, time, stress history, grain structure (i.e., flocculated or dispersed), environmental conditions (i.e., pore water, water table fluctuation, temperature), degree of saturation, conditions that formulate the formation of soil, capillary tension, and effective stresses in the direction normal to the plane of greatest shear distortion are some of the factors that may have significant effect, particularly for clay soils.

p–q Diagrams and Stress Paths

Instead of developing the strength envelope by plotting a series of Mohr circles, then drawing tangents to these circles, it is more convenient to plot the state of stress as a stress point whose coordinates are (p, q), defined as

$$p = \frac{\sigma_1 + \sigma_3}{2} \qquad q = \frac{\sigma_1 - \sigma_3}{2} \tag{2-47}$$

Actually the p and q values represent the coordinates of a point on Mohr's circle, with p representing the center of the circle (always located on the abscissa, or σ axis) and q representing the maximum shear stress, equal to the radius of the circle. The locus of p–q points for a test series is known as a *stress path* (13). Such a graphical depiction is known as the p–q diagram.

Figure 2.45 shows the Mohr envelope (ϕ line) developed from tangents to the circles at points 1, 2, and 3 and line K_f (K_f line), which passes through p–q points A, B, and C—points of maximum shear for the respective circles. Thus, the K_f line represents a limiting state of stress at impending failure. The relationship between the ϕ line and the K_f line may be better illustrated with the aid of Fig. 2.46 for cohesive ($c > 0$) and cohesionless ($c = 0$) soils.

From either Fig. 2.46a or Fig. 2.46b,

$$\tan \alpha_f = \sin \phi \tag{a}$$

But from Fig. 2.46a,

$$c \cot \phi = d \cot \alpha_f = \frac{d}{\tan \alpha_f} \tag{b}$$

Combining Eqs. (a) and (b), we get

$$c = \frac{d}{\cos \phi} \tag{2-48}$$

With the K_f line developed and d determined, the parameters c and ϕ may be either computed or scaled directly from the p–q diagram.

Figure 2.45 depicts the stress path for a series where $\sigma_3 < \sigma_3' < \sigma_3''$ and $\sigma_1 < \sigma_1' < \sigma_1''$. By contrast, Fig. 2.47 shows the stress path for a series where $\sigma_1 > \sigma_3$ and

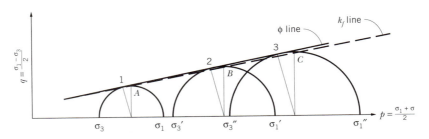

Figure 2.45 Stress path for a triaxial test series.

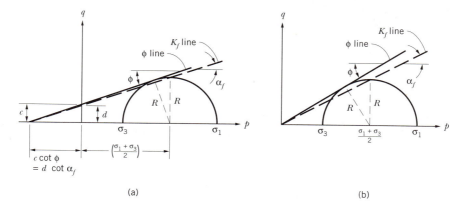

Figure 2.46 p–q diagrams and relationship between ϕ line, K_f line, and strength parameters ϕ and c. (a) Cohesive soil. (b) Cohesionless soil.

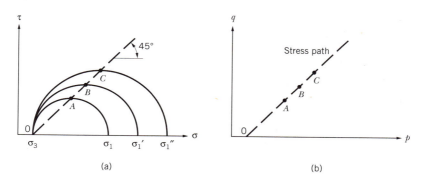

Figure 2.47 (a) Series of Mohr circles where σ_1 increases while σ_3 is held constant. (b) p–q diagram (stress path) corresponding to (a).

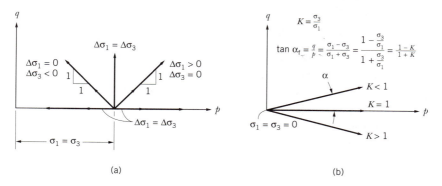

Figure 2.48 Stress paths for various combinations of stress conditions. (a) Stress paths starting from point where $\sigma_1 = \sigma_3$. (b) Stress paths starting from point where $\sigma_1 = \sigma_3 = 0$.

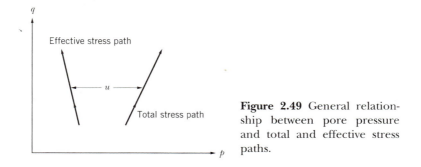

Figure 2.49 General relationship between pore pressure and total and effective stress paths.

σ_3 is held constant. In the same manner, we may hold σ_1 constant while varying σ_3, and so on. Thus, we may vary σ_1 and σ_3 in various ways to obtain any number of stress paths. Figure 2.48 shows a variety of such paths starting from points $\sigma_1 = \sigma_3$ and $\sigma_1 = \sigma_3 = 0$. Needless to say, the stress increments $\Delta\sigma_1$ or $\Delta\sigma_3$ do not have to vary linearly. Hence, the stress paths do not have to be straight lines, and they are not limited to a specific starting point.

Stress paths can be developed for either total or effective stresses. The horizontal distance between total and effective stress paths is the pore pressure u. The pore pressure is positive if the effective stress path is on the left of the total stress path, and vice versa, as shown in Fig. 2.49. Positive pore pressures are usually found in normally consolidated clays, while negative pore pressures may be expected in highly overconsolidated clays.

Problems

2.1   Describe the characteristics that make *soils* such a complex construction material.

2.2   What is encompassed by the term *soil* as used in engineering?

2.3   Define the following terms: (a) Settlement. (b) Differential settlement. (c) Stability as related to foundation design.

2.4   How does Geotechnical Engineering differ from other forms of mechanics, such as statics and dynamics, fluid mechanics, etc.?

Soil Formations and Deposits

2.5   What is the geological cycle? Give a brief description.

2.6   Differentiate between intrusive and extrusive rocks.

2.7   Name several factors that are directly tied to the disintegration of rocks.

2.8   Name and describe briefly the origin of three main groups of rocks, and give three examples of each.

2.9   How does mechanical weathering differ from chemical weathering?

2.10   What is residual soil? What is sedimentary deposit?

Density and Volume Relationships

For the following problems assume $\gamma_w = 1$ g/cm^3 = 9.807 kN/m^3 and, unless specified otherwise, $G = 2.7$. $G_m = G_{mass}$.

2.11 Why is the dry weight (weight of solids) rather than the total weight used in defining the water content? Can the water content exceed 100%? Explain.

2.12 A soil sample was determined to possess the following characteristics: $G = 2.74$, $e = 0.69$, and $w = 14\%$. Determine:
   (a) The degree of saturation.
   (b) The porosity.
   (c) The unit dry weight of the sample.

2.13 A moist soil sample was found to have a volume of 22.3 cm³ and to weigh 29.7 g. The dry weight of the sample was determined to be 23 g. Determine:
   (a) The void ratio.
   (b) The water content.
   (c) The porosity.
   (d) The degree of saturation of the sample.

2.14 Laboratory tests on a soil sample yielded the following information: $G = 2.71$, $G_m = 1.92$, $w = 13\%$. Calculate:
   (a) The void ratio.
   (b) The degree of saturation.
   (c) The porosity.

2.15 A soil sample was determined to have a water content of 8% and a degree of saturation of 42%. After adding some water, the degree of saturation was altered to 53%. Assuming no change in the volume of the voids, determine:
   (a) The void ratio.
   (b) The water content.
   (c) The specific gravity of the mass of the sample in the altered state.

2.16 Material for an earth fill was available from three different borrow sites. In the compacted state the fill measured 100,000 m³ at a void ratio of 0.70. The corresponding in-situ void ratio and cost (material and transportation) of the material for the three sites is as follows:

Borrow Site	Void Ratio	Total Cost per Cubic Meter
1	0.8	$6.40
2	1.7	$6.00
3	1.2	$5.15

Determine the most economical site.

2.17 A saturated soil sample weighs 1015 g in its moist, natural state and 704 g after drying. $G = 2.68$. Determine:
   (a) The water content.
   (b) The void ratio.
   (c) The porosity.

2.18 The mass specific gravity of an undisturbed soil sample was determined to be 1.96 at a water content of 14%. The void ratios in the loosest and densest states were determined to be 0.81 and 0.48, respectively. Determine the relative density of the mass.

2.19   An undisturbed soil mass has a weight of 788 g and a volume of 406 cm³. The weight of the dry sample was 670 g, $G = 2.65$. The void ratios in the loosest and densest states were found to be 0.77 and 0.52, respectively. Determine the relative density of the mass.

2.20   During compaction, a 1-m-thick stratum was consolidated a total of 3 cm via a vibrating roller. Initially the void ratio was determined to be 0.94, the water content 16%, and $G = 2.67$. Determine:

   (a) The void ratio.
   (b) The porosity.
   (c) The specific gravity of the mass in the compacted state.

2.21   A soil sample taken from a borrow pit has an in-situ void ratio of 1.15. The soil is to be used for a compaction project where a total of 100,000 m³ is needed in a compacted state with the void ratio predetermined to be 0.73. Determine how much volume is to be excavated from the borrow pit.

2.22   Laboratory tests determined the water content in a certain soil to be 14% at a degree of saturation of 62%. Determine:

   (a) The specific gravity.
   (b) The porosity.
   (c) The void ratio of the mass.

Soil Classification

Use the following data for Problems 2.23 through 2.28:

Percent Passing

Sieve		Soil Sample					
No.	Opening (mm)	A	B	C	D	E	F
4	4.76	100	90	100	100	94	100
8	2.38	97	64	100	90	84	100
10	2.00	92	54	96	77	72	98
20	0.85	87	34	92	59	66	92
40	0.425	53	22	81	51	58	84
60	0.25	42	17	72	42	50	79
100	0.15	26	9	49	35	44	70
200	0.075	17	5	32	33	38	63
Characteristics of −40 Fraction							
LL—ω_l		35	—	48	46	44	47
PL—ω_p		20	—	26	29	23	24

2.23   Draw the grain-size distribution curves for samples *A*, *C*, and *E*.

2.24   Draw the grain-size distribution curves for samples *B*, *D*, and *F*.

2.25   Classify soil samples *A, C,* and *E* by the unified classification system.

2.26   Classify soil samples *B, D,* and *F* by the unified classification system.

2.27   Classify soil samples *A, C,* and *E* by the AASHTO classification.

2.28   Classify soil samples *B, D,* and *F* by the AASHTO classification.

Soil–Water Phenomena

2.29   Determine the capillary rise of water in clean glass tubes of (a) $d = 0.04$ mm, (b) $d = 0.08$ mm, (c) $d = 0.15$ mm.

2.30   Determine the rise of water in Problem 2.29 if the contact angle is 30°.

2.31   The rise in a capillary tube is 52 cm above the free-water surface. Determine the surface tension if the radius of the tube is 0.03 mm.

2.32   What is the diameter of a clean glass tube if water rises to a total of 76 cm at approximately 20° C temperature?

2.33   Determine the water pressure in the capillary tube of Problem 2.31.

   (a) At 30 cm above the free surface.

   (b) At 40 cm above the free surface.

   (c) Just below the meniscus.

   (d) Just 1 cm above the free-water surface.

2.34   The effective size of a sandy silt is 0.04 mm and its void ratio is 0.7. Determine the approximate capillary rise for this soil. Assume $C = 0.32$.

2.35   A coarse-sand sample, 12 cm long and 7.3 cm in diameter, was tested in a constant-head permeameter under a head of 100 cm for 1 min 12 s. The quantity discharged was exactly 5 liters. Determine the coefficient of permeability.

2.36   How much water, per hour, would pass through a mass of soil described in Problem 2.35 if the mass is 60 cm long and 100 cm² in cross section, under a constant head of 2 m?

2.37   A constant-head permeability test is run on a soil sample 9.6 cm in diameter and 20 cm long. The total head at one end of the sample is 100 cm; at the other end, the head is 26 cm. Under these conditions the quantity of flow is determined to be 20 cm³/min. Determine the coefficient of permeability.

2.38   A falling-head permeability test was run on a soil sample 9.6 cm in diameter and 10 cm long. The head at the start of the test was 90 cm. Determine how much head was lost during the first 30 min if the coefficient of permeability of the soil was found to be 5×10^{-6} cm/s. The diameter of the standpipe was 1 cm.

2.39   A falling-head permeability test was run on a soil sample 7.3 cm in diameter and 18 cm long. The diameter of the standpipe was 1 cm. The water level in the standpipe dropped from 65 cm to 50 cm in 3 h 13 min. Determine the coefficient of permeability.

2.40   Determine the seepage per lineal meter of wall under the sheet pile wall shown in Fig. P2.1 for $h = 3$ m, $S = 6$ m, $d = 3$ m, and $k_x = k_z = 6 \times 10^{-4}$ cm/s.

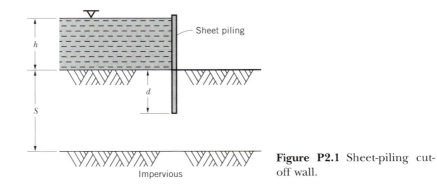

Figure P2.1 Sheet-piling cut-off wall.

2.41 For Problem 2.40 determine the pore pressures at each point where the equipotential lines meet the sheet piling on each side.

2.42 (a) Redraw the section in Fig. P2.2 to true scale and construct the flow net for this section.

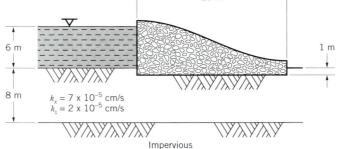

Figure P2.2 Concrete dam.

(b) Determine the seepage loss in cubic meters per day per meter of dam.

2.43 (a) Redraw the section in Fig. P2.3 to true scale and construct the flow net for this section.

(b) Determine the seepage loss in cubic meters per day per meter of dam.

Stresses in Soils

2.44 Calculate the vertical stress at a depth of 3 m due to the weight of the soil (geostatic stress) if $\gamma = 19.3 \text{ kN/m}^3$.

(a) The water table is below the 3 m depth.

(b) The water table is 1 m below the ground surface. Assume total saturation for the "wet" stratum.

2.45 Calculate the vertical stress from a concentrated surface load of 1000 kN at a point whose coordinates are $x = 1$ m, $y = 2$ m, and $z = 3$ m. Surface load coordinates are $(0, 0, 0)$.

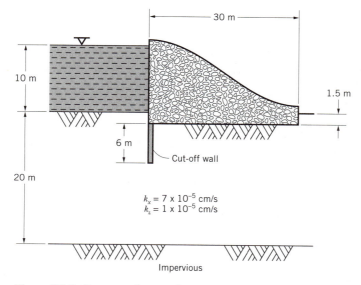

Figure P2.3 Concrete dam and cutoff wall.

2.46 A wall footing 10 m long induces a surface load of 300 kN/m. Assume this to be a line load parallel to the y axis and the one end of the footing to be located at coordinates $x = 0$, $y = 0$, and $z = 0$. Determine the vertical stress at points whose coordinates are $x_1 = 2$ m, $y_1 = 0$ m, $z_1 = 3$ m and $x_2 = 2$ m, $y_2 = 5$ m, and $z_2 = 3$ m.

2.47 A 700-kN load is supported by a circular footer (at surface) whose diameter is 2 m. Assume the surface pressure to be uniform over the area. (a) Determine the vertical stress at a depth of 3 m directly under the center and at radial distances of 1, 2, 3, and 4 m from the center. (b) Plot the values of σ_z from part (a) and connect these points with a smooth curve.

2.48 The footing shown in Fig. P2.4 exerts a uniform load to the soil of 300 kN/m². Determine the pressure at a point 4 m directly under point (a) A and (b) B.

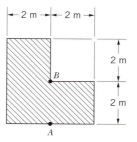

Figure P2.4 L-shaped footing.

2.49 A rectangular footing as shown in Fig. P2.5 (shaded) exerts a uniform pressure of 420 kN/m². Determine the vertical stress at point A for a depth of 3 m.

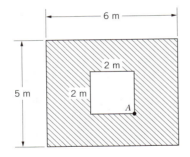

Figure P2.5 Composite footing.

2.50   Rework Problem 2.48 using Newmark's influence chart.

2.51   Rework Problem 2.49 using Newmark's influence chart.

2.52   A footing shaped as an equilateral triangle of 10 m sides is loaded with a uniform load of 500 kN/m. Determine the pressure at a point located 5 m directly under the middle of one of its sides.

2.53   Determine the total stress 8 m below point A due to the line pressures shown in Fig. P2.6.

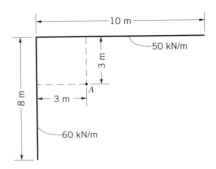

Figure P2.6 L-shaped wall footing.

2.54   Two bearing walls meet at a point A as shown in Fig. P2.7. Assume their loads to act as line loads. Determine the stress at a point 3 m below point A.

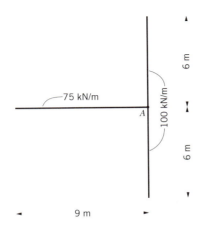

Figure P2.7 T-shaped wall footing.

Consolidation

2.55 Define: (a) normally consolidated clay, (b) overconsolidated clay, (c) over-consolidated ratio.

2.56 Define and explain the nature of (a) primary consolidation and (b) secondary compression.

2.57 The graph of e-log p shown in Fig. P2.8 represents the lab loading results of a one-dimensional consolidation test. Determine the compression index: (a) if $P_0 = P_c$, (b) if $P_0 = 90$ kN/m². Use Casagrande's method for determining P_c.

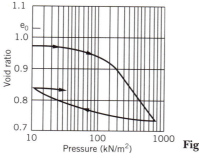

Figure P2.8 Consolidation curve.

2.58 Assume that the e–log p relationship shown in Fig. P2.8 is representative of the clay stratum shown in Fig. P2.9. Determine: (a) P_0 and P_c, (b) OCR.

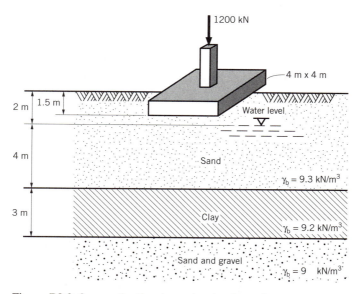

Figure P2.9 Square footing imposing load on clay.

2.59 Assume that the e–log p relationship shown in Fig. P2.8 is representative for the clay shown in Fig. P2.9. Determine:

   (a) C_c via Schmertmann's method.

   (b) An estimate of the settlement under the center of the footing.

   (c) An estimate of the settlement under a corner of the footing.

2.60 Assume that the e–*log* p relationship shown in Fig. P2.10 represents the laboratory results of a one-dimensional consolidation test. The clay stratum from which the test sample was extracted was 3.5 m thick, totally saturated. Assume $P_0 = P_c$. How much additional effective pressure can the clay withstand if the ultimate expected settlement is not to exceed 12 cm?

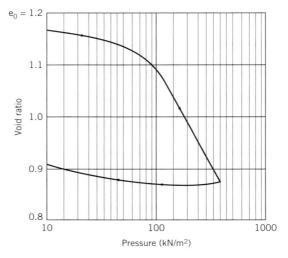

Figure P2.10 Consolidation test results.

The e–log p Test

2.61 (a) Repeat Problem 2.60 assuming $P_0 = 0.8P_c$.

   (b) For $P_0 = 0.8P_c$ determine how much additional effective pressure the clay can withstand if the expected settlement is not to exceed 24 cm. Is ΔP from part (b) twice that from (a)? Explain.

2.62 Refer to Fig. P2.11.

   (a) Determine the points of 0 and 100% consolidation.

   (b) Determine the percent consolidation scale from the results of part (a).

2.63 Determine the coefficient of consolidation in Fig. P2.11, assuming that the soil sample connected with this graph was 2.7 cm thick, tested with a top and a bottom porous stone.

2.64 Determine the time for 50% consolidation of the sample in Problem 2.63 for various thicknesses: (a) 1 m, (b) 3 m, (c) 8 m. State any assumptions.

2.65 (a) Repeat Problem 2.64 for 30, 40, 60, and 70% consolidation.

   (b) Draw a graph of percent consolidation versus time for a thickness of 3 m. Does the shape resemble that of Fig. 2.33? Should it? Explain.

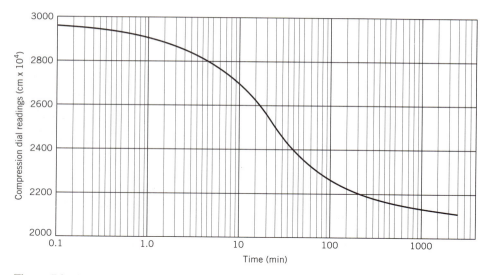

Figure P2.11 Time-deformation relationship.

BIBLIOGRAPHY

1. Atterberg, A., "Über die Physikalishe Bodenuntersuchung und über die Plastizität der Tone," *Int. Mitt. Boden.*, vol. 1, 1911.

2. Boussinesq, J., *Application des Potentiels à l'Etude de l'Equilibre et du Mouvement des Solides Elastiques*, Gauthier-Villars, Paris, 1885.

3. Casagrande, A., "The Determination of the Preconsolidation Load and Its Practical Significance," *1st Int. Conf. Soil Mech. Found. Eng.*, Cambridge, Mass., 1936.

4. Casagrande, A., "Research on the Atterberg Limits of Soils," *Public Roads*, Oct. 1932.

5. Casagrande, A., "Classification and Identification of Soils," *ASCE J. Geotech. Eng. Div.*, June 1947.

6. Casagrande, A., "Effect of Preconsolidation on Settlements," *ASCE J. Geotech. Eng. Div.*, vol. 90, Sept. 1964.

7. Cedegren, H. R., *Seepage, Drainage, and Flow Nets*, Wiley, New York, 1977.

8. D'Appolonia, D. J., E. E. D'Appolonia, and R. F. Brissette, "Settlement of Spread Footings on Sand," *ASCE J. Geotech. Eng. Div.*, vol. 94, May 1968.

9. Feld, J., "Tolerance of Structures to Settlement," *ASCE J. Soil Mech. Found. Eng.*, vol. 91, no. SM3, 1965.

10. Foster, C. R., and R. G. Ahlvin, "Stresses and Deflections Induced by a Uniform Circular Load," *Proc. Highw. Res. Board*, vol. 33, 1954.

11. Grant, R., J. T. Christian, and E. H. Vanmarcke, "Differential Settlement of Buildings," *ASCE J. Geotech. Eng. Div.*, vol. 100, Sept. 1974.

12. Hvorslev, M. J., "Uber die Festigkeitseigenschaften gestörter Bindiger Böden" (On the strength properties of remolded cohesive soils), Danmarks Naturvidenskabelige, Samfund, Copenhagen, 1937.

13. Lambe, T. W., "Stress Path Method," *ASCE J. Soil Mech. Eng. Div.*, vol. 93, 1967.

14. Newmark, N. M., "Influence Charts for Computation of Stresses in Elastic Foundations," *University of Illinois Bull.*, no. 338, 1942.

15. Newmark, N. M., "Influence Charts for Computation of Vertical Displacements in Elastic Foundations," *University of Illinois Eng. Exp. Stn. Bull.*, no. 367, 1947.

16. Pyke, R., H. B. Seed, and C. K. Chan, "Settlement of Sands under Multi-Directional Shaking," *ASCE J. Geotech. Eng. Div.,* vol. 101, Apr. 1975.

17. Scheidegger, A. E., *The Physics of Flow Through Porous Media,* 3rd ed., University of Toronto Press, Toronto, Canada, 1974.

18. Schmertmann, J. H., "The Undisturbed Consolidation Behavior of Clay," *Trans. ASCE,* vol. 120, 1955.

19. Skempton, A. W., R. B. Peck, and D. H. McDonald, "Settlement Analyses of Six Structures in Chicago and London," *Proc. Inst. Civ. Eng.,* July 1955.

20. Skempton, A. W., "The Colloidal Activity of Clays," *Proc. 3rd Int. Conf. Soil Mech. Found. Eng.,* vol. 1, 1953.

21. Taylor, D. W., *Fundamentals of Soil Mechanics,* Wiley, New York, 1948.

22. Terzaghi, K., and R. B. Peck, *Soil Mechanics in Engineering Practice,* 2nd ed., Wiley, New York, 1967.

23. Terzaghi, K., "Principles of Soil Mechanics II—Compressive Strength of Clay," *Eng. News Rec.,* vol. 94, 1925.

24. Westergaard, H. M., "A Problem of Elasticity Suggested by a Problem in Soil Mechanics: A Soft Material Reinforced by Numerous Strong Horizontal Sheets," *Mechanics of Solids,* S. Timoshenko, 60th Anniversary vol., Macmillan, New York, 1938.

C H A P T E R 3

Subsurface Exploration

3.1 INTRODUCTION

Perhaps all of us have heard it said that "one must understand the problem before one reaches a viable solution to that problem." So it is with geotechnical engineering problems: We need to know the characteristics of the formation on which the foundation rests before we design that foundation. Indeed, a well-planned subsurface investigation is a virtual prerequisite to a safe and economical design of the foundation components of a structure. This chapter introduces some of the methodologies and techniques employed in the more routine subsurface exploration efforts and field tests. Testing of soil is described in detail in many laboratory manuals and, thus, is outside the scope of this textbook.

Most subsurface exploration programs are coupled with viable laboratory testing; sometimes smaller projects forgo subsurface exploration programs, a practice not recommended. Usually the subsurface exploration program is tailored for the pertinent needs of a specific project and is frequently limited by budget restrictions. Generally, however, the overall investigation should be detailed enough to provide sufficient information for the geotechnical engineer to reach conclusions regarding the following phenomena:

Site Suitability

It goes without saying that not all locations are suitable building sites; that is, although one could generally say that all sites could be built upon if cost were not a factor, reality dictates that the feasibility of a site be judged in the context of economics as well as other criteria. Particularly in urban areas, good building sites may be scarce; many building sites in urban areas may be difficult and costly. Thus, the option of abandoning the site should always be reserved when confronted by unusually difficult and costly obstacles. Examples of problem sites may include: (1) sites over deep-coal mines; (2) particularly unstable formations (e.g., expansive shales or pyritic formations, highly compressible or highly expansive clays, landfill sites); or (3) unusual underground water problems that may pose costly foundation and construction problems.

Design Criteria

Most foundations will be classified as shallow or deep. Generally, shallow foundations include ordinary spread or isolated footings, combined footings, or mats. For such design, allowable bearing capacity (contact pressures) must be determined in order to determine the size and configuration and subsequent details of the foundation. Piles and caissons fall in the category of deep foundations. For the design of such components one needs information regarding friction factors (for friction piles) and end-bearing values for piles or caissons that derive their capacity from the end (tip) support media.

Construction Problems

Information regarding subsurface conditions is important to the geotechnical engineer for (1) formulating design parameters, (2) developing detailed plans and specifications, and (3) exposing and accounting for, as early as possible, potential construction problems. Knowing and accounting for such problems prior to bidding and construction greatly reduces the chances for construction delays, change-order costs, and general misunderstanding between the design professional, owner, and contractors. Typically, the unit cost for change-order work is significantly larger than work required under original specifications. Examples of problems associated with the foundation construction phase may include (1) large volume of water infiltration into the excavation and corresponding needs for a dewatering program (the effects of dewatering on an adjoining building are always a concern that has to be addressed by the geotechnical engineer); (2) rock excavation (typically, rock excavation is much more expensive than ordinary soil excavation, blasting of rock is not only expensive but potentially unacceptable if it is likely to adversely affect nearby structures or general environmental conditions such as water supply, noise, etc.); and (3) driving piles; this may not be an acceptable alternative to drilled, vibration-free piers owing to adverse effects driving vibrations may have on nearby structures, or the effects of noise on nearby inhabitants (e.g., hospitals or nursing home patients).

Environmental Impact

Society is becoming increasingly concerned about environmental impact that new construction will have on water quality, disposal techniques, and general hazards that may adversely affect humans, wildlife, and nature in general. For example, the construction of landfills is frequently perceived (sometimes correctly) by the public as a potential hazard to the water supply. Or the construction of an electricity-generating plant fueled by atomic power resting on a faulty strata (particularly if the site is subjected to possible earthquake effects) may be cause for alarm and subsequent opposition from the general public.

Generally a subsurface exploration provides a reasonable depiction of the stratigraphy (soil profile) and physical characteristics of the soil strata under the proposed building site; items typically included are

- *The nature of the underlying material* (soil or rock description, physical and index properties of the soil, etc.).
- *Stratification* (e.g., delineation of the formation including thickness, width, length, and stiffness of various formations).
- *Groundwater information* (e.g., water levels in various borings, potential seasonal fluctuation estimates, and contaminants, if any).
- *Pertinent bedrock information* (depth, stability, hardness, or the degree of fracture).
- *Samples* (for visual identification and laboratory testing).

In establishing the extent of the subsurface investigation, including the degree of sampling and laboratory analyses, a soil engineer usually takes into account the following:

1. The type, size, and function of the proposed structure
2. Soil (or rock) structure interaction
3. The effect the proposed structure or construction methodologies may have on adjoining structures or facilities

Generally, the soil engineer, in conjunction with the structural engineer, determines the spacing and depth of the borings, the type of exploration, the type of samples, and the frequency or degree or both of sampling, as well as the extent of laboratory testing.

For larger projects, it is generally advisable to run preliminary explorations prior to more detailed evaluations. Relatively speaking, a preliminary exploration is a rather inexpensive and expedient method for early or preliminary design estimates. It is also a rather sound basis on which to develop a more detailed exploration program. Depending on these early findings and specific project needs, the preliminary subsurface evaluations may be all that is needed for the project. However, most frequently a detailed exploration is necessary for a more reliable evaluation of the site condition.

3.2 SCOPE OF EXPLORATION PROGRAM

There are no hard and fast rules regarding the type and extent of the soil exploration program. For example, it may be apparent even to one not versed in geotechnical engineering details that the scope of exploration for a parking garage whose column loads may exceed 800 tons is likely to be significantly different from that for, say, an ordinary lightweight office building of one or two stories where the column loads may range between 10 and 30 tons. In all probability, the type of foundation for the garage column loads would be a deep foundation consisting of piles or caissons; for the lightweight office building ordinary spread footings are likely to be sufficient. Thus, the type and degree of testing is likely to focus on needs for a deep excavation in the case of high column loads versus the needs for a shallow foundation in the case of the office building example. The scope of the exploration program must be sufficient, however, to provide information of the nature and extent of the soil strata under a given site.

The exploration program usually encompasses conducting soil borings or test pits and extracting samples for visual and laboratory analysis. This is generally complemented, to various degrees, by a site reconnaissance and various field or in-situ tests. Although the scope of the program is usually formulated and established prior to the start of drilling and sampling, it is common practice (and advisable) to evaluate the findings as the work progresses and to make adjustments in the program as warranted. For example, the boring depth, the type and depth of samples, the number of borings, and so on may be adjusted as deemed necessary as the drilling progresses. On the other hand, this should be done under careful scrutiny, and in concert with other responsible parties (e.g., structural engineer, architect, owner), particularly if these changes reflect an increase in cost.

Depth of Exploration

In estimating the depth of exploration, a rule of thumb sometimes used for this purpose is to extend the borings to a depth where the additional load resulting from the proposed building is less than 10% of the average load of the structure, or less than 5% of the effective stress (described in Section 2.7) in the soil at that depth (1). The drilling should extend through any unsuitable material and into a stratum where the material of that formation is acceptably good to support the projected structural loads. Where piles are anticipated, the depth of exploration is to be at least to the anticipated pile tip; the author recommends that at least 2 or 3 borings be drilled to a depth of 20 to 30% beyond the anticipated tip of the pile. This generally not only provides information on some potentially relevant information but also facilitates potential changes in pile design (e.g., increase in pile length in exchange for a reduction in the number of piles). Figure 3.1 provides some guidelines in this regard. Generally, the cost for the geotechnical services is usually less than 1% of the total construction cost.

Artesian pressures and artesian flow may be a source of problems for low-lying areas or for deep excavations. This is sometimes the case in connection with sewage-treatment plants, since such sites are normally at low elevations, the water table is usually high (shallow depth), and some plant components are rather deep (e.g., pits, storage tanks). Dewatering and construction details (e.g., shoring, sheet-pile installations) and design considerations (e.g., hydrostatic pressures, uplift potential) demand fairly detailed information regarding the presence, depth, and magnitude of aquifers (water-bearing stratum) as well as pressure data and the type and character of the soil below the excavation. Heave, or boiling (see Eq. 2-17), is closely tied to these conditions. Again, as a rule of thumb, the depth of exploration is usually a minimum of 1.5 times the depth of excavation.

Rock exploration may be necessary for any number of possible reasons: the foundation rests directly on rock; excavation of the rock is necessary during construction (i.e., the cost for rock excavation may be many times that for ordinary soil); the rock formation may be subject to weathering effects (e.g., disintegration, expansion) during construction; variations in the rock stratum (e.g., flaws or fractures, variation

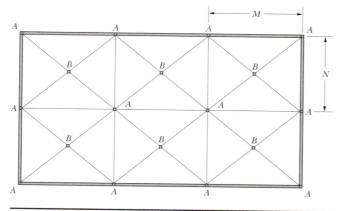

Column or Wall Loads	Depth $\lvert m(ft)\rvert$		Spacing $\lvert m(ft)\rvert$	
	Borings A	Borings B	M	N
Light	3–6 (10–20)	6–7.5 (20–25)	30 (100)	30 (100)
Medium	6–7.5 (20–25)	9–12 (30–40)	25–30 (80–100)	25–30 (80–100)
Heavy	9–12 (30–40)	15–25 (50–80)	15–25 (50–80)	15–25 (50–80)

Figure 3.1 General guidelines for boring layout and sampling used by the author.

in characteristics and formations, elevations). Rock sampling is commonly done via a core drill (discussed later in this chapter) for a minimum depth of 3 m. The depth may vary, however, and greater depths are common if much variation is detected in the rock formation within the site during the coring operations, or if voids (e.g., coal-mine voids, limestone sinkholes) may be present. (A giant limestone sinkhole about 200 m in diameter and 50 m deep swallowed homes, vehicles, etc., in Winter Park, Florida, in May 1981.) In such cases the strength and soundness of the rock strata above voids may be of paramount importance in the assessment of the ability of the rock formation to bridge over these voids and support the superstructure.

Spacing of Borings

It is normally difficult to determine the spacing and number of borings prior to the commencement of the drilling work because so much of such planning is tied to the underlying soil conditions, which are usually unknown at the time of planning. Hence, it is common practice to proceed with a rather skimpy preliminary investigation, and then follow up with a more structured and better planned effort.

The preliminary borings usually lack detailed sampling. Instead, drill cuttings or disturbed samples (see Section 3.5) and water table information usually suffice. The follow-up borings are planned as complements to the preliminary ones, but include specific requirements regarding the type, method, depth, and degree (or frequency) of sampling. For relatively light structures, the preliminary phase may be sufficient if the soil strata are good and appear uniform throughout the site. Frequently a comparison of the design data and performance history of an adjoining structure, if one is available, become useful tools for establishing the need for a more detailed investigation effort. On the other hand, for heavy or important structures, and for cases where the stratification information is doubtful or inconclusive, there is little choice but to proceed with the more comprehensive survey. This should be done with due care to extract only samples and perform only tests as required. For example, it is fruitless, if not irresponsible, to extract samples and perform tests that are not relevant to the design needs, or to perform tests if it becomes convincingly clear that the site is totally unsuitable for the intended purpose.

As is apparent by now, there are many factors related to the formulation of such a program. However, after a careful weighing of all the factors, the final decision on the part of the soil engineer is an empirical one, usually based on judgment and experience. Figure 3.1 shows a grid pattern for a typical boring layout used by the author on numerous projects. This particular layout assumes that the shape of the building is a rectangle, but the basic pattern could be improvised for nonrectangular buildings by merely adding grid blocks of M by N dimensions, as required by the shape of the building. This arrangement is convenient for plotting the stratification or profile of a series of borings in various directions (vertically, horizontally, diagonally). It facilitates evaluating the soil profile from a number of angles with a minimum of distortion of data. Again, it is merely a guide, intended to give the student an idea of some commonly used values for boring depth, spacing, and sampling for some general groups of structures. It depicts a wide range of values, provides only general ranges, and is very approximate. In this regard, the importance of judgment cannot be overemphasized in the overall program. Seldom are two sites identical, and even more seldom are two programs found to be completely identical in scope or needs.

3.3 SITE RECONNAISSANCE

Preliminary exploration is perhaps the most substantive basis for formulating the type and extent of a more detailed exploration and testing program. In fact, such "early" information may serve as the deciding factor for the feasibility of the site, or its rejection. For example, a preliminary subsurface exploration program conducted by the author, consisting of ten borings spread over several acres, revealed coal mines (voids) under the building site for a proposed new high school in a western Pennsylvania community. After some deliberation, and after reflecting on the preliminary findings, the local school board authorized a more detailed investigation and directed the design of a foundation as required to place the school over these mines. Obviously, in the judgment of the school board, the location of the school was important enough to warrant the additional expenditures not only for the much more detailed surface exploration, but also for the anticipated increase in design and foundation costs.

Sometimes the preliminary subsurface investigation may serve as the catalyst for a change in building design. For example, the author recalls a preliminary investigation that revealed a 2-m-thick layer of undesirable fill. The original design called for a basementless construction. However, after reviewing the cost of removing the undesirable fill and replacing it with controlled (compacted) fill, it was decided to incorporate a basement in the final design.

While indeed useful, a preliminary exploration should be regarded only as such. It is a useful intermediate step and one that is recommended whenever the time schedule permits. It is not intended to replace a more detailed and thorough evaluation of the site.

3.4 SUBSURFACE EXPLORATION

Within the efforts of evaluating a given site, one may reflect on one or any combination of the following considerations:

1. Environmental data
2. Relevant information on the behavior of adjoining structures
3. Electrical or geophysical testing
4. Soil drilling and sampling

In addition, the study is generally extended to include laboratory testing and evaluation and, perhaps, additional in-situ field tests (51).

Environmental Considerations

Geology The study of the geology of an area may provide useful information on the following points:

1. The general soil profile (e.g., groundwater conditions, flooding, erosion, metastable soil formation)
2. The state of the mass-rock formations (e.g., fractures or faults, formations, voids)
3. Areas of seismic activity

Sources for geological information may include the U.S. Geological Survey publications, state geological surveys of the Bureau of Mines, the U.S. Department of Agriculture, state highway departments, geological departments of universities, local well drillers, mining companies, and local libraries.

Seismic Zones Zones of potential seismicity should be identified in terms of both occurrence and intensity of seismic activity. This information is necessary if structures are to be properly designed and protected against potential damage from such occurrences, and the loss of life therefrom minimized.

Figure 3.2 depicts the probable seismic intensities, as a percent of **g**, for various zones in the United States. For example the likely bedrock intensity around the North Central region (e.g., Wisconsin, Minnesota, North Dakota) and some of the deep Southern region (e.g., Texas, Louisiana, Mississippi, Florida) is virtually nil, while in portions of California the intensity may be 0.6 **g**.

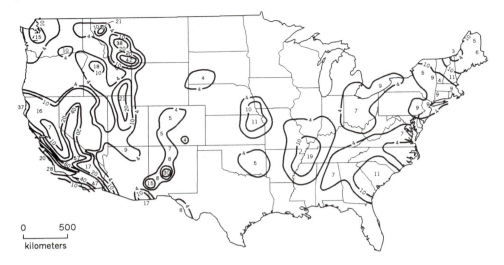

Figure 3.2 National probabilistic map of peak bedrock acceleration. The values of peak bedrock acceleration, in percent of **g**, have a 10% probability of being exceeded in a 50-year period. This map is the basis for all current building codes in the United States. (Courtesy United States Department of the Interior.)

Behavior of Adjoining Structures

Whenever available, information regarding the design criteria (e.g., type of foundations, design parameters), the behavioral history (e.g., settlement, problems during construction or later), and general soil information (e.g., available boring logs, conversations with persons acquainted with the general site or specific soil conditions) becomes an invaluable basis for formulating plans related to the new construction.

Underground facilities such as tunnels, pipes, or cables must be located and not interfered with either during the drilling and exploration or from the imposed building loads. Similarly, such installations or other buildings must be viewed in terms of the effects and changes in soil stresses and water conditions imposed by the new structure either during construction or during its life, or both. For example, some common questions may be: What is the effect of the additional building load on the settlement of the existing structure? What may be the effects on the existing structure from excavations, dewatering, pile driving, etc? What is the effect of the construction work or new building loads on an existing underground water main, sewer, or tunnel? In this regard, it is advisable that a careful inventory be taken of the conditions prior to construction of adjoining "structures" and a methodical assessment made of the potential effects the new conditions might impose on these structures. A final comparison should be undertaken to evaluate any subsequent changes of conditions and to ascertain potential damages, if any. Many lawsuits are instituted in this area of activity.

Geophysical Testing

Information regarding the topography of underlying rock and the water table is sometimes obtained by an indirect method known as *seismic refraction*. Briefly, the method consists of measuring the velocity of the compression wave that propagates through the soil. The waves are usually induced by an explosion (say, dynamite cartridge) at the ground surface and are picked up by detectors located at different known distances from the point of explosion. The velocity may vary (1) in soils from 150 to 2500 m/s, (2) in water, 1440 m/s, and (3) in rock from 1800 to 8000 m/s. Hence, some estimates of the stratification, the degree of weathering and hardness, and the depth of the hard rock may be made prior to a more reliable testing program (such as borings).

The *electric resistivity* method measures the change in the electric resistivity of the soil via electrodes placed at the surface. For example, hard and dense rocks have high resistivity, whereas softer rocks and soils are less resistant. Application of the method might also include the detection of cavities within the rock strata (e.g., mine voids, sinkholes, faults). It might further be a means for establishing the extent of a formation by comparative evaluations with quantities already established (say, from boring information), or to obtain groundwater information such as water tables and aquifers.

3.5 BORINGS

The drilling of holes, commonly referred to as *test borings*, and subsequent soil sampling at varying depths is one of the most widely used methods for subsurface investigations. The "drill" is usually advanced into the soil vertically, but may be inclined if one suspects faulty formations, say vertical rock fractures or joints, which might be missed by vertical drilling. The soil samples are taken at specified intervals or at changes of strata.

In nonrock formations (or in soft rock) the drilling is advanced by either augering or chopping the soil with a bit attached to a hollow drill rod through which water (sometimes mud, usually a bentonite clay mud) is injected into the hole under pressure and transports the loosened soil out. This is referred to as a *wash boring*. The use of water should be avoided in soils such as loess (wind-deposited soil), whose properties may be altered by water. A steel casing, pushed or pounded downward with the advancing depth of the boring, is normally used to prevent a potential collapse, particularly in cohesionless soils, of the hole during this operation. Samples may be obtained from the soil transported out, or via a sampling spoon or a thin-wall tube that is pushed or driven to the stratum at the bottom of the hole. (The latter two sampling devices are discussed in more detail below.)

Virtually all present-day augering is done by mechanically powered drills. The typical drilling rig, as shown in Figure 3.3, generally consists of a power unit, usually mounted on a truck, and is equipped with various types of augers, bits, core drills, hydraulic tools, and paraphernalia (e.g., pumps, reservoirs, hoses). Other accessories that may be needed for drilling, sampling, preparation and preservation of samples, and recording of information include split-spoon samplers, thin-wall tubes, sealing wax, sampling containers, and log sheets. Power units are sometimes mounted on skids to facilitate the drilling and sampling in places such as steep hills or inside low-clearance buildings, which are rather inaccessible to truck-mounted types.

Figure 3.3 (a) Drilling rig with mast in a down position—the typical position during transit. (b) The drilling rig positioned for drilling; note that the hydraulic jack lifted the front end in order to level the truck to a reasonable degree. (Courtesy of McKay & Gould Drilling, Inc., Darlington, Pennsylvania.)

Augers consist of flutes welded to a solid or hollow (pipe) stem. Their tips are usually sloped in order to facilitate the downward advancement of the auger. They are normally less than 20 cm (8 in.) in diameter for routine drilling, but sizes large enough for a person to enter and observe the formations within a drilled hole may be employed for special needs. Samples may be obtained from the flutes as the augers are extracted, or from the bottom of the hole. For samples from the bottom of the holes, the auger must be extracted in the case of solid-stem augers. Samples may be obtained through the stem without extracting the augers in the case of hollow-stem types.

Augering through soils that contain large cobbles, boulders, or rock may be quite difficult. In that case, the drilling is usually done by fracturing these boulders or rocks via percussion drills or by means of cable-tool drilling. In the latter case a steel rod, which is equipped with jagged teeth and weighs several hundred kilograms, is raised and dropped repeatedly by steel cables, thereby crushing the boulders or rock. Water and/or mud may be injected into the hole to facilitate the removal of the cuttings and, subsequently, extracted via a bailer at various intervals, as required.

Disturbed Samples

Disturbed samples encompass a wide array of soil specimens. Included are *wash-boring samples,* which are transported out by water and subsequently deposited in a tub or other container, sometimes on the ground. As mentioned previously, these samples have relatively limited value and are seldom kept for any laboratory analysis. The process obviously permits mixing of the various strata in the boring and, therefore, one may be provided with information of an average or general nature and texture of the total deposit. Continuous *auger boring samples* are more valuable. They provide reasonable data for stratum delineation and sample identification with depth, some information regarding moisture or the water table, and so on, but they are greatly lacking in information regarding the in-place characteristics of the soil such as stiffness, density, shear strength, or compressibility. Of those falling in the category of disturbed samples, the *split-spoon sample* is by far the most reliable. These samples are obtained by driving a steel tube into the undisturbed stratum and extracting the sample. The sampler is shown in Fig. 3.4. Briefly, the assembly consists of a short tube with a cutting edge (cutting shoe) on one end and threads on the other. A split tube threads to the shoe and to a head assembly, which is attached to the drilling rod, as detailed in Fig. 3.5. When unscrewed from the shoe and head assembly, the split barrel can actually be opened into two equal

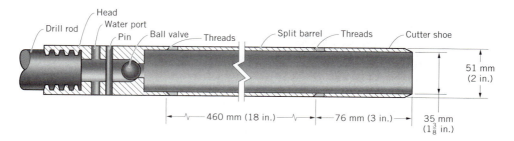

Figure 3.4 Split-barrel (split-spoon) sampler.

Figure 3.5 (a) Split-spoon sampler; on left, assembled and ready to be screwed on to extension rod; opened into two halves in the center; cutting head on lower right-hand corner and rod splicing unit at upper point. (b) Cutting head being screwed on to the two sections. (c) A "retention basket" sometimes used inside the split-spoon sampler to keep very loose soil from sliding out during the extraction process. (Courtesy of Lininger Drilling & Pumps, Inc., Greenville, Pennsylvania.)

94

segments for visual inspection of the sample or for removing part of the sample for preservation or future analysis.

The split-spoon is obtained by driving the sampler a total of 450 mm (18 in.) with a 64-kg (140-lb) hammer falling for a distance of 760 mm (30 in.). A record is made of the number of blows required to drive each 150-mm (6-in.) segment. The number of blows required to drive the sampler for the last 300 mm (12 in.) is an indication of the relative density of the material and is generally referred to as the *standard penetration test* (SPT). A more detailed description of such sampling and the relevant factors is given by ASTM D-1586-67 and other sources (1, 12, 18, 23, 25, 41, 44, 49, 60, 63).

Split-spoon samples are generally taken at every change of soil stratum or at specified intervals of depth, usually every 1.5 m (approx. 5 ft) or at every change of stratum detected by the driller. The samples are preserved in properly labeled sample bottles (e.g., project, date, boring number, sample depth). Occasionally, some of these samples are coated with a paraffin wax for the purpose of preserving the moisture in the material as well as the sample shape, however deformed it may be.

The blow count from the standard penetration test is frequently used as a measure of the relative density of sand or of the stiffness of the stratum in which the split-spoon sample is taken. The method is not recommended to measure comparable characteristics of formations that contain gravels, particularly large sizes, or for cohesive soils.

Undisturbed Samples

An undisturbed sample is a somewhat more expensive and more time-consuming sample, but it is considerably more valuable (31, 54, 56, 70, 75). As might be expected, the grain structure of particle arrangement of many soils may be sensitive to disturbance. Hence, if the samples have experienced considerable disturbance during extraction, the test resuls and the predictions based on such results may be seriously in error. Yet total duplication of in-situ conditions is virtually impossible. That is, the mere extraction of a sample changes the pressure of the sample from in-place conditions to atmospheric, for instance (33, 34, 39, 46, 48, 65). Quite obviously, therefore, due care should be exercised to obtain samples least disturbed and to account for whatever disturbances have occurred during the interpretation of the test results. A thin-walled tube, sometimes referred to as a *Shelby tube* (Shelby Tube Company was among the first to manufacture thin-wall tubing) is one of the most widely used devices for in-situ or "undisturbed" soil sampling.

Figure 3.6 shows a typical Shelby-tube sampler. The tube has a thin wall, about 1.6 mm thick, and anywhere from 50 to 100 mm in diameter. It is recommended, however, that the diameter be 76 mm or larger. In order to minimize friction between soil and tube, the cutting tip of the sample is slightly beveled inward (about 0.5 mm), as shown in Fig. 3.6. Furthermore, it is recommended that the ratio of the peripheral cross section (encompassed by the wall thickness and beveling) to that of the soil not exceed 10%. For example, if we assume the diameter of the specimen to be 73 mm (2.875 in.) and a beveled thickness (wall thickness plus inward protrusion) of 1.8 mm, the ratio is $(74.8)\pi \times 1.8 \times 4/\pi(73)^2 = 0.10$ (or 10%).

The samples are usually obtained by pushing, smoothly and continuously, the

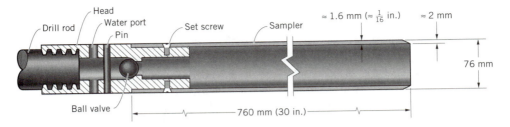

Figure 3.6 Thin-wall Shelby-tube sampler.

Shelby tube into the soil. (Some engineers claim less disturbance by driving the sample into some of the more plastic soils.) Once extracted, the sample is left in the tube, with both ends of the tube waxed to prevent moisture escape. The tube with the sample inside is taken to the laboratory, where it is cut and the sample is extracted and tested. The procedure for performing the various tests may be found in the ASTM standards or in most laboratory manuals dealing with the subject.

Core Boring in Rock

For most rock drilling and sampling, the same drilling rig could be used as for soil sampling. On the other hand, most of the sampling tools and implements used for drilling in the soil are not normally adequate for rock sampling. For example, a split-spoon sampler or auger bit may penetrate some soft rock, but usually these tools are limited to a relatively negligible portion of the upper segment of the massive rock formation. Coring the rock layer is perhaps the most reliable method of sampling. The typical tool used for rock coring is the *core barrel*, shown in Fig. 3.7b. In essence, it is a hardened steel tubing, 5 to 10 cm in diameter and 60 to 300 cm long (2 to 4 in. in diameter and 2 to 10 ft long), equipped with a cutting bit, which contains tungsten carbide or commercial diamonds at its cutting end. During sampling, the bit and core barrel rotate, while a steady stream of water or air is pushed down through the hollow rods and barrel into the bit. The water and air serve as coolants and as transporting agents in the process of bringing the cuttings up to the surface. A number of such tools are available commercially, and a more detailed description of their characteristics is deemed unnecessary at this time.

The typical rock sample consists of a cylindrical core cut from the rock formation. Through careful analysis, one may extract much useful and relevant information about the rock. Figure 3.8 depicts the typical manner of workers during the removal and logging of core samples.

1. Type (e.g., shale, sandstone, limestone)
2. Texture (e.g., fine or coarse grain, mixtures)
3. Compressive strength (e.g., compression test results)
4. Orientation of formation (e.g., bedding planes, vertical, horizontal variations)
5. Degree of stratification (e.g., laminations)
6. Soundness (e.g., weathering, fissures, faults, degree of fracture)
7. Miscellaneous (clay seams, coal formations, stability)

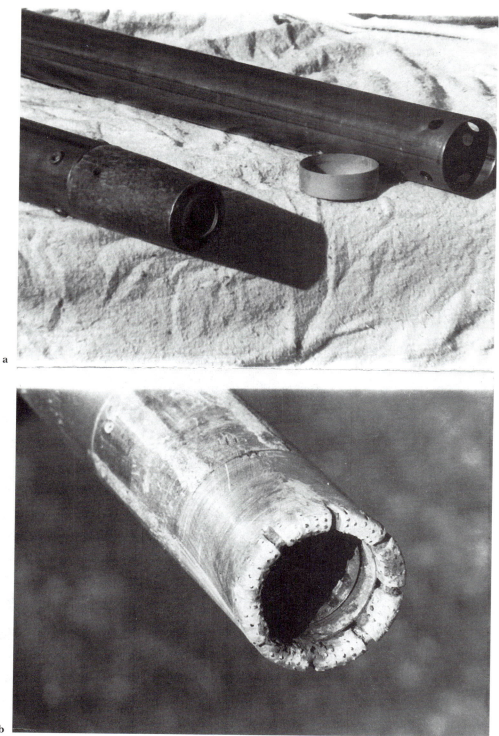

Figure 3.7 (a) A thin-wall, Shelby tube sampler; lower photo shows the tube secured to rod splicer via set screws; upper photo is a view of the upper end of the tube that connects to rod splicer; next to the tube is a plastic cap, which is typically placed over each end after extraction in order to contain the soil in the tube during transport to lab. Along with wax, which is melted in the field on each end immediately after extraction, the plastic cap may also prevent moisture evaporation. (b) Diamond-studded cutting tip of rock coring barrel used to obtain rock core samples. (Courtesy of Lininger Drilling & Pumps, Inc., Greenville, Pennsylvania.)

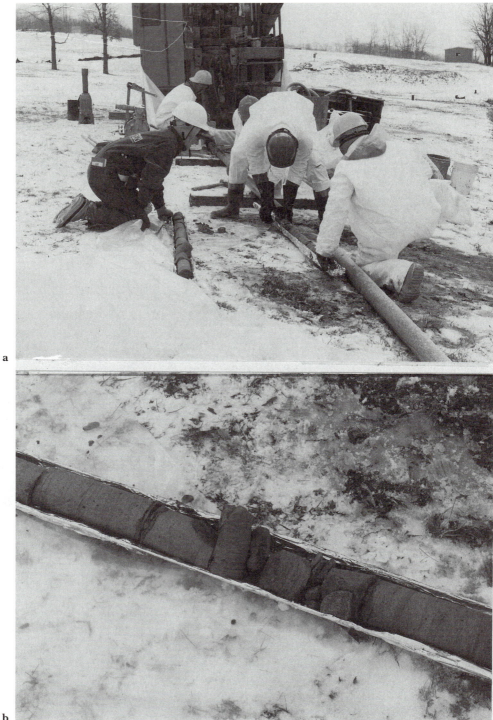

a

b

Figure 3.8 (a) Workers removing rock cores from core barrel. (b) Core sample is preserved in order of extraction (i.e., not mixed); the rock quality designation (RQD) is the percentage of rock core samples that are 4 in. or longer in a typical 5-ft core length. (Courtesy of McKay & Gould Drilling, Inc., Darlington, Pennsylvania.)

By noting the ratio of the length of core obtained to the distance drilled, commonly referred to as *core recovery*, one is able to formulate an opinion regarding the consistency and soundness of the rock. For example, a shale stratum laminated with a number of clay seams will generally result in a low percentage recovery since much of the clay may be "washed away" by the water or air during the drilling process. Similarly, a layer of broken sandrock may show a relatively small percentage recovery, with a percentage of the rock stratum being carried away by the water or air.

The small-diameter core samples are generally sufficient for the exploration of the rock strata for most jobs. Occasionally, larger diameter cores are taken for evaluating the rock strata not only from the extracted cores, but also from more detailed observations via actual visual inspection of the natural formation, perhaps through mirrors or personal inspection from within the hole. Such explorations and investigations are frequently coupled with coring for caisson installations.

Strengths as determined from compression tests of core samples are of rather limited value in predicting the strength of the in-situ rock. Strength and deformation characteristics of the in-situ rock are greatly affected by discontinuities (e.g., joints, bedding planes, fractures, weathered seams), which are too complex to account for during the strength evaluation from the rock core compression test.

3.6 TEST PITS

Another method of subsurface exploration is via an *open pit,* usually dug with a backhoe or power shovel. An ordinary construction backhoe with a reach of 3 to 4 m (10 to 13 ft) is usually adequate for this type of exploration. It is one of the most dependable and informative methods of investigation, since it permits a most detailed examination of the soil formation for the entire depth. For example, the stiffness of the strata (via penetrometers), the texture and grain size of the soil, detailed sampling, in-situ testing, moisture evaluations, and the like, are among some of the items of information that can be conveniently and reliably obtained from such explorations.

For deeper holes, the excavation is frequently shored to protect against collapse, so that an observer may be able to view the stratification from within the hole or take samples of the undisturbed soil from the walls or bottom of the hole. One may obtain relatively undisturbed samples by entering the pit itself. One sampling method is to press a thin-wall steel tube into the bottom or sides of the excavation and extract a portion of the stratum by means of the tube. Another sampling method consists of carving a relatively undisturbed soil sample from the sides or bottom of the pit. The sample is then waxed to preserve its moisture during the interim between sampling and laboratory testing.

The open-pit method of exploration has both advantages and disadvantages. Some of the positive features are

1. It provides a vivid picture of the stratification.
2. It is relatively fast and inexpensive.
3. It permits rather reliable in-place testing and sampling.

Some of the disadvantages associated with the open-pit method of exploration are

1. It is usually practical for only relatively shallow depths, generally 4 to 5 m.
2. A high water table may prohibit or at least limit the depth of excavation.
3. If shoring or extraordinary safety requirements are confronted during excavation, the cost of such exploration may be unacceptably high.
4. The backfilling of the holes, generally under controlled compaction conditions, may produce a series of nonuniform stratum characteristics over the site.

3.7 FIELD TESTS

The reliability of laboratory results to represent the in-situ soil properties remains a significant concern to the geotechnical engineer. The change in the environmental conditions (e.g., pressure, moisture) and the disturbance that the soil sample undergoes during extraction and subsequent handling and testing may greatly influence the test data. Furthermore, some valuable data may be obtained only through actual "field" testing. Several of the more commonly used in-situ tests are discussed in this section.

Penetrometer Tests

A number of penetrometer-type tests have been used by engineers to determine relative density, stiffness, strength, or bearing capacity of the soil strata. Figure 3.9 shows three common hand-operated penetrometers used for that purpose. In essence, they all relate the force or energy required to push or drive a probe a certain distance through a soil. A cone-type penetrometer has been developed in Holland and is widely used in Europe. The Standard Penetration Test (SPT) is more widely used in the United States, although the cone penetrometer has experienced increased use in the United States as well.

The details for performing the Standard Penetration Test (ASTM D-1586-67) have already been briefly mentioned in Section 3.5 and will therefore not be repeated here. The test is used as a measure of the relative density of sands and noncohesive soils, excluding cobbles, boulders, or very coarse gravels; it is not recommended for cohesive soils. The blow count increases with depth and with increasing coarseness of the cohesionless material. Figure 3.10 shows the relationship between the relative density of sand or gravelly sands, and the corresponding blow counts for various effective overburden pressures (or confining stresses).

Cone penetrometers come in a variety of designs (58). Basically they all have a cone-shaped tip, which is used to facilitate the downward advancement. Some are solid rods, others are a composite of rod and casing. Some are pushed down (static), whereas others are driven (dynamic) into the soil. Generally, they are limited in accuracy and should be used relative to and in conjunction with information from other methods and evaluations. The Begemann friction cone (45, 67) is a sophisticated, widely used tool, which appears relatively reliable for measuring the relative density and bearing capacity of strata in a continuous fashion.

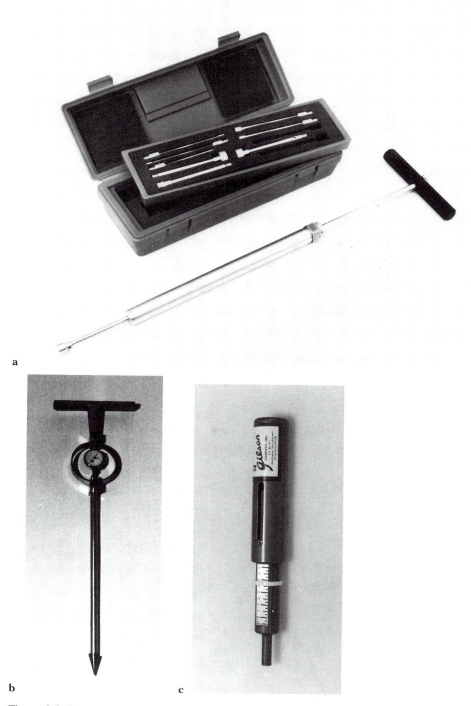

Figure 3.9 Penetrometers provide a measure of soil bearing/unconfined compressive strength. They are pushed into the soil to the designated mark. (a) The cylindrical-shaped head type has a sliding ring, which moves over a calibrated scale as load is applied. (b) The cone-shaped penetrometer measures the load via a calibrated dial gage–proving ring arrangement. (Courtesy of Soiltest, Inc., Lake Bluff, Illinois.) (c) Hand penetrometer consists of a cylindrical-shaped head, which is pushed into the ground to the designated mark of $\frac{1}{4}$ in. Load is indicated via the sliding ring, which slides over the calibrated scale as the load is applied. (Courtesy of Gilson Lab Equipment, Worthington, Ohio.)

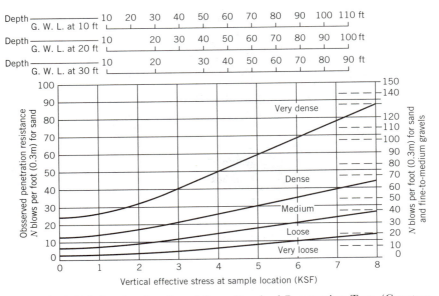

Figure 3.10 Relative density of sand from Standard Penetration Test. (Courtesy of ASCE.)

Vane Shear Tests

Vane shear tests (ASTM D-2573-72) are used with increasing acceptability to determine the shear strength of soils in situ. Although reportedly first used in Sweden almost 60 years ago, they did not gain significant acceptance until their extensive use and development from 1947 to 1949 by the Royal Swedish Geotechnical Institute. There are a number of different versions of the instrument, based on essentially the same principle. In essence, the instrument consists of a rod with radial vanes, as shown in Figs. 3.11 and 3.12a. Once pushed into the soil, the vane is rotated by an applied torque. The resistance to the applied torque T in Fig. 3.12a is provided by the shearing forces on the two ends of the vane and on the circumferential plane. Figure 3.12b shows a commonly used ratio of the vane. Figure 3.12c shows an assumed shear stress distribution on the ends of the vane, similar to that used in round bars subjected to torsion. Figure 3.12d shows a uniform shear stress around the vertical cylindrical surface of the rotating vane.

The torque for the dimension shown in Fig. 3.12b is resisted by T_1 and T_2 (Figs. 3.12c and 3.12d, respectively). If both ends of the vane are "submerged" in the soil stratum, and if the maximum shear stress is τ for all shear surfaces, then

$$T = 2T_1 + T_2 = 2\tau(\pi R^2)\left(\tfrac{2}{3}R\right) + \tau(2\pi R)(4R)R$$

or

$$T = \tfrac{28}{3}\tau\pi R^3$$

from which

$$\tau = \frac{3T}{28\pi R^3} \tag{3-1a}$$

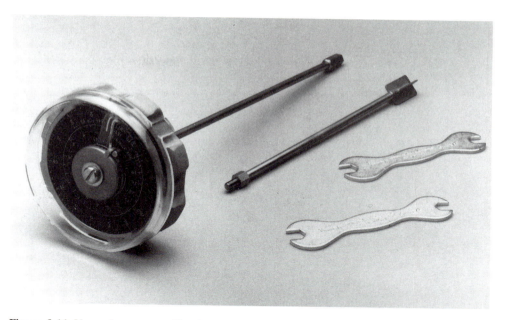

Figure 3.11 Vane shear tester. Heads are pushed into the ground. Torque is measured via a calibrated scale, as torque is applied to soil failure. (Courtesy of Soiltest, Inc., Lake Bluff, Illinois.)

If the vane penetration is only to the top of the vane, only one end plane develops shear resistance. Therefore,

$$T = \tau(\pi R^2)(\tfrac{2}{3}R) + \tau(2\pi R)(4R)R$$

or

$$\tau = \frac{3T}{26\pi R^3} \tag{3-1b}$$

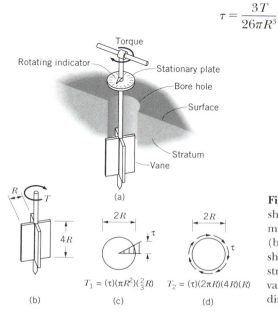

Figure 3.12 Basic features of a shear vane tester. (a) Schematic of shear vane apparatus. (b) Common dimensions of shear vane. (c) Assumed shear stress distribution at ends of vane. (d) Assumed shear stress distribution on vertical surface.

The vane shear test is adapted to clay soils, particularly to soft sensitive clays. Generally speaking, for such clays these tests give results that are somewhat superior to those obtained from unconfined compression or direct shear tests.

Load Tests

Load tests are sometimes run at the surface or on the bottom of an excavation in order to determine the behavior of the soil stratum under load, for example, shear strength or settlement. However, as desirable as it may be to have the test conditions simulate the "actual" loads and footer sizes, this approach is impractical. That is, the footer sizes, loads, duration of loading, and so on are normally not reasonable to duplicate. Instead, loads are generally transmitted through relatively rigid bearing plates, usually less than 1 m² in area, bearing on the soil stratum at the anticipated footer elevations. The resulting pressures from the plates to the soil may be anywhere from 100 to 300% of the expected pressures from the footing. Likewise, the time of the sustained pressures may vary from a few hours to perhaps several weeks. Deformation readings (rough approximations of settlement and/or shear strength) during the time of the sustained load are made at regular intervals. Also, a reading of the net settlement with the load removed is made after completion of the test. Load-settlement readings may be used to determine the *coefficient of subgrade reaction.* Figure 3.13 is a schematic of such a load test.

Observation Wells

Special installations are sometimes used for the purpose of determining the ground-water location and seasonal fluctuations. Although the level of the water table is generally recorded during the drilling of bore holes, the information may be misleading; the reading may represent water from capillary saturation, surface infiltration, or a perched water table. And, of course, such short-time observations do not indicate the seasonal fluctuations that the water table may undergo.

A commonly used approach for this purpose, serving for both granular and fine-grained strata, consists of using perforated plastic pipe. The perforated pipe is lowered into the bore hole, which is held open by a steel casing. The space between the perforated pipe and the steel casing is filled with a sand and gravel mix, and then the steel casing is pulled out. In order to minimize surface infiltration, the

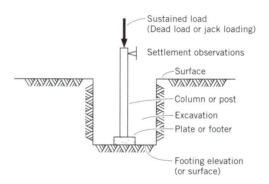

Figure 3.13 Schematic of a field load test.

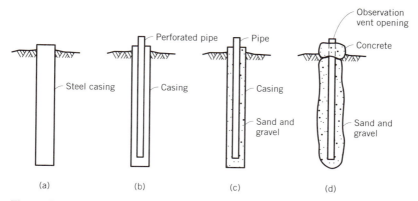

Figure 3.14 Schematic of a common, long-term observation well.

top of the bore hole is sometimes sealed with a concrete cap, with only a small opening left to serve as a vent. Figure 3.14 shows a schematic installation of such a well. Note that the purpose of the sand and gravel is to act as a filter; that is, it is easy to plug the perforations via fine-grained particle infiltration, thereby reducing or perhaps completely destroying the effectiveness of the well.

In-Situ Permeability Tests

A rough estimate of in-situ permeability can be obtained by pumping or bailing water out of a boring. Another way is to observe the rate of infiltration of water poured into a hole. (Percolation tests run in connection with septic systems employ shallow and small holes dug near the surface.) Another method entails observations of the time required for an identifiable liquid (e.g., dye) injected into the soil to reach a well at a given distance fom the point of injection. Still another method is described in Section 5.6. *The Earth Manual* of the U.S. Bureau of Reclamation provides procedures and sample calculations for this purpose.

Pile Load Tests

The purpose of the pile load test is to verify the capacity of the pile. Pile tests provide acceptably reliable information regarding load capacity, but are not generally reliable for long-term settlement data since the pile loads are normally sustained for only relatively short periods (e.g., 3 to 4 days). The procedure for performing the tests is described in Section 11.10, as well as in ASTM D-1143.

3.8 BORING LOGS

Data obtained from drilling the bore holes (or test pits or core samples) must be recorded accurately, completely, and at the time the data become available—not later. That is, as the drilling progresses and information regarding the strata becomes available, either through visual observations of the material brought up by the auger, air, or water, or from samples taken by the split-spoon or Shelby-tube samplers, the information is immediately recorded. Do not depend on memory to record this information at completion of drilling. Samples that are saved for future evaluations in the laboratory (Shelby-tube samples, split-spoon, or rock cores) are likewise properly labeled on the container in which they are preserved (a jar, a Shelby tube, or a core box). Simultaneously, that information is also recorded in the boring log.

Typically, a driller's boring log will have the following information:

1.   Name, address, and telephone number of the drilling company
2.   Name and address of the project

Driller's Name _____
Address _____
Phone Number _____

Name _____   Address _____   Location _____
Boring No. _____   Page No. _____ of _____
Type Boring _____   Date _____
Rig _____   Elevation _____
Casing Size _____   Driller _____

Standard Penetration Test
Correlation Chart
Sampler 51 mm OD; 35 mm ID
Hammer 64 kg; 760 mm fall

Soil–Clay	Blows	Sand/Silt	Blows
Very Soft	2	Loose	0-10
Soft	3-5	Medium	11-30
Medium	6-15	Dense	31-50
	16-25	Very Dense	Over 50
Hard	Over 25	Stiff	

Description
Soil Type, Water, Firmness, Drive Notes, Remarks

Sample Number	Sample Depth	Blows Each 150 mm	Casing Depth	I.D. of Sampler	Length of Sampler	Shelby	S.S.	Depth in Feet or Meters	
								0	X X X X X X X X X X X X X X X X
								1	
								2	
								3	
								4	
								5	
								6	
								7	
								8	
								9	
								10	
								11	
								12	
								13	
								14	
								15	
								16	
								17	
								18	
								19	
								20	
								21	
								22	
								23	
								24	
								25	
								26	
								27	
								28	
								29	
								30	

Water Level
Time
Date

Figure 3.15 Typical boring log form.

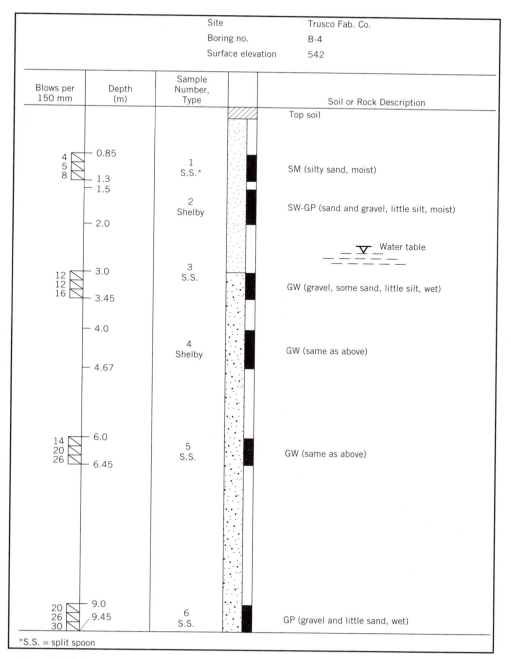

Figure 3.16 Typical boring log.

3. Type and number of boring
4. The date of drilling
5. The type of drilling rig and equipment
6. Ground elevation at the boring location
7. The driller's name

8. The water level at the time of drilling and at a given later time (frequently 24 hours)
9. The sample number
10. The sample depth
11. The resistance to penetration of the split-spoon sampler
12. The description of the sampler
13. A complete description of the strata, coupled with pertinent notes that the driller or field technician deemed appropriate

Figure 3.15 shows a typical boring log the driller used to record the above-mentioned data. The correlation chart in the upper right-hand corner of the figure is an empirical guide, for the sake of uniformity and classification of the strata from job to job. It is not a universally accepted standard, but it is widely accepted as a reasonable guide for general classification purposes in the field.

From the information provided from the field log and that obtained from laboratory samples, a new log is drawn, which facilitates the stratification interpretation. Figure 3.16 shows a typical log.

Problems

3.1 Name several sources for information and preliminary evaluation of a building site.

3.2 Explain briefly the difference between a *preliminary* exploration and a *detailed* exploration program; suggest relevant features for each of the two scopes.

3.3 Name several pieces of information that are obtained from a typical subsurface investigation.

3.4 Name several factors that are relevant to the planning for a well-balanced exploration program.

3.5 How does one go about planning the depth of the boring, the boring layout, and the type of samples?

3.6 What is a test pit? Give some of the negative and positive aspects of a test pit.

3.7 What is a test boring? How does it differ from a test pit?

3.8 What is a disturbed soil sample? What is an undisturbed soil sample? How is each obtained?

3.9 Describe the features of a Shelby-tube sampler. How does this differ from the split-spoon sampler?

3.10 What is a core sample? How is it obtained? What information can be obtained by evaluating this sample?

3.11 What is a vane shear test? Describe the apparatus.

3.12 What are boring logs?

3.13 Describe the basic construction and features of observation wells.

3.14 Can a split-spoon sampler penetrate a typical rock formation? Can a flight auger penetrate a rock formation? Explain.

3.15   Shelby tubes are usually pushed into the strata. However, some practitioners regard driving the tube an acceptable approach. How might the disturbance be affected by the two methods for each type listed below?

   (a) Very soft clay

   (b) Hard clay

   (c) Fine sandy silt

   (d) Sandy soil

3.16   What is drilling mud? What is bentonite?

BIBLIOGRAPHY

1.   American Society of Civil Engineers, "Subsurface Investigation for Design and Construction of Condition of Building," *Manual,* no. 56, 1976.

2.   Adam, J., discussion of V. F. B. deMello, "The Standard Penetration Test," *4th PanAm. Conf. Soil Mech. Found. Eng.,* vol. III, 1971.

3.   Alperstein, R., and S. A. Leifer, "Site Investigation with Static Cone Penetrometer," *ASCE J. Geotech. Eng. Div.,* vol. 102, May 1976.

4.   Begemann, H.K. S., "A New Method for Taking of Samples of Great Length," *5th Int. Conf. Soil Mech. Found. Eng.,* vol. 1, 1961.

5.   Begemann, H. K. S., "The New Apparatus for Taking a Continuous Soil Sample," *LGM Meded.,* vol. 10, no. 4, 1966.

6.   Begemann, H. K. S., "Soil Sampler for Taking Undisturbed Sample 66 mm in Diameter and with a Maximum Length of 17 m," *4th Asian Conf. Int. Soc. Soil Mech. Found. Eng.,* Specialty Session Quality in Soil Sampling, 1971.

7.   Begemann, H. K. S., "The Delft Continuous Soil Sampler," *Eng. Geol.,* no. 10, 1974.

8.   Berre, T., K. Schjetne, and S. Sollie, "Sampling Disturbance of Soft Marine Clays," *7th Int. Conf. Soil Mech. Found. Eng.,* Specialty Session 1, Mexico, 1969; also *Norw. Geotech. Inst. Publ.* no. 85.

9.   Bishop, A. W., "A New Sampling Tool for Use in Cohesionless Soils below the Water Level," *Geotechnique,* vol. 1, 1948.

10.   Bishop, A. W., D. L. Webb, and P. I. Lewin, "Undisturbed Samples of London Clay from the Ashford Common Shaft, Strength Effective Stress Relationships," *Geotechnique,* vol. 15, 1965.

11.   Broms, B. B., and A. Hallen, "Sampling of Sand and Moraine with the Swedish Foil Sampler," *4th Asian Conf. Int. Soc. Soil Mech. Found. Eng.,* Specialty Session Quality in Soil Sampling, 1971.

12.   Broms, B. B., "Soil Sampling in Europe, State-of-the-Art," *ASCE J. Geotech. Eng. Div.,* Jan. 1980.

13.   Brown, R. E., "Drill Rod Influence on Standard Penetration Test," *ASCE J. Geotech. Eng. Div.,* vol. 103, no. GTII, Proc. Paper 13313, Nov. 1977.

14.   Burghignoli, A., and G. Calabresi, "A Large Sampler for the Evaluation of Soft Clay Behavior," *9th Int. Conf. Soil Mech. Found. Eng.,* papers presented at Specialty Session 2, *Soil Sampling,* vol. 1, 1977.

15.   Cass, J. R., "Subsurface Explorations in Permafrost Areas," *ASCE J. Geotech. Eng. Div.,* vol. 85, Oct. 1959.

16.   Christian, J. T., and F. Swiger, "Statistics of Liquefaction and SPT Results," *ASCE J. Geotech. Eng. Div.,* vol. 101, no. GTII, Proc. Paper 11701, Nov. 1975.

17.   Cooling, L. F., discussion of "Site Investigation Including Boring and Other Methods of Sub-Surface Exploration," *J. Inst. Civ. Eng.,* vol. 32, 1949.

18. De Mello, V. F., "The Standard Penetration Test," *4th PanAm. Conf. Soil Mech. Found. Eng.*, San Juan, Puerto Rico (published by ASCE), vol. 1, 1971.

19. Drnevich, V. P., and K. R. Massarsch, "Effect of Sample Disturbance on Stress Strain Behavior of Cohesive Soils," presented at the 1978 ASCE Speciality Session on Soil Sampling and Its Importance in Dynamic Laboratory Testing, Chicago, Ill., 1978.

20. Ducker, A., "Method for Extraction of Undisturbed Frozen Cores," *7th Int. Conf. Soil Mech. Found. Eng.*, Specialty Session I, Soil Sampling, Mexico, 1969.

21. Durante, V. A., J. L. Kogan, V. I. Ferronsky, and S. I. Nosal, "Field Investigations of Soil Densities and Moisture Contents," *Int. Conf. Soil Mech. Found. Eng.*, London, England, vol. 1, 1957.

22. Fahlquist, F. E., "New Methods and Technique in Subsurface Explorations," *Contribution in Soil Mechanics 1941–1953*, Boston Soc. Civ. Eng., 1941.

23. Fletcher, G. F., "Standard Penetration Test, Its Uses and Abuses," *ASCE J. Soil Mech. Found. Eng. Div.*, vol. 91, 1965.

24. Friis, J., "Sand Sampling," *5th Int. Conf. Soil Mech. Found. Eng.*, vol. 1, 1961.

25. Gibbs, H. J., and W. G. Holtz, "Research on Determining the Density of Sands by Spoon Penetration Testing," *4th Int. Conf. Soil Mech. Found. Eng.*, London, England, vol. 1, 1957.

26. Helenelund, K. V., and C. Sundman, "Influence of Sampling Disturbances of the Engineering Properties of Peat Samples," *4th Int. Peat Congr.*, Helsinki, Finland, vol. 2, 1972.

27. Holm, G., and R. D. Holtz, "A Study of Large Diameter Piston Samplers," *9th Int. Conf. Soil Mech. Found. Eng.*, papers presented at Specialty Session 2, *Soil Sampling*, vol. 1, 1977.

28. Hryeiw, R. C., S. Vitton, and T. G. Thomann, "Liquefaction and Flow Failure During Seismic Exploration," *ASCE J. Geotech. Eng. Div.*, vol. 116, Dec. 1990.

29. Hvorslev, M. J., "Subsurface Exploration and Sampling of Soils for Civil Engineering Purposes," Rep. Comm. Sampling and Testing, ASCE Soil Mech. Found. Div., 1948.

30. Hvorslev, M. J., "Subsurface Exploration and Sampling of Soils for Civil Engineering Purposes," Rep. Res. Project ASCE, U.S. Army Eng. Exp. Stn., Vicksburg, 1949.

31. Idel, K. H., H. Muhs, and P. von Soos, "Proposal for Quality-Classes in Soil Sampling in Relation to Boring Methods and Sampling Equipment," *7th Int. Conf. Soil Mech. Found. Eng.*, Specialty Session I, Soil Sampling, Mexico, 1969.

32. Jakobson, B., "Influence of Sampler Type and Testing Method on Shear Strength of Clay Samples," *Swed. Geotech. Inst.*, no. 8, 1954.

33. Kallstenius, T., and W. Kjellman, "A Method of Extracting Long Continuous Cores of Undisturbed Soil," *2nd Int. Conf. Soil Mech. Found. Eng.*, vol. 1, 1948.

34. Kallstenius, T., "Mechanical Disturbances in Clay Samples Taken with Piston Samplers," *Swed. Geotech. Inst.*, no. 16, 1958.

35. Kallstenius, T., "A Standard Piston Sampler Prototype," *Swed. Geotech. Inst.*, no. 19, 1961.

36. Kallstenius, T., "The Soil Mechanics Aspects of Soil Sampling in Organic Soils," *7th Int. Conf. Soil Mech. Found. Eng.*, Specialty Session I, Soil Sampling, Mexico, 1969.

37. Kezdi, A., I. Kabai, E. Biezok, and L. Marczal, "Sampling Cohesive Soils," *Period. Polytech. Civ. Eng.*, Budapest, Hungary, vol. 18, no. 4, 1974.

38. Kirkpatrick, W. M., and I. A. Rennie, "Stress Relief Effects in Deep Sampling Operations," *Underwater Constr. Conf.*, University College, Cardiff, Wales, Apr. 1975.

39. Kjellman, W., "A Method of Extracting Long Continuous Cores of Undisturbed Soil," *2nd Int. Conf. Soil Mech. Found. Eng.*, vol. 1, 1948.

40. Kjellman, W., T. Kallstenius, and O. Wager, "Soil Sampler with Metal Foils," *Swed. Geotech. Inst.*, no. 1, 1950.

41. Kovacs, W. D., et al., "A Comparative Investigation of the Mobile Drilling Company's Safe-T-Driver with the Standard Cathead with Manila Rope for the Performance of the Standard Penetration Test," School of Civ. Eng., Purdue University, Lafayette, Ind., 1975.

42.   Kovacs, W. D., and L. A. Salomone, "SPT Hammer Energy Measurements," *ASCE J. Geotech. Eng. Div.,* vol. 108, Apr. 1982.

43.   Kovacs, W. D., J. C. Evans, and A. H. Griffith, "Towards a More Standardized SPT," *9th Int. Conf. Soil Mech. Found. Eng.,* vol. 2, 1977.

44.   Kovacs, W. D., "Velocity Measurements of Free-Fall SPT Hammer," *ASCE J. Geotech. Eng. Div.,* vol. 105, Jan. 1979.

45.   Landva, A., "Equipment for Cutting and Mounting Undisturbed Specimens of Clay in Testing Devices," *Norw. Geotech. Inst. Publ.,* no. 56, 1964.

46.   La Rochelle, P., and G. Lefebvre, "Sampling Disturbance in Champlain Clays," *ASTM Spec. Tech. Publ.* no. 483, Am. Soc. Test Mater., Philadelphia, Pa., 1970.

47.   Lundstrom, R., "Influence of Sample Diameter in Consolidation Tests," *Swed. Geotech. Inst.,* no. 19, 1961.

48.   McGown, A., L. Barden, S. H. Lee, and P. Wilby, "Sample Disturbance in Soft Alluvial Clyde Estuary Clay," *Can. Geotech. J.,* vol. 11, no. 4, 1974.

49.   McLean, F. G., A. G. Franklin, and T. K. Dahlstrand, "Influence of Mechanical Variables on the SPT," *Specialty Conf. In-Situ Meas. Soil Prop.,* ASCE, vol. 1, 1975.

50.   Mohr, H. A., "Exploration of Soil Conditions and Sampling Operations," *Harvard Bull.,* no. 208, 1962.

51.   Mulilis, J. P., C. K. Chan, and H. B. Seed, "The Effects of the Method of Sample Preparation on the Cyclic Stress-Strain Behavior of Sands," Rep. 75–18, Earthquake Eng. Res. Center, University of California, Berkeley, Calif., July 1975.

52.   Noorany, I., and I. Poorinand, "Effect of Sampling on Compressiblity of Soft Clay," *ASCE J. Soil Mech. Found. Eng. Div.,* vol. 99, no. SM 12, Dec. 1973.

53.   Palmer, D. J., and J. G. Stuart, "Some Observations on the Standard Penetration Test and a Correlation of the Test with a New Penetrometer," *4th Int. Conf. Soil Mech. Found. Eng.,* vol. 1, 1957.

54.   Proctor, D. C., "Requirements of Soil Sampling for Laboratory Testing," *Offshore Soil Mechanics,* Cambridge University and Lloyd's Register of Shipping, 1976.

55.   Raymond, G. P., D. L. Townsend, and M. J. Lojkasch, "The Effect of Sampling on the Undrained Soil Properties of a Leda Clay," *Can. Geotech. J.,* vol. 8, no. 4, 1971.

56.   Rutledge, P. C., "Relation of Undisturbed Sampling to Laboratory Testing," *Trans. ASCE,* vol. 109, 1944.

57.   Sandegren, E., "Sample Transportation Problems," *Swed. Geotech. Inst.,* no. 19, 1961.

58.   Sanglerat, G., *The Penetrometer and Soil Exploration,* Elsevier, Amsterdam, 1972.

59.   Schjetne, K., "The Measurement of Pore Pressure during Sampling," *4th Asian Conf. Int. Soc. Soil Mech. Found. Eng.,* Specialty Session 1, Quality in Soil Sampling, 1971.

60.   Schmertmann, J., discussion of V. F. B. deMello, "The Standard Penetration Test," *4th Pan Am. Conf. Soil Mech. Found. Eng.,* vol. III, 1971.

61.   Schmertmann, J., "Use the SPT to Measure Dynamic Soil Properties?—Yes, But . . . !" *Dynamic Geotechnical Testing, ASTM Spec. Tech. Publ* 654, 1978.

62.   Schmertmann, J. H., "Statics of SPT," *ASCE J. Geotech. Eng. Div.,* vol. 105, May 1979.

63.   Schmertmann, J. H., and A. Palacios, "Energy Dynamics of SPT," *ASCE J. Geotech. Eng. Div.,* Aug. 1979.

64.   Seed, H. B., S. Singh, and C. K. Chan, "Considerations in Undisturbed Sampling of Sands," *ASCE J. Geotech. Eng. Div.,* vol. 108, Feb. 1982.

65.   Serota, S., and R. A. Jennings, "Undisturbed Sampling Techniques for Sands and Very Soft Clays," *4th Int. Conf. Soil Mech. Found. Eng.,* vol. 1, 1957.

66.   Skempton, A. W., and V. A. Sowa, "The Behavior of Saturated Clays during Sampling and Testing," *Geotechnique,* vol. 13, no. 4, 1963.

67.   Sone, S., H. Tsuchiya, and Y. Saito, "The Deformation of a Soil Sample during Extrusion from a Sample Tube," *4th Asian Conf. Int. Soc. Soil Mech. Found. Eng.,* Specialty Session Quality in Soil Sampling, 1971.

68. Sowers, G. F., "Modern Procedures for Underground Investigations," ASCE, separate no. 435, 1954.

69. Swedish Committee on Piston Sampling, "Standard Piston Sampling," *Swed. Geotech. Inst.*, no. 19, 1961.

70. Thornburn, T. H., and W. R. Larsen, "A Statistical Study of Soil Sampling," *ASCE J. Geotech. Eng. Div.*, vol. 85, Oct. 1959.

71. Underwood, L. B., "Classification and Identification of Shales," *ASCE J. Soil Mech. Found. Eng. Div.*, vol. 93, no. SM 6, Nov. 1967.

72. van Bruggen, J. P., "Sampling and Testing Undisturbed Sands from Boreholes," *1st Int. Conf. Soil Mech. Found. Eng.*, vol. 1, 1936.

73. Ward, W. H., "Some Field Techniques for Improving Site Investigation and Engineering Design," *Roscoe Memorial Symp. Stress-Strain Behavior of Soils*, Cambridge, 1971.

74. Wineland, J. D., "Borehole Shear Device," *6th Pan Am. Conf. Soil Mech. Found. Eng.*, vol. 1, 1975.

75. Yoshimi, Y., M. Hatanaka, and H. Oh-Oka, "A Simple Method for Undisturbed Sand Sampling by Freezing," *9th Int. Conf. Soil Mech. Found. Eng.*, papers presented at Specialty Session 2, *Soil Sampling*, vol. 1, 1977.

CHAPTER 4

Bearing Capacity—
Shallow Foundations

4.1 INTRODUCTION

It is common practice to separate structures supported by the earth (e.g., buildings, bridges, dams) into two categories: The upper part is designated as the *superstructure,* and the components that interface the superstructure to the adjacent zone of soil or rock as the *foundation.*

The function of the foundation is to transfer the load of the superstructure to the underlying soil formation without overstressing the soil. Hence, a safe foundation design provides for a suitable factor against (1) shear failure of the soil and (2) excessive settlement. The soil's limiting shear resistance is referred to as the *ultimate bearing capacity,* q_u, of the soil. For design, one uses an *allowable bearing capacity,* q_a, obtained by dividing the ultimate bearing capacity by a suitable safety factor (i.e., $q_a = q_u/F$). Interrelated to the bearing capacity used for design is the magnitude and rate of settlement (soil deformation) and related effects on the superstructure. Indeed, the geotechnical–foundation engineer is always faced with the dual requirement: (1) select a suitable bearing pressure that is compatible with settlement restrictions (i.e., provide a safe foundation) and (2) keep the cost at a minimum (economic considerations are almost always an indispensable part of the decision making, since budget limitations are frequently common restrictions).

Depending on the importance of the proposed structure, the process of selection of a suitable bearing capacity may include extensive subsurface investigation and laboratory analysis, or it may be a matter of merely complying with the minimum standard of a building code. As so typical of engineering, the "relative importance" of the structure is a dilemma that the engineer has to resolve. Building codes and handbooks are frequently relied on as the sole basis for the selection of a bearing capacity value (a design number) for ordinary buildings such as homes and other light structures. Although this could prove a dangerous practice even for light buildings, it is highly inadvisable as a general practice, particularly for heavy or tall structures. Hence, the more reliable approach consists of field exploration and testing, laboratory analysis, and subsequent geotechnical evaluation.

Sometimes the foundation loads are transmitted directly to a rock formation. Except for possibly poor-quality rock (such as potentially expansive shales, pyrite formations, or rock heavily laminated with clay seams or highly fractured), rock formations generally have more than ample strength for a safe bearing support. In such a case, ordinary spread footings or wall footings are the routine foundation components. Indeed, such footings are common for residential, commercial, and even industrial structures of light to moderate loads bearing directly on soil. Such foundations are referred to as *shallow foundations,* discussed in this chapter. On the other hand, if the stratum directly underneath the building foundation consists of soil that has low shear strength or is highly compressible, the foundation loads may have to be transmitted to a greater depth, to a stiffer stratum or to rock, perhaps by means of piles or drilled piers. These are generally referred to as *deep foundations,* and are discussed in Chapter 11.

Much investigation on the subject of bearing capacity has been carried out during the past century. Subsequently, numerous proposals have been advanced regarding considerations, criteria, and procedures for the evaluation of the bearing capacity of soils (57). In fact, the development of a bearing capacity equation is still undergoing some degree of evolution. Among the very early contributors were Prandtl, Terzaghi, and Taylor; somewhat more recent refinements were made by Meyerhof, Vesic, Hanson, Cuquot, Kerisel, and many, many others. Indeed, it is perhaps exhaustive and counterproductive to attempt to delineate the work of all of the numerous contributors to the subject. Hence, an attempt to provide a merely sequential overview of development of current bearing capacity equations will be described in several of the following sections.

4.2 LOAD–SOIL DEFORMATION RELATIONSHIP

As we know from basic mechanics, all materials will deform if subjected to load; so will soil. However, unlike the more homogeneous materials such as steel or other metals, the load-deformation relationship in soils is not nearly as well defined or predictable; also, time is frequently a factor in soil deformation. Nevertheless, a certain degree of elastic behavior could be detected even in soils. Most fairly dense or stiff soils will depict a somewhat elastic relationship between load and deformation up to a significant percentage of their ultimate strength. This is somewhat analogous to the stress–strain relationship for materials such as concrete or steel, although the degree of linearity in soils is not comparable to that of steel. In soil, the early phase of deformation is primarily attributed to the densification of the stratum as a result of the reduction in voids within a soil mass. As the load is increased, a further increase in the deformation takes place, but at a somewhat more rapid rate. This increased rate of yielding is deemed to be due partially to an additional decrease in the void ratio, as well as to high lateral displacement coupled with vertical deformation. With further increase in load, excessive vertical penetration and ultimate shear failure of the soil stratum will take place (again, somewhat analogous to the stress–strain behavior in other construction materials).

Figure 4.1 depicts a general pattern of load–deformation relationship in a soil subjected to a spread or isolated footing. Within the elastic range, the amount of the footer penetration in Fig. 4.1a is relatively small (e.g., the soil shifting under and around the footing is small). With an increase in load, the soil under the footing becomes increasingly compressed, while that around the footing has a tendency to bulge out laterally and upward, as shown in Fig. 4.1b. This phenomenon becomes more apparent with increasing footer penetration, as shown in Fig. 4.1c.

Figure 4.2 depicts a rather typical load–settlement relationship for a case of spread (isolated) footing on a soil stratum. In general, no part of the load–settlement curve of soils is a truly straight line (hence, the modulus of elasticity, E, is an elusive quantity). Furthermore, the yield and ultimate strengths are not at well-defined levels. In most instances, such a load–deformation relationship (diagram) is not even available for determining yield or ultimate strengths. The lack of such information is usually attributed to cost and time. Occasionally, load-bearing tests are run in the field on relatively small plates, say $\frac{1}{4}$ to $\frac{1}{2}$ square meters, using dead loads. That is, due to the very large loads required for prototype-size loading tests, only bearing plates of relatively small size are deemed practical.

When load–deformation relationships are available, the ultimate bearing capacity q_u is taken at pressures associated with large vertical penetration, as indicated in

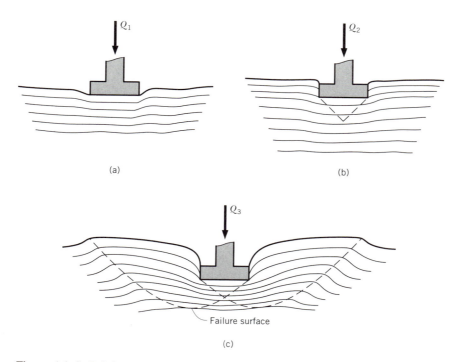

Figure 4.1 Soil deformation under and around isolated footing at various stages of increasing load.

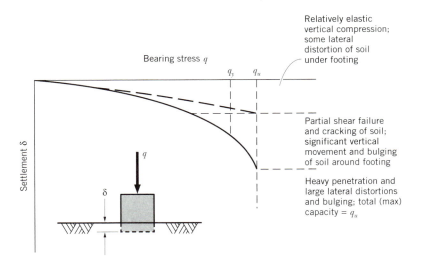

Figure 4.2 General load intensity—settlement in soils.

Figure 4.2. In lieu of such field tests, the ultimate bearing capacity is determined via equations, as described in the following sections.

4.3 BEARING FAILURE PATTERNS

Bearing failure of a foundation usually results because the soil supporting the foundation fails in shear. Depending on the type of soil and soil density, bearing failures are usually accompanied by rather deep penetration and side bulging, perhaps resembling one's foot sinking in soft mud and bulging of surrounding soil.

The modes of shear failure are commonly separated into three categories: (1) *general shear failure,* (2) *local shear failure,* and (3) *punching shear failure.* Based on his personal research and observations, and on a comprehensive summary of findings by others, Vesic (60) notes that the failure load is linked to several factors. In general, it depends on the relative compressibility of the soil in the particular geometrical and loading conditions.

Vesic depicts the basic features of these failure modes as shown in Fig. 4.3 for tests conducted in Chattahoochee River sand (North Carolina). For the case of general shear failure, usually associated with dense soils of relatively low compressibility, the slip surface is continuous from the edge of the footing to the soil surface, and full shear resistance of the soil is developed along the failure surface. The load–deformation curve has a fairly constant slope for a substantial percentage of the ultimate load. A gradual deviation from a straight line and subsequent yielding are observed as the load intensity approaches q_u. The slope becomes virtually vertical as q_u is reached. Under stress-controlled conditions (e.g., dead loads, building loads), an increase in load past q_u may result in significant continuous and cumulative vertical or tilting deformation, with likely sudden and total failure. Under strain-controlled conditions (e.g., constant rate of penetration), deformation is likely even

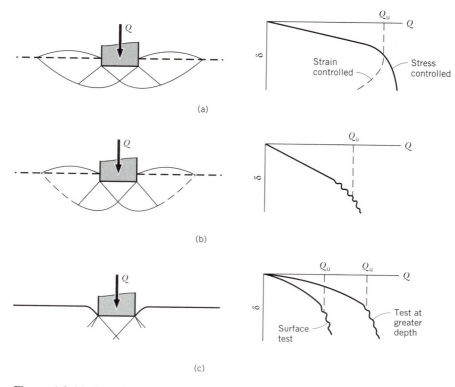

Figure 4.3 Modes of bearing failures. (a) General shear. (b) Local shear. (c) Punching shear. [After Vesic (60).]

under a reduced level of load. Bulging of the soil near the footing is usually apparent throughout most of the loading cycle, as shown in Fig. 4.3a.

For the local shear failure, the failure surface extends from the edge of the footing to approximately the boundary of the Rankine passive state. The shear resistance is fully developed over only part of the failure surface (solid segment of the line). There is a certain degree of bulging on the sides and considerable vertical compression under the footing. This is not usually apparent until significant vertical penetration occurs. As shown in Fig. 4.3b, the load–deformation curve displays a lesser degree of linearity, a steeper slope, and a smaller q_u than the load–deformation relationship for general shear, represented in Fig. 4.3a.

In the case of punching shear failure, a condition common for loose and very compressible soils, the pattern of failure is not easily detected. Generally, some vertical shear deformation is visible around the periphery of the footing, for there is little horizontal strain and no apparent bulging of the soil around the footing. Test results for footings at the surface (zero depth) indicate that the ultimate load is significantly lower than for the case where the footing is placed at a greater depth. Furthermore, the load–deformation curve has a steeper slope for the surface than it does for the greater depth tests, as shown in Fig. 4.3c. Although deformation magnitudes are considerable, sudden collapse or tilting failures are not common.

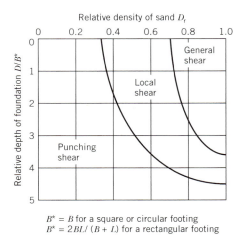

$B^* = B$ for a square or circular footing
$B^* = 2BL/(B + L)$ for a rectangular footing

Figure 4.4 Modes of failure of model footings in Chattahoochee sand. [After Vesic (60).]

Vesic developed Fig. 4.4, which shows the mode of failure that may be expected within the relative density ranges indicated for the various relative depth of penetration. For example, a footing on a very loose sand may fail in punching shear, while the same footing on the surface of a very dense sand would fail in general shear. On the other hand, a footing on a dense sand may fail in general shear near the surface; it may fail in local shear or even punching shear if it is located at greater depths.

4.4 PRANDTL'S THEORY FOR ULTIMATE BEARING CAPACITY— TAYLOR'S CONTRIBUTION

Many of the current-day fundamental principles, however limited or incomplete, regarding bearing capacity determination appear to have had their beginning with Prandtl's theory (50) of plastic equilibrium. Subsequent to Prandtl's findings in the early 1920s, extensive investigations by numerous individuals have pointed to some deficiencies in Prandtl's theory when used to predict bearing capacities of soils. These observations led to modifications and improvements in the Prandtl equation. Still, we have not reached the point, and possibly never will, where the bearing capacity of the soil stratum can be predicted with total confidence. Instead, we may be limited to further improving on the work started by Prandtl and amplified and improved by others (see bibliography at the end of this chapter).

Prandtl's theory of plastic equilibrium reflects on the penetration (deformation) effects of hard objects into much softer material. Within the context of Prandtl's assumed conditions, we make use of this theory to assume a rigid (i.e., concrete) footing penetrating into a relatively soft soil. Unlike some of Prandtl's assumptions, however, the typical soil is not isotropic and homogeneous. Furthermore, the typical footing assessed in terms of practical design limitations is not infinitely long, not smooth at the interface of footing and soil, and quite probably never applied at the very surface of the soil—conditions assumed in Prandtl's work.

Figure 4.5 shows three zones developed within the soil stratum as a long footing (say, $L/B > 5$) subjected to increasing load, resulting in bearing failure within the

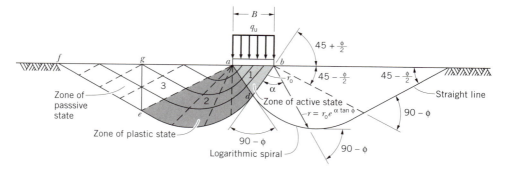

Figure 4.5 Prandtl's theory of plastic equilibrium.

soil. The load is transmitted through the soil wedge, zone 1, which is assumed to remain intact during failure. Zone 2 undergoes considerable plastic flow, with the dashed lines depicting arbitrary failure planes developed throughout this zone. The curved boundaries of this zone are approximately logarithmic spirals with the radius of curvature being represented by $r = r_0 e^{\alpha \tan \phi}$. Zone 3 represents the passive state as developed by Rankine. The failure surface of zone 3 is a straight line, as indicated in Fig. 4.5.

At impending failure, zones 2 and 3 are pushed aside by the penetrating wedge. Subsequently, a shear resistance to such movement is developed along the logarithmic spiral and the straight-line segment. Assuming a constant value for cohesion, the shear strength along the plane of failure could be expressed as $s = c + \sigma \tan \phi$. On this premise the ultimate bearing capacity of a soil based on Prandtl's theory is given by:

$$q_u = c \cot \phi \left[\tan^2 \left(45 + \frac{\phi}{2} \right) e^{\pi \tan \phi} - 1 \right] \qquad (4\text{-}1)$$

From Eq. 4-1 we observe that if the cohesion of the soil were zero ($c = 0$), the bearing capacity would also equal zero. In fact, this is not the case. Quite to the contrary; anyone familiar with foundation design recognizes that a dense granular (say, $c = 0$) stratum may provide a most desirable foundation base when considering not only the bearing capacity but also consolidation and predictable behavior. Taylor (56) added a term to Eq. 4-1 to account for the shear strength induced by the overburden pressure. Taylor's term equals to $[(\gamma B/2) \tan (45 + \phi/2)]$. Taylor recommends that the term fit into Prandtl's ultimate bearing capacity as given by Eq. 4-2.

$$q_u = \left[c \cot \phi + \frac{\gamma B}{2} \tan \left(45 + \frac{\phi}{2} \right) \right] \left[\tan^2 \left(45 + \frac{\phi}{2} \right) e^{\pi \tan \phi} - 1 \right] \qquad (4\text{-}2)$$

The introduction of Taylor's term into Prandtl's bearing capacity equation removed the limitation, indeed the misleading inference, inherent in Prandtl's equation as depicted by Eq. 4-1 (i.e., that the equation is not applicable to cohesionless soils). Nevertheless, with time, it became apparent that even with Taylor's term the equation falls short as a reasonable formula for predicting a q_u that gives results consistent with observations. In fact, Prandtl's equation is never used in current practice; but it was a beginning.

4.5 BEARING CAPACITY BASED ON RANKINE WEDGES

Although the following analysis will not yield the bearing capacity equation employed in the present-day analysis, it is a basic, dimensionally correct expression. Subsequent sections in this chapter will discuss additional elements and refinements to this expression proposed by some investigators in an attempt to account for a number of variables in a way that is consistent with their findings and observations (experimental).

Figure 4.6a shows a long, narrow footing (i.e., L/B is very large) at a depth D into a $(c - \phi)$ soil. Figure 4.6b depicts the Rankine wedges used in this analysis. Wedge I is assumed to be an active Rankine wedge, which is pushed down and slides to the right during the failure sequence; wedge II is assumed to be a passive wedge, which is pushed to the right and upward in the process. The horizontal (lateral) resistances are designated by P and are assumed to act at the interface of the two wedges as shown; they have the same magnitude but opposite directions. However, the P associated with wedge I represents the active pressure resultant, whereas the P for wedge II is the passive thrust. Thus, for the active case, wedge I, from Eq. 8.15 we have*

$$P = \tfrac{1}{2}\gamma K_a H^2 - 2cH \sqrt{K_a} + q_u K_a H \tag{a}$$

where the last term reflects the confining influence of the surcharge q_u, and $K_a = \tan^2(45 - \phi/2)$. For the passive case, wedge II, from Eq. 8.16 we have

$$P = \tfrac{1}{2}\gamma K_p H^2 + 2cH \sqrt{K_p} + q K_p H \tag{b}$$

where $K_p = \tan^2(45 + \phi/2)$. The two resultants are assumed to have the same magnitude. Hence,

$$\tfrac{1}{2}\gamma K_a H^2 - 2cH \sqrt{K_a} + q_u K_a H = \tfrac{1}{2}\gamma K_p H^2 + 2cH \sqrt{K_p} + q K_p H$$

Solving for q_u, we have

$$q_u = \tfrac{1}{2}\gamma H \left(\frac{1}{K_a}\right)(K_p - K_a) + \frac{2c}{K_a}(\sqrt{K_p} + \sqrt{K_a}) + q K_p^2 \tag{c}$$

But $K_p = 1/K_a$; also, from Fig. 4.6b,

$$H = \frac{B}{2\tan(45 - \phi/2)} = \frac{B}{2\sqrt{K_a}}$$

Hence, Eq. (c) becomes

$$q_u = \tfrac{1}{4}\gamma B K_p^{3/2}(K_p - K_p^{-1}) + 2cK_p(K_p^{1/2} + K_p^{-1/2}) + q K_p^2$$

or

$$q_u = \tfrac{1}{4}\gamma B(K_p^{5/2} - K_p^{1/2}) + 2c(K_p^{3/2} + K_p^{1/2}) + q K_p^2 \tag{d}$$

Let

$$N_\gamma = \tfrac{1}{2}(K_p^{5/2} - K_p^{1/2})$$
$$N_c = 2(K_p^{3/2} + K_p^{1/2})$$
$$N_q = K_p^2$$

---

* The active and passive pressure phenomena are covered in Chapter 8.

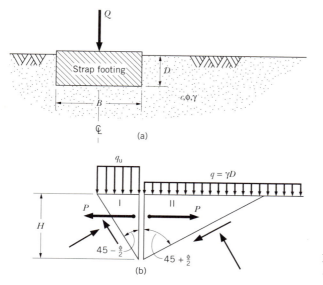

Figure 4.6 (a) Footing in ($c - \phi$) soil. (b) Rankine wedges.

Thus Eq. (d) becomes

$$q_u = cN_c + qN_q + \tfrac{1}{2}\gamma B N_\gamma \qquad (4\text{-}3)$$

Equation 4-3 is the basic form of the general bearing capacity expression used in the soil mechanics field. Variations in the values for N_c, N_q, and N_γ have been proposed over the years by different investigators, as have been factors to account for footing shapes, depth, inclination, ground, and base variations, and the like. These will be discussed in later sections.

The above derivation is based on less-than-accurate asumptions: (1) the shear at the interface of the two wedges was neglected and (2) the failure surfaces are not straight lines as was indicated for the two wedges. Hence, subsequent changes in these assumptions prompted some changes in the N_c, N_q, and N_γ values.

4.6 TERZAGHI'S BEARING CAPACITY THEORY

Terzaghi (58) improved on the wedge analysis described in the preceding section by working with trial wedges of the type assumed by Prandtl (50). However, he expanded on Prandtl's theory to include the effects of the weight of the soil above the footer (bottom) level, an aspect that Prandtl omitted in his work. Terzaghi assumed the general shape of the various zones to remain unchanged, as illustrated in Fig. 4.7. Terzaghi assumed the angle that the wedge faced forms with the horizontal to be ϕ, rather than the $(45 + \phi/2)$ assumed in Prandtl's and most other theories.

Figure 4.7 provides the basic elements in the development of Terzaghi's theory. As did Prandtl, Terzaghi assumed a strip footing of infinite extent and unit width. Unlike Prandtl, however, Terzaghi assumed a rough instead of a smooth base surface. Furthermore, although he neglected the shear resistance of the soil above the base of the footing (segment *gf* in Fig. 4.7), he did account for the effects of

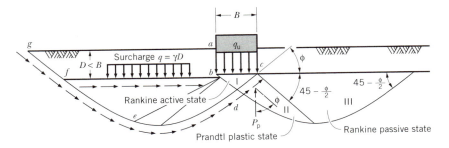

Figure 4.7 Terzaghi's bearing-capacity theory.

the soil weight by superimposing an equivalent surcharge load $q = \gamma D$. Otherwise, the shape of the failure surface is similar to that assumed by Prandtl.

Figure 4.8 shows the penetrating wedge, including equilibrium, where the downward load is resisted by the forces of the inclined faces of the wedge. These forces consist of the cohesion and the resultant of the passive pressure. Thus, assuming

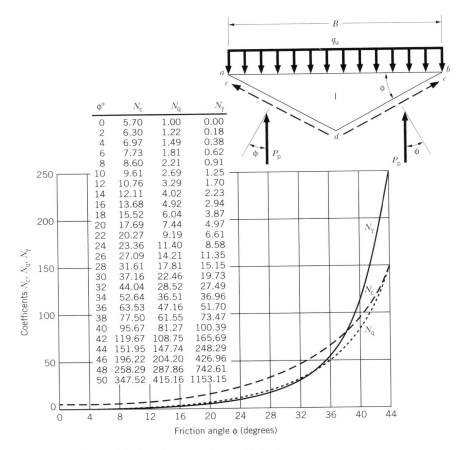

$\phi°$	N_c	N_q	N_γ
0	5.70	1.00	0.00
2	6.30	1.22	0.18
4	6.97	1.49	0.38
6	7.73	1.81	0.62
8	8.60	2.21	0.91
10	9.61	2.69	1.25
12	10.76	3.29	1.70
14	12.11	4.02	2.23
16	13.68	4.92	2.94
18	15.52	6.04	3.87
20	17.69	7.44	4.97
22	20.27	9.19	6.61
24	23.36	11.40	8.58
26	27.09	14.21	11.35
28	31.61	17.81	15.15
30	37.16	22.46	19.73
32	44.04	28.52	27.49
34	52.64	36.51	36.96
36	63.53	47.16	51.70
38	77.50	61.55	73.47
40	95.67	81.27	100.39
42	119.67	108.75	165.69
44	151.95	147.74	248.29
46	196.22	204.20	426.96
48	258.29	287.86	742.61
50	347.52	415.16	1153.15

Figure 4.8 Terzaghi's bearing-capacity coefficients.

a unit length of the footing normal to the page, we obtain q_u as follows: From $\Sigma F_y = 0$, we have, from Figure 4.8

$$q_u B = 2P_p + 2(bd) c \sin \phi$$

But $bd = B/2 \cos \phi$. Thus,

$$q_u B = 2P_p + Bc \tan \phi \tag{4-4}$$

Terzaghi represented the value of P_p as the vector sum of the three components: (1) that from cohesion, (2) that from surcharge, and (3) that resulting from the weight of the soil (*bdef* in Fig. 4.7). With the addition for shaped factors in the cohesion and base terms, Terzaghi obtained expressions for the ultimate bearing capacity for general shear conditions as:

Long footings: $q_u = cN_c + \gamma D N_q + \frac{1}{2}\gamma B N_\gamma$ (4-5)

Square footings: $q_u = 1.3 cN_c + \gamma D N_q + 0.4 \gamma B N_\gamma$ (4-6)

Circular footings: $q_u = 1.3 cN_c + \gamma D N_q + 0.3 \gamma B N_\gamma$ (4-7)

where

$$N_c = \cot \phi \left[\frac{a^2}{2 \cos^2(45 + \phi/2)} - 1 \right]$$

$$N_q = \frac{a^2}{2 \cos^2(45 + \phi/2)}$$

$$N_\gamma = \frac{1}{2} \tan \phi \left(\frac{K_{p\gamma}}{\cos^2 \phi} - 1 \right)$$

with $a = e^{(3\pi/4 - \phi/2)\tan\phi}$

Figure 4.8 gives the values for the various bearing capacity factors recommended for Eqs. 4-5 through 4-7. The value of $K_{p\gamma}$ used to compute N_γ is a factor Terzaghi used in determining N_γ but one that he did not thoroughly explain. However, the curves from Fig. 4.8 were provided by Terzaghi, thereby circumventing the need for specific values for $K_{p\gamma}$.* Also, although the N_γ factor varies rather widely with different authors for ranges of ϕ, particularly for ϕ values exceeding 40°, the term itself is not contributing significantly to q_u, thereby minimizing the significance of such wide variation on q_u.

Although Terzaghi's equations account for the shape factors in the "cohesion" and "base" terms, they lack other factors that will be described in subsequent sections. The equations are limited to concentrically loaded horizontal footings; they are not suitable for footings that support eccentrically loaded columns or to tilted footings. Furthermore, they are regarded as somewhat overly conservative. Yet, although not as widely used as when first proposed, they still have wide appeal because of the greater ease for not having to compute the many factors developed and proposed since Terzaghi's time. Indeed, many practitioners view the effort for computing some of these factors as counterproductive since the ultimate answer

---

* A close approximation is given by $K_{p\gamma} = 3 \tan^2 \left[45 + \left(\frac{\phi + 33}{2} \right) \right]$. (After S. Husain, Professor, Youngstown State University.)

is divided by frequently arbitrary safety factors, thereby negating the merit for refinements and related efforts. Indeed, the results are quite within acceptable limits for shallow footings (e.g., $D/B \leq 1$) subjected to only vertical loads—the frequently encountered case.

4.7 MEYERHOF'S BEARING CAPACITY EQUATION

Meyerhof proposed a bearing capacity equation similar to that of Terzaghi but added shape factors, s, depth factors, d, and inclination factors, i (39, 44). Meyerhof's expressions are presented via Eqs. 4-8 and 4-9. His expressions for the shape, depth, and inclination factors, and the N values are shown in Table 4.1 and in Fig. 4.9, respectively.

Vertical load: $q_{\mathrm{u}} = cN_c s_c d_c + \overline{q}N_q s_q d_q + 0.5\gamma BN_\gamma s_\gamma d_\gamma$ (4-8)

Inclined load: $q_{\mathrm{u}} = cN_c s_c d_c i_c + \overline{q}N_q s_q d_q i_q + 0.5\gamma BN_\gamma s_\gamma d_\gamma i_\gamma$ (4-9)

$$N_q = e^{\pi\tan\phi}\tan^2\left(45 + \frac{\phi}{2}\right)$$

$$N_c = (N_q - 1)\cot\phi$$

$$N_\gamma = (N_q - 1)\tan(1.4\phi)$$

Table 4.1 Meyerhof's Factors

For ϕ	Shape	Depth	Inclination
Any ϕ	$s_c = 1 + 0.2K_p\dfrac{B}{L}$	$d_c = 1 + 0.2\sqrt{K_p}\dfrac{D}{B}$	$i_c = i_q = \left(1 - \dfrac{\alpha}{90°}\right)^2$
For $\phi = 0°$	$s_q = s_\gamma = 1.0$	$d_q = d_\gamma = 1.0$	$i_\gamma = 1$
For $\phi \geq 10°$	$s_q = s_\gamma = 1 + 0.1\,K_p\dfrac{B}{L}$	$d_q = d_\gamma = 1 + 0.1\sqrt{K_p}\dfrac{D}{B}$	$i_\gamma = \left(1 - \dfrac{\alpha}{\phi}\right)^2$

$K_p = \tan^2\left(45 + \dfrac{\phi}{2}\right)$

α = angle of resultant measured from vertical axis

When triaxial ϕ is used for plane strain, adjust ϕ to obtain $\phi_{ps} = \left(1.1 - 0.1\dfrac{B}{L}\right)\phi_{\text{triaxial}}$

Meyerhof suggested that footing dimensions $B' = B - 2e_y$ and $L' = L - 2e_x$ be used in determining the total allowable load eccentrically applied in the x and y directions, respectively (i.e., $Q_{\mathrm{u}} = q_{\mathrm{u}}B'L'$), and in the corresponding terms in the ultimate bearing capacity equations and in the various correction factors for shape and inclination.

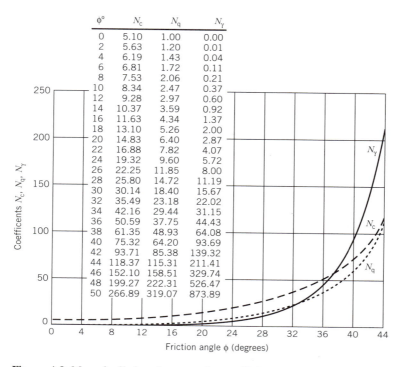

$\phi°$	N_c	N_q	N_γ
0	5.10	1.00	0.00
2	5.63	1.20	0.01
4	6.19	1.43	0.04
6	6.81	1.72	0.11
8	7.53	2.06	0.21
10	8.34	2.47	0.37
12	9.28	2.97	0.60
14	10.37	3.59	0.92
16	11.63	4.34	1.37
18	13.10	5.26	2.00
20	14.83	6.40	2.87
22	16.88	7.82	4.07
24	19.32	9.60	5.72
26	22.25	11.85	8.00
28	25.80	14.72	11.19
30	30.14	18.40	15.67
32	35.49	23.18	22.02
34	42.16	29.44	31.15
36	50.59	37.75	44.43
38	61.35	48.93	64.08
40	75.32	64.20	93.69
42	93.71	85.38	139.32
44	118.37	115.31	211.41
46	152.10	158.51	329.74
48	199.27	222.31	526.47
50	266.89	319.07	873.89

Figure 4.9 Meyerhof's bearing-capacity coefficients.

4.8 HANSEN'S BEARING CAPACITY EQUATION—VESIC'S FACTORS

J. Brinch Hansen (23) proposed what is referred to as the *general bearing-capacity equation*, given by Eq. 4-10. To a great extent, Hansen's equation is an extension of Meyerhof's proposed equations. The N_c and N_q coefficients are identical. The N_γ coefficient recommended by Hansen is almost the same as Meyerhof's for ϕ values up to about 35°. There are some deviations or differences for higher values of ϕ; Hansen's values are slightly more conservative than Meyerhof's. Both Meyerhof and Hansen provided expressions for the shape, depth, and inclination factors; however, Hansen also added what he called ground factors and base factors to include conditions for a footing on a slope.

$$q_u = cN_c s_c d_c i_c b_c g_c + \overline{q}N_q s_q d_q i_q b_q g_q + \tfrac{1}{2}\overline{\gamma}BN_\gamma s_\gamma d_\gamma i_\gamma b_\gamma g_\gamma \tag{4-10}$$

In the special case of a horizontal ground surface,

$$q_u = -c \cot \phi + (\overline{q} + c \cot \phi)N_q s_q d_q i_q b_q + \tfrac{1}{2}\overline{\gamma}BN_\gamma s_\gamma d_\gamma i_\gamma b_\gamma \tag{4-10a}$$

Where $\overline{q}$ is the effective overburden pressure at base level.

Figure 4.10 provides the relationships between N_c, N_q, N_γ, and the ϕ values, as proposed by Hansen. Table 4.2 has some expressions for inclination, shape, depth, base, and ground inclination expressions proposed by Hansen.

With regard to the load-inclination factors, Hansen recommends that if the quantity inside the bracket for the expression for i_q becomes negative, the equation is not applicable; the bearing capacity in such a case will be negligible. Also, i_γ must

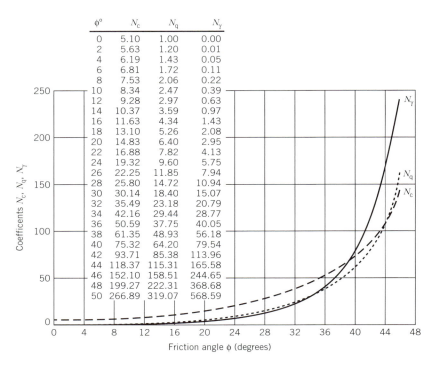

$\phi°$	N_c	N_q	N_γ
0	5.10	1.00	0.00
2	5.63	1.20	0.01
4	6.19	1.43	0.05
6	6.81	1.72	0.11
8	7.53	2.06	0.22
10	8.34	2.47	0.39
12	9.28	2.97	0.63
14	10.37	3.59	0.97
16	11.63	4.34	1.43
18	13.10	5.26	2.08
20	14.83	6.40	2.95
22	16.88	7.82	4.13
24	19.32	9.60	5.75
26	22.25	11.85	7.94
28	25.80	14.72	10.94
30	30.14	18.40	15.07
32	35.49	23.18	20.79
34	42.16	29.44	28.77
36	50.59	37.75	40.05
38	61.35	48.93	56.18
40	75.32	64.20	79.54
42	93.71	85.38	113.96
44	118.37	115.31	165.58
46	152.10	158.51	244.65
48	199.27	222.31	368.68
50	266.89	319.07	568.59

Figure 4.10 Hansen's bearing-capacity coefficients.

be modified if the foundation base is inclined. Hansen also notes that his expression for i_c is a rather simple formula, for a special case of $\phi = 0$. He notes that exact formulas for i_c and i_q have been derived by several authors but with rather complicated results.

Noting that theoretical values of the shape factors require a three-dimensional theory of plasticity evaluation, Hansen proposed an empirical set of inclination factors. Since failure can take place either along the long side or along the short side, he recommended two sets of inclination factors, as indicated in Table 4.2. He recommends that a value exceeding 0.6 should always be used in connection with the expression for $s_{\gamma B}$ and $s_{\gamma L}$.

Regarding the depth factors, Hansen indicates that the formulas presented in the table are valid for the usual case of failure along the long side L of the base. For the investigation of possible failure along the short side B, he recommends that B be replaced by L in the two expressions; see Example 4.3.

For the base and ground inclination factors, reference is made to the figure in Table 4.2. V depicts the foundation load normal to the base and H the load parallel to the base, with ν as the angle between base and the horizontal line. He notes that the above formulas should be used only for positive values of ν and β, the latter being smaller than ϕ. Also, $\nu + \beta$ must not exceed 90°. D is measured vertically, as indicated on the figure.

Vesic's (60) approach is much the same as Hansen's. The N_γ is somewhat higher than Hansen's for ϕ values up to approximately 40°, and slightly lower for values exceeding 45°. Also, Vesic's expressions for inclination, base, and ground factors are somewhat less conservative than Hansen's.

Table 4.2 Hansen's Factors

Shape Factors

Since failure can take place either along the long sides, or along the short sides, Brinch Hansen proposed two sets of shape factors.

$$s_{cB}^a = 0.2\, i_{cB}^a\, B/L$$

$$s_{cL}^a = 0.2\, i_{cL}^a\, L/B$$

$$s_{qB} = 1 + \sin\phi \cdot B i_{qB}/L$$

$$s_{qL} = 1 + \sin\phi \cdot L i_{qL}/B$$

$$s_{\gamma B} = 1 - 0.4\,(B i_{\gamma B}) : (L i_{\gamma L})$$

$$s_{\gamma L} = 1 - 0.4\,(L i_{\gamma L}) : (B i_{\gamma B})$$

For the last two factors the special rule must be followed, that the value exceeding 0.6 should always be used.

Base and Ground Inclination Factors

$$b_c^a = \frac{2\nu}{\pi + 2} = \frac{\nu^\circ}{147^\circ} \qquad b_q = e^{-2\,\nu\tan\phi} \qquad b_\gamma = e^{-2.7\,\nu\tan\phi}$$

$$g_c^a = \frac{2\beta}{\pi + 2} = \frac{\beta^\circ}{147^\circ} \qquad g_q = [1 - 0.5\tan\beta]^5 = g_\gamma$$

Load Inclination Factors

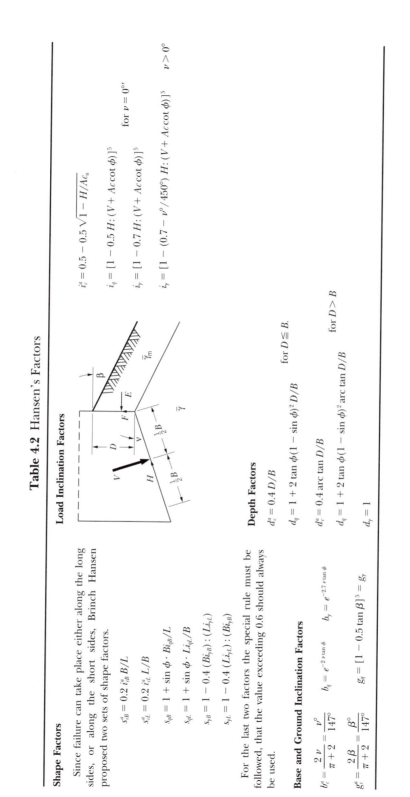

$$i_c^a = 0.5 - 0.5\,\sqrt{1 - H/Ac_u}$$

$$i_q = [1 - 0.5\, H : (V + Ac\cot\phi)]^5$$

$$i_\gamma = [1 - 0.7\, H : (V + Ac\cot\phi)]^5 \qquad \text{for } \nu = 0^{o'}$$

$$i_\gamma = [1 - (0.7 - \nu^\circ/450^\circ)\, H : (V + Ac\cot\phi)]^5 \qquad \nu > 0^\circ$$

Depth Factors

$$d_c^a = 0.4\, D/B$$

$$d_q = 1 + 2\tan\phi(1 - \sin\phi)^2\, D/B \qquad \text{for } D \leqq B.$$

$$d_c^a = 0.4\arctan D/B$$

$$d_q = 1 + 2\tan\phi(1 - \sin\phi)^2 \arctan D/B \qquad \text{for } D > B$$

$$d_\gamma = 1$$

4.9 A COMPARATIVE OVERVIEW OF TERZAGHI'S, MEYERHOF'S, AND HANSEN'S EQUATIONS

Of the bearing capacity equations previously discussed, perhaps the three most widely used are Terzaghi's, Meyerhof's, and Hansen's. For expediency, they are reproduced below, followed by a comparative assessment of their merits. The respective "N" values are given in Table 4.3.

Terzaghi's equations were and are still widely used, perhaps because they are somewhat simpler than Meyerhof's or Hansen's; one need not compute the shape, depth, inclination, base, and ground factors. Of course, they are suitable for a concentrically loaded horizontal footing but are not applicable for columns with moment or tilted forces. Generally, they are somewhat more conservative than Hansen's or Meyerhof's. Practitioners frequently use Terzaghi's for very cohesive soil where the ratio $D/B \lesseqqgtr 1$.

Table 4.3 Bearing-Capacity Coefficients

$\phi°$	Terzaghi			Meyerhof			Hansen		
	N_c	N_q	N_γ	N_c	N_q	N_γ	N_c	N_q	N_γ
0	5.70	1.00	0.00	5.10	1.00	0.00	5.10	1.00	0.00
2	6.30	1.22	0.18	5.63	1.20	0.01	5.63	1.20	0.01
4	6.97	1.49	0.38	6.19	1.43	0.04	6.19	1.43	0.05
6	7.73	1.81	0.62	6.81	1.72	0.11	6.81	1.72	0.11
8	8.60	2.21	0.91	7.53	2.06	0.21	7.53	2.06	0.22
10	9.61	2.69	1.25	8.34	2.47	0.37	8.34	2.47	0.39
12	10.76	3.29	1.70	9.28	2.97	0.60	9.28	2.97	0.63
14	12.11	4.02	2.23	10.37	3.59	0.92	10.37	3.59	0.97
16	13.68	4.92	2.94	11.63	4.34	1.37	11.63	4.34	1.43
18	15.52	6.04	3.87	13.10	5.26	2.00	13.10	5.26	2.08
20	17.69	7.44	4.97	14.33	6.40	2.37	14.83	6.40	2.95
22	20.27	9.19	6.61	16.88	7.82	4.07	16.88	7.82	4.13
24	23.36	11.40	8.58	19.32	9.60	5.72	19.32	9.60	5.75
26	27.09	14.21	11.35	22.25	11.85	8.00	22.25	11.85	7.94
28	31.61	17.81	15.15	25.80	14.72	11.19	25.80	14.72	10.94
30	37.16	22.46	19.73	30.14	18.40	15.67	30.14	18.40	15.07
32	44.04	28.52	27.49	35.49	23.18	22.02	35.49	23.18	20.79
34	52.64	36.51	36.96	42.16	29.44	31.15	42.16	29.44	28.77
36	63.53	47.16	51.70	50.59	37.75	44.43	50.59	37.75	40.05
38	77.50	61.55	73.47	61.35	48.93	64.08	61.35	48.93	56.18
40	95.67	81.27	100.39	75.32	64.20	93.69	75.32	64.20	79.54
42	119.67	108.75	165.69	93.71	85.38	139.32	93.71	85.38	113.96
44	151.95	147.74	248.29	118.37	115.31	211.41	118.37	115.31	165.58
46	196.22	204.20	426.96	152.10	158.51	329.74	152.10	158.51	244.65
48	258.29	287.86	742.61	199.27	222.31	526.47	199.27	222.31	368.68
50	347.52	415.16	1153.15	266.89	319.07	873.89	266.89	319.07	568.59

Terzaghi's

Long footings:	$q_u = cN_c + \gamma D N_q + \frac{1}{2}\gamma G N_\gamma$
Square footings:	$q_u = 1.3 cN_c + \gamma D N_q + 0.4\gamma B N_\gamma$
Circular footings:	$q_u = 1.3 cN_c + \gamma D N_q + 0.3\gamma B N_\gamma$

Meyerhof's

Vertical load:	$q_u = cN_c s_c d_c + \bar{q} N_q s_q d_q + \frac{1}{2}\gamma B N_\gamma s_\gamma d_\gamma$
Inclined load:	$q_u = cN_c s_c d_c i_c + \bar{q} N_q s_q d_q i_q + \frac{1}{2}\gamma B N_\gamma s_\gamma d_\gamma i_\gamma$

Hansen's

$$q_u = cN_c s_c d_c i_c b_c g_c + \bar{q} N_q s_q d_q i_q b_q g_q + \frac{1}{2}\bar{\gamma} B N_\gamma s_\gamma d_\gamma i_\gamma b_\gamma g_\gamma$$

In the special case of a horizontal ground surface,

$$q_u = -c \cot \phi + (\bar{q} + c \cot \phi) N_q s_q d_q i_q b_q + \frac{1}{2}\bar{\gamma} B N_\gamma s_\gamma d_\gamma i_\gamma b_\gamma$$

where $\bar{q}$ is the effective overburden pressure at base level.

Currently, Meyerhof's and Hansen's equations are perhaps more widely used than Terzaghi's. Both are viewed as somewhat less conservative and applicable to more general conditions. Hansen's, however, is used when the base is tilted or when the footing is on a slope and for $D/B > 1$.

Terzaghi developed his bearing capacity equations assuming a general shear failure in a dense soil and a local shear failure for a loose soil. For the local shear failure he proposed reducing the cohesion and ϕ as:

$$c' = 0.67c$$
$$\phi' = \tan^{-1}(0.67 \tan \phi)$$

One notes from the N_c factor that cohesion is the predominant parameter in cohesive soil and the N_q is the predominant factor in cohesionless soil. Also, where the soil is not uniform (not homogeneous), judgment is necessary as to the appropriate bearing capacity value. Generally, as indicated previously, Terzaghi's equation is easier to use and has significant appeal to many practitioners. This is particularly acceptable in view of the fact that the ultimate bearing capacity q_u is reduced in practice by dividing by a safety factor. Indeed, the argument is frequently made that while the ultimate bearing capacity is determined via rather laborious and detailed steps and calculations of various factors, the effort is greatly negated when the result q_u is divided by what is frequently an arbitrary safety factor in the final step to obtain q_a.

As so frequently is the case in engineering, and particularly in geotechnical engineering, judgment still remains a decisive ingredient in the use of the bearing-capacity equations.

Example 4.1 _____

Given The data in Fig. 4.11.

Find q_u via Terzaghi's, Meyerhof's, and Hansen's equations.

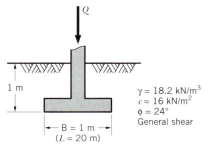

$\gamma = 18.2$ kN/m^3
$c = 16$ kN/m^2
$\phi = 24°$
General shear

Figure 4.11 A given isolated footing.

Procedure Via Terzaghi's equation:

$$q_u = cN_c + \gamma DN_q + \tfrac{1}{2}\gamma BN_\gamma$$

For $\phi = 24°$, $N_c = 23.36$; $N_q = 11.4$; $N_\gamma = 8.58$. Hence,

$$q_u = 16(23.36) + 18.2(1)11.4 + \tfrac{1}{2}(18.2)(1)8.58$$
$$q_u = 373.76 + 207.48 + 78.08 = \underline{\underline{659.3}} \text{ kN/m}$$

Via Meyerhof's equation:

$$q_u = cN_c s_c d_c + \gamma DN_q s_q d_q + \tfrac{1}{2}\gamma BN_\gamma s_\gamma d_\gamma$$

For $\phi = 24°$, $N_c = 19.32$; $N_q = 9.60$; $N_\gamma = 5.72$.

$$K_p = \tan^2\left(45 + \frac{\phi}{2}\right) = 2.37$$

Hence,

$$s_c = 1 + 0.2\ K_p\left(\frac{B}{L}\right) = 1 + 0.2(2.37)\left(\tfrac{1}{20}\right) = 1.02$$

$$s_q = s_\gamma = 1 + 0.1\ K_p\frac{B}{L} = 1 + 0.1(2.37)\left(\tfrac{1}{20}\right) = 1.01$$

$$d_c = 1 + 0.2\sqrt{K_p}\left(\frac{D}{B}\right) = 1 + 0.2\sqrt{2.37}\left(\tfrac{1}{1}\right) = 1.31$$

$$d_q = d_\gamma = 1 + 0.1\sqrt{K_p}\left(\frac{D}{B}\right) = 1 + 0.1\sqrt{2.37}\left(\tfrac{1}{1}\right) = 1.15$$

$$i_c = i_q = i_\gamma = 1$$

Thus,

$$q = 16(19.32)(1.02)(1.31) + 18.2(1)(9.6)(1.01)(1.15)$$
$$+ \tfrac{1}{2}(18.2)(1)(5.72)(1.01)(1)$$

or

$$q_u = 412.62 + 202.94 + 52.57 = \underline{\underline{668.13}} \text{ kN/m}^2$$

Via Hansen's equation: For horizontal ground surface and horizontal footing base, ground, and inclination factors $= 1$

$$q_u = -c\cot\phi + (\overline{q} + c\cot\phi)\,N_q s_q d_q + \tfrac{1}{2}\overline{\gamma}BN_\gamma s_\gamma d_\gamma$$

For $\phi = 24°$, $N_c = 19.32$; $N_q = 9.6$; $N_\gamma = 5.75$.

$$s_q = 1 + \sin \phi \left(\frac{B}{L}\right) = 1 + 0.41\left(\tfrac{1}{20}\right) = 1.02$$

$$s_\gamma = 1 - 0.4 \left(\frac{B}{L}\right) = 1 - 0.4\left(\tfrac{1}{20}\right) = 0.98$$

$$d_c = 0.4 \tan^{-1} \left(\frac{D}{B}\right) = 0.4 \tan^{-1} \left(\tfrac{1}{1}\right) = 0.31$$

$$d_q = 1 + 2 \tan \phi (1 - \sin \phi)^2 \left(\frac{D}{B}\right) = 1 + 0.89(0.353)\left(\tfrac{1}{1}\right) = 1.31$$

$$d_\gamma = 1$$

$$q_u = -c \cot \phi + (\bar{q} + c \cot \phi) N_q s_q d_q + \tfrac{1}{2}\bar{\gamma} B N_\gamma s_\gamma d_\gamma$$

$$= -16(2.25) + (18.2 \times 1 + 35.94)(9.6)(1.02)(1.31)$$
$$+ \tfrac{1}{2}(18.2)(5.75)(0.98)(1)$$
$$= -35.94 + 694.44 + 51.84 = \underline{710.34} \text{ kN/m}^2$$

Hence, q_u results are shown below.

Answer

Method	Terzaghi	Meyerhof	Hansen
q_u, kN/m^2	659.3	668.13	710.34

Example 4.2

Given   The results of a full-scale test conducted by H. Muhs in Berlin and reported by J. B. Hansen (24). The pertinent data associated with this test, as reported by Hansen are: Footing dimension $L = 2$ m; $B = 0.5$ m ($A = 1$ m^2); $D = 0.5$ m. Dense sand, $\bar{\gamma} = 0.95$ t/m^3; $c = 0$. Failure load, $Q = 190$ t; $\phi_{pl} = 47°$ ($\phi_{tr} \approx 40°$; Muhs' measured $\phi_{tr} \approx 40°$).

Find q_u via Terzaghi's, Meyerhof's, and Hansen's equations.

Procedure   Via Terzaghi's equation: Note $L/B = 4$; use of long footing equation is perhaps acceptable. For $c = 0$.

$$q_u = \gamma D N_q + \tfrac{1}{2}\gamma B N_\gamma$$

for $\phi \approx 47°$, $N_q \approx 246$; $N_\gamma = 585$. Thus,

$$q_u = (0.95)(0.5)(246) + \tfrac{1}{2}(0.95)(0.5)(585)$$
$$q_u = 116.85 + 138.9 = \underline{255.78} \text{ t/m}^2$$

Via Meyerhof's equation:

$$q_u = \bar{q} N_q s_q d_q + \tfrac{1}{2}\gamma B N_\gamma s_\gamma d_\gamma$$

For $\phi = 47°$, $N_q \approx 190$; $N_\gamma = 429$ (via linear interpolation).

$$K_p = \tan^2(45 + \tfrac{47}{2}) = 6.44; \sqrt{K_p} = 2.54$$

$$s_q = s_\gamma = 1 + 0.1\,K_p\frac{B}{L} = 1 + 0.1(6.44)\left(\frac{0.5}{2}\right) = 1.16$$

$$d_q = d_\gamma = 1 + 0.1\sqrt{K_p}\left(\frac{D}{B}\right) = 1 + 0.1\sqrt{6.44}\left(\frac{0.5}{0.5}\right) = 1.25$$

$$\therefore q_u = (0.95)(0.5)(190)(1.16)(1.25) + \tfrac{1}{2}(0.95)(0.5)(429)(1.16)(1.25)$$
$$q_u = 130.86 + 147.74 = \underline{\underline{278.60}} \text{ t/m}^2$$

Via Hansen's equation (for $c = 0$, and i, b, and g terms $= 0$):

$$q_u = \overline{q}N_q s_q d_q + \tfrac{1}{2}\overline{\gamma}BN_\gamma s_\gamma d_\gamma$$

For $\phi = 47°$, $N_q = 190$; $N_\gamma = 300$ (from Hansen's calculations).

$$s_q = 1 + \sin\phi\frac{B}{L} = 1 + 0.73\left(\frac{0.5}{2}\right) = 1.18$$

$$s_\gamma = 1 - 0.4\frac{B}{L} = 0.9$$

$$d_q = 1 + 2\tan\phi(1 - \sin\phi)^2\frac{D}{B}$$

$$d_q = 1 + 2(1.07)(1 - 0.731)^2\left(\frac{0.5}{0.5}\right) = 1.15$$

$$d_\gamma = 1 + i_{qB} = i_{qB} = 1$$

Hence,

$$q_u = (0.95)(0.5)(190)(1.15)(1.18) + \tfrac{1}{2}(0.95)(0.5)(300)(0.9)(1)$$
$$q_u = 122.47 + 64.13 = 186.6 \text{ t/m}^2$$

Summary of q_u (t/m²) values

Method	Terzaghi	Meyerhof	Hansen	Measured
q_u	255.78	278.6	186.6	190

Answer

The larger q_u values obtained via Terzaghi's and Meyerhof's method is attributed to the larger values of N_γ (see Section 4.9).

Example 4.3

Given The result of a full-scale test conducted by H. Muhs in Berlin and reported by B. Hansen (24). The pertinent data, as reported by Hansen are: Footing dimension was $L = 2$ m; $B = 0.5$ m $(A = 1$ m$^2)$; $D = 0.5$ m. Dense sand, $\overline{\gamma} = 0.95$ t/m^3; $c = 0$; $\phi_{pe} = 47°$. The 2 m $\times$ 0.5 m block was centrally loaded, but the load was inclined in the direction of the long side L. At failure the forces $Q_u = V = 108$ t and $H_L = 39$ t $(q_u = Q/A$, etc.$)$.

Find q_u via Meyerhof's and Hansen's equations. (Note that Terzaghi's equations are not applicable for inclined loads.)

Procedure   Via Meyerhof Eq. 4-9

$$q_u = \overline{q}N_q s_q d_q i_q + \tfrac{1}{2}\gamma\, Bs_\gamma d_\gamma i_\gamma \qquad \text{since } c = 0$$

For $\alpha \approx \tan^{-1}\left(\dfrac{H}{V}\right) = \tan^{-1}(\tfrac{39}{108}) = 19.85°$

$$K_p = \tan^2\left(45 + \frac{\phi}{2}\right) = 6.44$$

$$d_c = 1 + 0.2\sqrt{K_p}\left(\frac{D}{B}\right) = 1 + 0.2\sqrt{6.44}\left(\frac{0.5}{0.5}\right) = 1.51$$

$$d_q = d_\gamma = 1 + 0.1\sqrt{K_p}\left(\frac{D}{B}\right) = 1.15$$

$$i_c = i_q = \left(1 - \frac{\alpha}{90}\right)^2 = \left(1 - \frac{19.85}{90}\right)^2 = 0.61$$

$$i_\gamma = \left(1 - \frac{\alpha}{\phi}\right)^2 = \left(1 - \frac{19.85}{47}\right)^2 = 0.33$$

$$s_c = 1 + 0.2\,K_p\left(\frac{B}{L}\right) = 1 + 0.2(2.37)\left(\frac{0.5}{0.5}\right) = 1.47$$

$$s_q = s_\gamma = 1 + 0.1K_p\left(\frac{B}{L}\right) = 1.24$$

$N_q = 190$; $N_\gamma = 429$ (N_c is not needed since $c = 0$)
$q_u = (0.5)(0.95)(190)(1.24)(1.15)(0.61)$
     $+ (0.5)(0.95)(0.5)(429)(1.24)(1.15)(0.33)$
$q_u = 78.50 + 45.04 = \underline{123.54}$ t/m^2

Via Hansen's equation:
   Where the horizontal force has both a component H_B parallel with the short sides B, and a component H_L parallel with the long sides L of the equivalent effective rectangle, Hansen recommends the following formulas:

$$q_u = -c\cot\phi + (\overline{q} + c\cot\phi)\,N_q d_{qL} s_{qL} i_{qL} b_q + \tfrac{1}{2}\overline{\gamma}N_\gamma L s_{\gamma L} i_{\gamma L} b_\gamma$$

or

$$q_u = -c\cot\phi + (\overline{q} + c\cot\phi)\,N_q d_{qB} s_{qB} i_{qB} b_q + \tfrac{1}{2}\overline{\gamma}N_\gamma B s_{\gamma B} i_{\gamma B} b_\gamma$$

According to Hansen, these formulas should be used in the following way: Of the two possibilities for the γ term, the upper one should be used when $Bi_{\gamma B} \leqq Li_{\gamma L}$; whereas the lower one should be used when $Bi_{\gamma B} \geqq Li_{\gamma L}$. A check on the right choice is that $s_\gamma \geqq 0.6$. Of the two possibilities for the q term, we must always choose the one giving the smallest numerical value.

The factors (i.e., depth, shape, inclination) in one direction are (note that $H_B = 0$, in B direction; also see Example 4.2).

$$d_{qB} = 1 + 2 \tan \phi (1 - \sin \phi)^2 \frac{D}{B} = 1.15$$

$$d_{\gamma B} = 1$$

$$s_q = 1 + \sin \phi \left(\frac{B}{L} \right) = 1.18$$

$$i_{\gamma B} = i_{qB} = 1$$

Hence, $Bi_{\gamma B} = (0.5)(1) = 0.5$; $d_{qB} s_{qB} i_{qB} = (1.15)(1.18)(1) = 1.32$.
 In the other direction, the factors are:

$$d_{qL} = 1 + 2 \tan \phi (1 - \sin \phi)^2 \frac{D}{L} = 1 + 2(1.07)(1 - 0.731)^2 \left(\frac{0.5}{2} \right) = 1.04$$

$$i_{\gamma L} = (1 - 0.7H/V)^5 = (1 - 0.7 \times 39/108)^5 = 0.233$$
$$i_{qL} = (1 - 0.5H/V)^5 = 0.368$$
$$Li_{qL} = 2(0.368) = 0.74$$
$$Li_{\gamma L} = 2(0.233) = 0.47 < Bi_{\gamma B} = 0.5$$

Therefore, use the second q_u and the lower γ term in the above equation.

$$s_{\gamma L} = 1 - 0.4 Bi_{\gamma B}/Li_{\gamma L} = 1 - 0.4(0.47)/0.5 = 0.624$$
$$s_{qL} = 1 + \sin \phi (Li_{qL}/B) = 1 + \sin 47°(0.74/0.5) = 2.08$$
$$d_{qL} s_{qL} i_{qL} = (1.04)(2.08)(0.368) = 0.796$$

Note, since

$$d_{qB} s_{qB} i_{qB} = 1.32 > d_{qL} \cdot s_{qL} \cdot i_{qL} = 0.796$$

Use the lower q term in equation. Thus,

$$q_u = \overline{\gamma} D N_q d_{qL} s_{qL} i_{qL} + \tfrac{1}{2} \gamma L N_\gamma d_{\gamma L} s_{\gamma L} i_{\gamma L}$$
$$q_u = (0.95)(0.5)(190)(1.04)(2.08)(0.368) + \tfrac{1}{2}(0.95)(2)(300)(1)(0.624)(0.233)$$
$$q_u = 71.84 + 41.44 = \underline{\underline{113.3}} \ \text{t/m}^2$$

Hence, q_u results are (t/m^2)

	Method	Meyerhof	Hansen	Measured
Answer	q_u (vertical)	123.54	113.3	108

Note: The measured value for $V = 108$ and $H = 39$ were used in establishing α; otherwise, the value of α would need to be known.

4.10 EFFECT OF WATER TABLE ON BEARING CAPACITY

The soil's unit weight used in the second and third term (the γ in N_q and N_γ terms) of the bearing capacity equations presented in the preceding sections are the **effective** unit weights. Of course, if a dry subsoil becomes saturated with a rising of the water table, the unit weight of the submerged soil is reduced to perhaps half the weight for that soil of the water table; obviously, we have to account for the buoyant effect of the water. A reduction in the unit weight results in a decrease in the ultimate bearing capacity of the soil. Indeed, a rise in the water table may result in swelling of some fine-grain soils, possible loss of apparent cohesion, a reduction of the angle of internal friction and decrease in the shear strength of the soil.

Figure 4.12 depicts three cases of water levels. When the water level is at a distance of B or below the bottom of the footing, no adjustment in the γ value is deemed needed; the γ_e in the second and third terms of the bearing capacity equations is merely the unit weight of the soil. However, adjustment is recommended when the

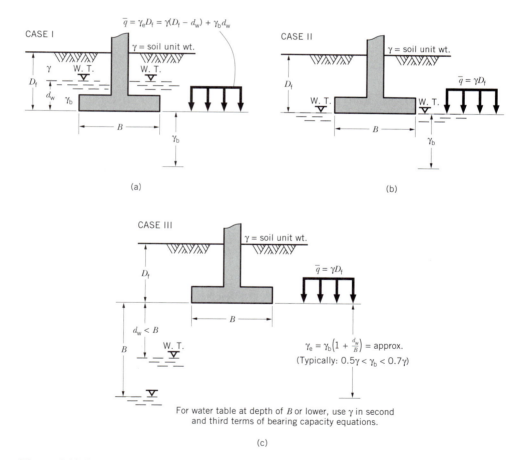

Figure 4.12 Water table above, at, or below the bottom of footing.

water level ranges between the ground surface and a distance *B* below the base of the footing, as follows:

Water Table Above the Base of the Footing

Figure 4.12a depicts a case of the water table located between the ground surface and base of the footing. This condition is not frequently encountered; where practical, designers will circumvent such conditions by relocating the foundation to higher elevations since, typically, high water tables create construction problems, require dewatering, lower bearing capacity, and so on. However, when this condition is encountered, both the second and third terms of the bearing-capacity equations are affected by a lower value of γ. The second term has an effective $\bar{q}$ as indicated in Fig. 4.12a; the third term has a γ_e equal to approximately γ_b. For all practical purposes, γ_b could be considered as $\frac{1}{2}\gamma$. Indeed, some practitioners advocate the complete elimination of the third term in such an instance for a more conservative solution.

Water at the Base of Footing.

For this case, depicted in Fig. 4.12b, the γ in the second term (N_q) requires no adjusting. The third term will be γ_b (as indicated above, γ_b is approximately $\frac{1}{2}$ of γ).

The Water Table Below the Base of Footing

This case is depicted in Fig. 4.12c. For this case, the second term requires no adjusting. The γ_e for the third term may be approximated via the formula submitted under Fig. 4.12c. If the water depth, designated as d_w, is equal to or larger than the footing dimension *B*, no adjustment is needed for any of the terms.

4.11 BEARING CAPACITY BASED ON STANDARD PENETRATION TESTS

The following coverage focuses on some of the current-day expressions for determining the **allowable bearing capacity,** q_a, based on standard penetration tests (SPT). SPT tests are described in Chapter 3 and are designated under ASTM D-1586-67.

Terzaghi and Peck (59) proposed a series of curves for estimating the allowable soil pressure for footings on sand on the basis of SPT test data. Although widely used for a time after their introduction, subsequent field data have shown these curves to be too conservative and, therefore, they are not used to any extent by current-day practitioners.

Meyerhof (42, 46) proposed equations for determining the allowable bearing capacity from SPT values for a one-inch (2.54-cm) settlement. Meyerhof's expressions are presented in Tables 4.4 and 4.5 for both FPS and SI units. Figure 4.13 represents a graphical depiction of these equations; these curves are similar to those of Terzaghi and Peck.

Bowles (5) makes the observation that, based on additional data, Meyerhof's equations are also conservative. Hence, Bowles proposes a modification of Meyerhof's equations, reflecting a substantial increase in the allowable bearing capacity

Table 4.4 Suggested Expressions for q_a Based on SPT Values

Units	Meyerhof	Bowles	For
FPS	$q_a = \dfrac{N}{4} k_d$	$q_a = \dfrac{N}{2.5} k_d$	$B \leq 4$ ft
	$q_a = \dfrac{N}{6}\left(\dfrac{B+1}{B}\right)^2 k_d$	$q_a = \dfrac{N}{4}\left(\dfrac{B+1}{B}\right)^2 k_d$	$B > 4$ ft
SI	$q_a = 12\,Nk_d$	$q_a = 20N\,k_d$	$B \leq 1.22$ m
	$q_a = 8N\left(\dfrac{B+0.305}{B}\right)^2 k_d$	$q_a = 12.5N\left(\dfrac{B+0.305}{B}\right)^2 k_d$	$B > 1.22$ m

Note: $k_d = 1 + 0.33\left(\dfrac{D}{B}\right) \leq 1.33$ suggested by Meyerhof

q_a units: kips/ft² @ FPS system

kN/m² @ SI system

from SPT data. The adjusted Meyerhof equations as proposed by Bowles are presented in Table 4.4.

The author is of the opinion that both Meyerhof's and Bowle's equations are most viable and only reliable in formations of sand, silty sand, or mixtures of silt, sand, and fine gravel (say one-inch or less in size). Thus, careful scrutiny should be used in establishing a q_a from SPT tests in fine-grain soils such as silt and particularly clay, since silt and clay may be softened or stiffened with an increase or decrease in the moisture content. Correspondingly, the SPT results may vary in

Table 4.5 Allowable Bearing Capacity in kN/m² (kips/ft²) for Surface-Loaded Footings for 2.54-cm (1-in.) Settlement, Based on Meyerhof Equations

				B			
N	0–4	6	8	10	12	14	16
5	60	54.3	50.5	48.3	46.83	45.82	45.06
	(1.25)	(1.134)	(1.055)	(1.008)	(0.978)	(0.957)	(0.941)
10	120	108.6	101	96.6	93.66	91.64	90.12
	(2.5)	(2.268)	(2.109)	(2.016)	(1.956)	(1.913)	(1.882)
15	180	162.9	151.5	144.8	140.50	137.46	135.18
	(3.75)	(3.403)	(3.164)	(3.024)	(2.934)	(2.871)	(2.823)
20	240	217.2	202	193.1	187.32	183.28	180.24
	(5)	(4.537)	(4.220)	(4.032)	(3.912)	(3.827)	(3.764)
25	300	271.5	252.5	241.4	234.15	229.10	225.30
	(6.25)	(5.671)	(5.275)	(5.040)	(4.89)	(4.785)	(4.705)
30	360	325.8	303	289.7	281	274.92	270.36
	(7.5)	(6.805)	(6.328)	(6.048)	(5.868)	(5.742)	(5.646)
35	420	378.2	353.5	338	327.8	320.74	315.42
	(8.75)	(7.940)	(7.385)	(7.056)	(6.846)	(6.699)	(6.587)

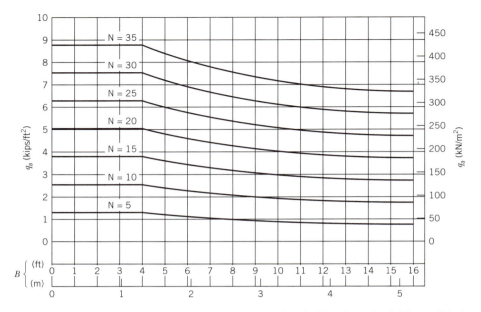

Figure 4.13 Allowable bearing capacity for surface-loaded footings for 2.54-cm (1-in.) settlement, based on Meyerhof's equations; $k_d = 1.0$.

the same silt or clay formations if the moisture conditions change. The author recalls a situation related to a one-story school building that was designed using a q_a based on a very high blow count (large N values) obtained during a dry season (and low water table). The SPT information was used as the sole basis for determining q_a. Gradually, but cumulatively, over a three-year span from construction, cracks exceeding 5 cm (2 in.) developed in some of the masonry walls, and significant cracking and heaving occurred in the on-grade concrete slab. A subsequent evaluation revealed a clay stratum vulnerable to significant shrinkage and swelling from notable changes in the water content. In the same context, SPT numbers may be misleading if the formation should contain large-size gravel. The large size gravel may wedge itself into a split-spoon sampler (remember that the I.D. of the sampler is $1\frac{3}{8}$ in.) thereby resulting in a large, misleading N value.

Perhaps buoyed by their expediency (SPT testing is done at the time of drilling) and relative simplicity, these expressions are widely used and increasing in popularity. Indeed, frequently they are used as the sole basis for determining the design bearing capacity of the soil, especially for more ordinary and less important buildings. However, they should always be used with discreet judgment and common sense, and in conjunction with available data on the strata characteristics whenever possible.

4.12 GENERAL OBSERVATIONS ON q_u and q_a EXPRESSIONS

By now the student may have noticed a number of phenomena related to the development of the ultimate bearing capacity equations: (1) the integrated expressions for q_u, such as the general bearing capacity formula, are essentially the culmination and product of theory, experimentation, and some empirical considerations; (2) soil properties such as unit weight, cohesion, and angle of internal friction are shown to be key ingredients in the q_u equations; and (3) footing data such as width, depth, shape, and so on are also accounted for in the q_u expressions.

When judged in the context of the laborious and diligent effort put forth by so many in the development of the q_u expressions, the results must be regarded as indeed impressive and worthy of a great deal of merit. On the other hand, they are long and cumbersome, and their use require extensive calculations. Indeed, the extensive effort required toward a solution is a notable deterrent to their use and is frequently unwarranted when so often the "final" q_u values are divided by a somewhat arbitrary and not always rational factor of safety in the process of obtaining a practical allowable bearing capacity to be used in design (i.e., $q_a = q_u/F$).

The expressions based on standard penetration data have particular appeal in their simplicity and expediency. This is further enhanced by the fact that q_a could be determined with relative ease at various and many depths (generally, the split-spoon sample is taken at 5.0-ft intervals or at any change in the stratum) and laboratory testing costs are virtually eliminated.

The cost savings realized from eliminating laboratory tests may also be viewed as a negative factor since potentially pertinent physical and index properties of the soil are thus unknown. Similarly, the footing factors such as shape, depth, and inclination are ignored. Also, the reliability of SPT data obtained in very fine-grain soils (clay, silt) or strata containing large gravels is indeed questionable, and, thus, so is the use of such q_a expressions for such formations.

Perhaps a most significant limitation to both approaches (for determining q_u or q_a) is that both methodologies are reflective of only shear strength parameters and do not take into account the potential deformation of the soil. That is, particularly for clay, consolidation or expansion may indeed be significant and frequently governing factors in the foundation design.

Example 4.4 _____

Given Footing: 4 ft × 4 ft (1.22 m × 1.22 m); $D = 4$ ft, $N = 12$

Find q_a via (1) Meyerhof's, (2) Bowles' equations

Procedure
$$k_d = 1 + 0.33 \left(\frac{D}{B}\right) = 1 + 0.33 \left(\frac{4}{4}\right) = 1.33$$

FPS Units

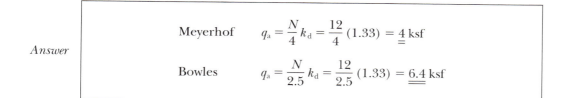

Meyerhof $q_a = \dfrac{N}{4} k_d = \dfrac{12}{4} (1.33) = \underline{\underline{4}} \text{ ksf}$

Answer

Bowles $q_a = \dfrac{N}{2.5} k_d = \dfrac{12}{2.5} (1.33) = \underline{\underline{6.4}} \text{ ksf}$

SI Units

Answer

$$\text{Meyerhof} \qquad q_a = 12Nk_d = 12(12)(1.33) = \underline{\underline{191.5 \text{ kN/m}^2}}$$

$$\text{Bowles} \qquad q_a = 20Nk_d = 20(12)(1.33) = \underline{\underline{319.2 \text{ kN/m}^2}}$$

Comparison: Bowles' values are roughly 60% higher than Meyerhof's.

Example 4.5 ───

Given   Footing: 6 ft × 6 ft (1.83 m × 1.83 m); $D = 4$ ft; $N = 12$

Find q_a via (1) Meyerhof's and (2) Bowles' equations.

Procedure
$$k_d = 1 + 0.33 \left(\frac{D}{B}\right) = 1 + 0.33 \left(\frac{4}{6}\right) = 1.22$$

FPS Units

Answer

$$\text{Meyerhof} \qquad q_a = \frac{N}{6}\left(\frac{B+1}{B}\right)^2 k_d = \frac{12}{6}\left(\frac{6+1}{6}\right)^2 (1.22) = \underline{\underline{3.32 \text{ ksf}}}$$

$$\text{Bowles} \qquad q_a = \frac{N}{4}\left(\frac{B+1}{B}\right)^2 k_d = \frac{12}{4}\left(\frac{6+1}{6}\right)^2 (1.22) = \underline{\underline{5 \text{ ksf}}}$$

SI Units

Answer

$$\text{Meyerhof} \qquad q_a = 8N\left(\frac{B+0.305}{B}\right)^2 k_d = 8(12)\left(\frac{1.83+0.305}{1.83}\right)^2 (1.22)$$

$$= \underline{\underline{159.4 \text{ kN/m}^2}}$$

$$\text{Bowles} \qquad q_a = 12.5N\left(\frac{B+0.305}{B}\right)^2 k_d = \underline{\underline{249.1 \text{ kN/m}^2}}$$

Comparison: Bowles' values are approximately 60% higher than Meyerhof's.

4.13 FOUNDATIONS IN CHALLENGING SOIL

The bearing capacity expressions for q_u and q_a, discussed in the preceding sections, are based primarily on the strength parameters ϕ and c and are suitable for soils that fit into a general-shear-failure mode; only a limited adjustment is made for a loose soil (see Section 4.9). These expressions are not suitable for soils prone to big volume changes that may ensue from large compression, expansion, or collapse. Indeed, in such cases, the governing criteria shift from shear strength to deformation. Such situations require some special assessment, which subsequently is likely to call for some special foundations. At this point, we shall briefly identify some situations that may pose challenging foundation conditions. Methodologies for improving difficult soil formations are presented in Chapter 5; the design of special foundations follows in subsequent chapters.

Highly Compressible Formations

Figure 4.14a is a schematic diagram intended to illustrate the $q - \Delta H$ features for two different soil strata. Both curves depict a reasonably smooth, continuous shape, similar to a typical stress–strain curve, but they show a significant difference in the stiffness (e.g., E) of the two strata. Curve A represents a case of a rather dense or stiff soil. Curve B illustrates a much more compressible character (e.g., smaller E); this is a feature that may be associated with soils from the following groups:

Loose Sands The formation is characterized by grains perched on top of one another, inherently unstable, and a low relative density (e.g., $D_r < 50\%$).

Organic Clays These are predominantly organic colloids produced by decaying vegetative matter.

Sensitive Clays They soften upon remolding.

High-Plasticity Clays Typically, they display large shrinkage upon drying and swelling when saturated.

Uncompacted Fills Generally these fills settle for a prolonged period, due mostly to their own weight and water that causes a breakdown of the soil structure.

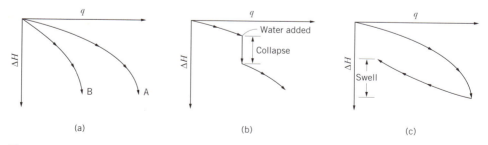

Figure 4.14 Examples of formations that may be subject to unusual or large deformation due to various factors.

Lowering of Water Table   The effect is an increase in the effective stress and significant consolidation of cohesive soils.

Sanitary Landfills   They feature organic decomposition, rubbish debris, and biochemical decay.

Permafrost Formation   Surface subsidence occurs as heat, induced by buildings or environmental warm-up, melts the frozen surface.

Subsurface Erosion   This represents a condition of water-transported soil from under structures (e.g., river scour under bridge foundations).

Collapsible Formations

Figure 4.14b is intended to illustrate the typical $q - \Delta H$ behavior of a soil formation prone to a large and sudden collapse; as shown, the culprit in this case is water, but other phenomena may have a similar effect.

Loess Deposits   These, mostly wind-deposited soils, also known as *metastable* soils, are characterized by high-void, low-unit weight, reasonably hard and incompressible when dry. However, they are subject to sudden and large collapse if wetted, as indicated in Fig. 4.14b, or when subjected to dynamic loads, shock waves from earthquakes, explosions, and the like.

Deep-Mine Sites   Sites underlain with coal, clay, or salt mines may be subject to sudden and large subsidence when the formation above such cavities, usually rock, shears and collapses.

Limestone Formations   Although limestone formations are generally very stable, limestone is soluble in acids and may be slowly dissolved by water, resulting in crack-type crevasses that gradually enlarge into large cavities known as sinkholes.

Slopes   Failure of slopes may cause sudden and large movement. Such movements may affect not only the sliding mass but also adjoining formations as a result of changing confining support, changes in the effective stress, and so on.

Water Leakage   Sewers, leaking pipes, and drains may develop a piping or undermining conditions in the stratum, thereby posing the danger of a sudden subsidence similar to a deep-mine collapse.

Expansive Formations

In Fig. 4.14c, the "rebound" segment of the $q - \Delta H$ curve typifies the behavior of compressed soil upon removal of pressure (the average slope of the rebound portion is generally about 20–40% of the compression index C_c). Introduction of water into expansive soil produces a similar "swell" effect.

Expansive Clays   Some clays (e.g., montmorillonites) swell when saturated. An indicator of such potential is its activity (see Section 2.4). Another test, commonly referred to as a *free-swell* test, consists of placing a sample of clay in a consolidometer, flooding to total saturation, and measuring the swell—analogous to a reverse consolidation test procedure.

Compaction Shales   Unlike *cementation* shale, which is generally stable, *compaction* shales frequently disintegrate upon the removal of a load (e.g., overburden), or when exposed to water and air, and subsequently expand as they turn back into clay. A *free-swell* test (or *pressure-swell* if a "restrictive" load is used), as described above, is deemed a reasonable measure of the swell potential of shales.

Frost-Susceptible Soils   Water retained in fine-grained soils will form ice-lenses with freezing temperatures and thus expand, lift structures, and so on. Cold-storage buildings are vulnerable to such effects, even in warm regions.

Problems

4.1   A continuous-wall footing, 1.5 m wide, is supported on a soil whose physical properties are as follows: $\phi = 18°$; $\gamma = 18.9$ kN/m^3; $c = 22$ kN/m^2. The water table is 3.2 m below the surface. Determine the ultimate bearing capacity by both Eqs. 4-5 and 4-8, assuming the base of the footing to be:

   (a) At the surface.

   (b) 1.3 m below the surface.

4.2   A continuous-wall footing, 2 m wide, is placed on a silty sand stratum whose physical properties are as follows: $c = 19$ kN/m^2; $\gamma = 18.3$ kN/m^3; $\gamma_b = 9.5$ kN/m^3; $\phi = 15°$. The base of the footing is 1.8 m below the surface, and the water table is 0.8 m below the surface. Calculate the ultimate bearing capacity using Eq. 4-8:

   (a) If the effect of the water table was neglected.

   (b) If the water table was accounted for.

4.3   Using Terzaghi's bearing capacity formula and the bearing capacity factors given by Fig. 4.8, determine the ultimate bearing capacity for Problem 4.1 when the base of the footing is assumed to be:

   (a) At the surface.

   (b) 0.8 m below the surface.

   (c) 1.3 m below the surface.

   (d) 2.0 m below the surface.

   (e) Plot the relationship between q_u and D for the four cases. Is there a trend?

4.4   Determine the ultimate bearing capacity using Terzaghi's formula and the corresponding bearing capacity factors for the data given in Problem 4.2 when the water table is:

   (a) At the surface.

   (b) 0.8 m below the surface.

   (c) 1.6 m below the surface.

   (d) 2.4 m below the surface.

(e) 3.2 m below the surface.

(f) Plot the relationship between q_u and the water table depth by plotting the results of parts (a)–(e). Is there an apparent trend?

4.5   Assuming that the footing in Problem 4.1 is 10 times as long as it is wide, determine the ultimate bearing capacity for the conditions given in Problem 4.1 using Eq. 4-8 and the corresponding bearing capacity factors given in Fig. 4.10 when the base of the footing is:

(a) At the surface.

(b) 1.3 m below the surface.

4.6   Determine the ultimate bearing capacity via Eq. 4-8 for the conditions given in Problem 4.2, assuming that the ratio of length to width of the footing is 10.

4.7   For the data given in Problem 4.1 determine the ultimate bearing capacity of the soil using Eq. 4-10 for a footing whose ratio $L/B = 10$ if the footing is to be placed at 2.5 m below the surface.

4.8   Determine the ultimate bearing capacity for the conditions given in Problem 4.2 via Eq. 4-10, assuming a ratio of $L/B = 5$, if the base of the footing is:

(a) At the surface.

(b) 1 m below the surface.

(c) 2 m below the surface.

(d) 3 m below the surface.

4.9   Determine the ultimate bearing capacity for the rectangular footing shown in Fig. P4.9. (a) Use Terzaghi's equation and the corresponding bearing capacity factors given by Fig. 4.8. (b) Use the general bearing capacity equation (Eq. 4-10) for the data shown. $L = 3.5$ m; $B = 2$ m; $D = 2$ m; $D_w = 3$ m; $\gamma = 18.6$ kN/m³; $c = 12.5$ kN/m²; $\phi = 24°$.

4.10   Rework Problem 4.9 for the condition where the water table is at the surface.

4.11   Determine the total Q that the footing shown in Fig. P4.9 can support with a safety factor of 5 against ultimate failure for the following conditions: $L = B = 3.0$ m; $D = 2$ m; $D_w = 1$ m; $\gamma = 19.1$ kN/m³; $c = 14.5$ kN/m²; $\phi = 20°$. Use the general bearing capacity formula, i.e. Eq. 4-10.

4.12   Rework Problem 4.11 if the water table reaches the ground surface.

4.13   Rework Problem 4.9 if D_f is equal to 4 m.

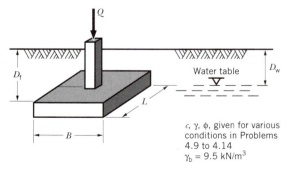

Figure P4.9 Isolated footing.

4.14 Determine the ultimate bearing capacity for the footing shown in Fig. P4.9 for the following data: $L = 3.5$ m; $B = 2$ m; $D = 4$ m; $\gamma = 19.1$ kN/m³; $c = 0$; $\phi = 33°$; and the soil is virtually dry.

4.15 A square footing, 3 m × 3 m, rests on a dry sand and gravel stratum with the following physical properties: $\gamma = 20$ kN/m³; $\phi = 35°$; $c = 0$. If the base of the footing is at 2.5 m below the surface, determine the total load the footing could support with a safety factor of 4 via the general capacity equation.

4.16 The blow count from a standard penetration test in a relatively dense sand stratum intermixed with relatively small gravel was 20 blows per 30.5 cm, taken at an average depth of 2.4 m. In a moist state the unit weight of the soil was determined to be 18.6 kN/m³. The water table was approximately 4 m below the surface. A 2-m × 2.5-m footing is anticipated to be placed at a base depth of 2 m below the surface. Determine:
   (a) The allowable bearing capacity.
   (b) The total load the footing can be expected to support safely for an anticipated 2.5-cm settlement.

4.17 Rework Problem 4.15 if the water table is expected to occasionally reach the surface. In that case a submerged unit weight of the soil of 9.1 kN/m³ is to be assumed. The blow count from a standard penetration test in a relatively dense sand was 37 blows per 30.5 cm at a depth of 4.8 m. A 3-m × 3-m footing is anticipated to be placed at a 4.5-m depth. Normally, and at the time of drilling, the water table was determined to be at 4.5 m below the surface. The unit weight of the moist soil was 18.7 kN/m³. Determine the allowable bearing capacity for an anticipated 2.5-cm maximum settlement.

4.18 Plate-load tests were run on two square plates whose sides were 0.305 and 0.61 m, respectively. The loads were 25 kN for each plate. The observed settlements (average) at the corner of the plates were 1.6 and 0.9 cm, respectively. The Poisson ratio for the soil was estimated at 0.3. Determine:
   (a) The stress–strain modulus of the soil.
   (b) The expected elastic settlement of a 2-m × 3-m footing subjected to a contact pressure comparable to the small plate.

4.19 For the data of Problem 4.18, what might be the elastic settlement at a corner of the 2-m × 3-m footing when loaded with a 1500-kN column load, based on Eq. 4-11?

BIBLIOGRAPHY

1. Badie, A., M. C. Wang, "Stability of Spread Footing Above Void in Clay," *ASCE J. Geotech. Eng. Div.*, vol. 110, Nov. 1984.
2. Balla, A., "Bearing Capacity of Foundations," *ASCE J. Soil Mech. Found. Div.*, vol. 88, no. SM5, Oct. 1962.
3. Baus, R. L., and M. C. Wang, "Bearing Capacity of Strip Footing Above Void," *ASCE J. Geotech. Eng. Div.*, vol. 109, Jan. 1983.
4. Bazaraa, A. R., "Use of the Standard Penetration Test for Estimating Settlements of Shallow Foundations on Sand," Ph.D. thesis. Univeristy of Illinois, Urbana, 1967.
5. Bowles, J. E., *Foundation Analysis and Design*, McGraw-Hill, New York, 1988.

6.   Caquot, A., and J. Kérisel, *Tables for the Calculation of Passive Pressure, Active Pressure and Bearing Capacity of Foundations* (transl. by M. A. Bec, London), Gauthier-Villars, Paris, France, 1948.

7.   Caquot, A., and J. Kérisel, *Traité Méchanique des Sols,* Gauthier-Villars, Paris, France, 1949.

8.   Caquot, A., and J. Kérisel, "Sur le Terme de Surface dans le Calcul des Fondations en Milieu Pulvérulent," *3rd Int. Conf. Soil Mech. Found. Eng.,* Zurich, Switzerland, vol. 1, 1953.

9.   Caquot, A., and J. Kérisel, *Traité de Mécanique des Sols,* 3rd ed., Gauthier-Villars, Paris, France, 1956.

10.   Chummar, A. V., "Bearing Capacity Theory from Experimental Results," *ASCE J. Soil Mech. Found. Div.,* vol. 98, no. SM12. Dec. 1972.

11.   Clemence, S. P., and A. O. Finbarr, "Design Considerations for Collapsible Soils," *ASCE J. Geotech. Eng. Div.,* vol. 107, Mar. 1981.

12.   Davis, H. E., and R. J. Woodward, "Some Laboratory Studies of Factors Pertaining to the Bearing Capacity of Soils," *Proc. Highw. Res. Board,* vol. 29, 1949.

13.   Davis, H. E., and J. T. Christian, "Bearing Capacity of Anisotropic Cohesive Soil," *ASCE J. Geotech. Eng. Div.,* vol. 97, May 1971.

14.   De Beer, E. E., "Bearing Capacity and Settlement of Shallow Foundations on Sand," *Proc. Symp. Soil Mech.,* Duke University, Durham, N.C., 1965.

15.   De Beer, E. E., "The Scale Effect on the Phenomenon of Progressive Rupture in Cohesionless Soils," *6th Int. Conf. Soil Mech. Found. Eng.,* Montreal, Canada, vol. II, 1965.

16.   De Beer, E. E., and A. Vesic, "Etude Experimentale de la Capacité Portante de Sable sous des Foundations Directes Etablies en Surface," *Ann. Trav. Publiques Belg.,* vol. 59, no. 3, 1958.

17.   De Beer, E. E., "Experimental Determination of the Shape Factors and the Bearing Capacity Factors of Sand," *Geotechnique,* vol. 20, no. 4, Dec. 1970.

18.   Eden, W., and M. Bozozuk, "Foundation Failure of a Silo on Varved Clay," *Eng. J.,* Montreal, Canada, 1962.

19.   Feda, J., "Research on Bearing Capacity of Loose Soil," *5th Int. Conf. Soil Mech. Found. Eng.,* Paris, France, vol. 1, 1961.

20.   Hansen, B., "The Bearing Capacity of Sand, Tested by Loading Circular Plates," *5th Int. Conf. Soil Mech. Found. Eng.,* Paris, France, vol. 1, 1961.

21.   Hansen, B., "Bearing Capacity of Shallow Strip Footings in Clay," *7th Int. Conf. Soil Mech. Found. Eng.,* Mexico City, Mexico, vol. 2, 1969.

22.   Hansen, B., and N. H. Christensen, discussion of A. Larkin, "Theoretical Bearing Capacity of Very Shallow Footings," *ASCE J. Soil Mech. Found. Div.,* vol. 95, no. SM6, Proc. Paper 6258, Nov. 1969.

23.   Hansen, J. Brinch, "A General Formula for Bearing Capacity," Dan. Tech. Inst., Copenhagen, Denmark, Bull. no. 11, 1961.

24.   Hansen, J. Brinch, "A Revised and Extended Formula for Bearing Capacity," Dan. Geotech. Inst., Copenhagen, Denmark, Bull. no. 28, 1970.

25.   Heller, L. W., "Failure Modes of Impact-Loaded Footings on Dense Sand," U.S. Naval Civ. Eng. Lab., Port Hueneme, Calif., Tech. Rep. R-281, 1964.

26.   Hough, B. K., "Compressibility as the Basis for Soil Bearing Value," *ASCE J. Geotech. Eng. Div.,* vol. 85, Aug. 1959.

27.   Hough, B. K., *Basic Soils,* 2nd ed., Ronald Press, New York, 1969.

28.   Houston, S. L., W. N. Houston, and D. J. Spadola, "Prediction of Field Collapse of Soils Due to Wetting," *ASCE J. Geotech. Eng. Div.,* vol. 114, Jan. 1988.

29.   Hvorslev, M. J., "The Basic Sinkage Equations and Bearing Capacity Theories," U.S. Army Waterways Exp. Stn., Vicksburg, Miss., Tec. Rep. M-70-1, Mar. 1970.

30.   Ingra, T. S., and G. B. Baecher, "Uncertainty in Bearing Capacity of Sands," *ASCE J. Geotech. Eng. Div.,* vol. 109, July 1983.

31.   Jennings, J. E., and Knight, K., "A Guide to Construction on or with Materials Exhibiting Additional Settlements Due to 'Collapse' of Grain Structure." *Proceedings,* Sixth Regional Conference for Africa on Soil Mechanics and Foundation Engineering, Johannesburg, pp. 99–105, 1975.

32.   Knodel, P. C., "Construction of Large Canal on Collapsing Soils," *ASCE J. Geotech. Eng. Div.,* vol. 107, Jan. 1981.

33.   Ko, H. Y., and L. W. Davidson, "Bearing Capacity of Footings in Plane Strain," *ASCE J. Soil Mech. Found. Div.,* vol. 99, no. SM1, Jan. 1973.

34.   Krizek, R. J., "Approximation for Terzaghi's Bearing Capacity Factors," *ASCE J. Geotech. Eng. Div.,* vol. 91, Mar. 1965.

35.   Lambe, T. W., and R. V. Whitman, *Soil Mechanics,* Wiley, New York, 1969.

36.   Leary, D. J., and B. F. Langen, "Shallow Foundations for Tall Structures in Florida," *ASCE J. Geotech. Eng. Div.,* vol. 108, Mar. 1982.

37.   Lundgren, H., and K. Mortensen, "Determination by the Theory of Plasticity of the Bearing Capacity of Continuous Footings on Sand," *3rd Int. Conf. Soil Mech. Found. Eng.,* Zurich, Switzerland, vol. 1, 1953.

38.   Meyerhof, G. G., "An Investigation of the Bearing Capacity of Shallow Footings on Dry Sand," *2nd Int. Conf. Soil Mech. Found. Eng.,* Rotterdam, The Netherlands, vol. 1, 1948.

39.   Meyerhof, G. G., "The Ultimate Bearing Capacity of Foundations," *Geotechnique,* vol. 2, 1951.

40.   Meyerhof, G. G., "The Bearing Capacity of Foundations under Eccentric and Inclined Loads," *3rd Int. Conf. Soil Mech. Found. Eng.,* Zurich, Switzerland, vol. 1, 1953.

41.   Meyerhof, G. G., "Influence of Roughness of Base and Ground Water Conditions on the Ultimate Bearing Capacity of Foundations," *Geotechnique,* vol. 5, no. 3, 1955.

42.   Meyerhof, G. G., "Penetration Tests and Bearing Capacity of Cohesionless Soils." *ASCE J. Soil Mech. Found. Div.,* vol. 82, no. SM1, 1956.

43.   Meyerhof, G. G., "The Ultimate Bearing Capacity of Wedge-Shaped Foundations," *5th Int. Conf. Soil Mech. Found. Eng.,* Paris, France, vol. 2, 1961.

44.   Meyerhof, G. G., "Some Recent Research on the Bearing Capacity of Foundations," *Can. Geotech. J.,* vol. 1, no. 1, 1963.

45.   Meyerhof, G. G., "Shallow Foundations," *ASCE J. Soil Mech. Found. Div.,* vol. 91, no. SM2, 1965.

46.   Meyerhof, G. G., "Ultimate Bearing Capacity of Footings on Sand Layer Overlying Clay," *Can. Geotech. J.,* vol. 11, no. 2, May 1974.

47.   Meyerhof, G. G., and T. Koumoto, "Inclination Factors for Bearing Capacity of Shallow Footings," *ASCE J. Geotech. Eng. Div.,* vol. 113, Sept. 1987.

48.   Mitchell, J. K., and W. S. Gardner, "In-Situ Measurements of Volume Change Characteristics," *ASCE J. Soil Mech. Found. Div.,* 1975.

49.   Muhs, H., "On the Phenomenon of Progressive Rupture in Connection with the Failure Behavior of Footings in Sand," discussion, *6th Int. Conf. Soil Mech. Found. Eng.,* Montreal, Canada, vol. 3, 1965.

50.   Prandtl, L., "Über die Eindringungsfestigkeit plastischer Baustoffe und die Festigkeit von Schneiden," *Z. Angew. Math. Mech.,* Basel, Switzerland, vol. 1, no. 1, 1921.

51.   Reddy, A. S., and R. J. Srinvasan, "Bearing Capacity of Footings on Layered Clays," *ASCE J. Soil Med. Found. Div.,* vol. 93, no. SM2, Mar. 1967.

52.   Schleicher, F., "Zur Thèorie des Baugrundes," *Bauingenieur,* vol. 7, 1926.

53.   Sherif, M. A., I. Ishibashi, and B. W. Medhin, "Swell of Wyoming Montmorillonite and Sand Mixtures," *ASCE J. Geotech. Eng. Div.,* vol. 108, Jan. 1982.

54.   Skempton, A. W., "The Bearing Capacity of Clays," *Proc. Build. Res. Congr.,* London, England, 1951.

55.   Steinbrenner, W., "Tafeln zur Setzungsberechnung," Die Strasse, vol. 1, Oct. 1934.

56.   Taylor, D. W., *Soil Mechanics,* Wiley, New York, 1948.

57.   Teng, W. C., *Foundation Design,* Prentice-Hall, Englewood Cliffs, N.J., 1962.

58. Terzaghi, K., *Theoretical Soil Mechanics*, Wiley, New York, 1943.

59. Terzaghi, K., and R. B. Peck, *Soil Mechanics in Engineering Practice*, 2nd ed., Wiley, New York, 1967.

60. Vesic, A. S., "Analysis of Ultimate Loads of Shallow Foundations," *ASCE J. Soil Mech. Found. Div.*, vol. 99, Jan. 1973.

61. Wang, M. C., and A. Badie, "Effect of Underground Void on Foundation Stability," *ASCE J. Geotech. Eng. Div.*, vol. 111, Aug. 1985.

CHAPTER 5

Site Improvement

5.1 INTRODUCTION

An ideal building site is one whose foundation soils provide for a safe as well as an economical design, be it for a building, pavement, or dam. Ideally, the foundation soils will possess the following properties: (1) have adequate shear strength; good bearing capacity, (2) will undergo minimum deformation; minimum consolidation under the imposed loads, (3) will undergo minimum volume change from swelling, shrinkage, or dynamic loading, (4) will retain strength and resist deformation with time, and (5) possess special qualities that may be desired for a particular construction (e.g., favorable water table, permeability, minimal construction problems).

Some sites possess most, if not all, of the above qualities. Of course, many building sites do not. Indeed, as the more "suitable" sites are built on, the problems associated with the poor sites appear to become more prevalent. Thus, the engineer may frequently be faced with the choice of one of the following: (1) adapt the design details to be compatible with the soil conditions (e.g., use piles, increase footing dimensions to compensate for low bearing capacity), (2) alter or improve the soil properties toward a designated goal (e.g., increase strength, reduce permeability, reduce compressibility), and (3) abandon the site in favor of one with more favorable soil characteristics.

The process of altering the soil properties of the given site is broadly referred to as *soil stabilization*. Encompassed in stabilization are a number of techniques:

- Densification of soil via compaction, precompression, drainage, vibrations, or a combination of these

- Mixing or impregnation of the soil formation with chemicals or grouting, or using geofabrics to develop a more stable base for compaction
- Replacement of an undesirable soil with a suitable one under controlled conditions

Not all methods and techniques mentioned are practical for all sites and for all types of soils. For example, the impregnation of a clay strata with a grout chemical is likely to be appreciably more difficult than injection of such materials into a granular layer; permeability of stratum to be impregnated is of obvious importance. Also, the construction details and equipment (e.g., type of equipment, field control) are generally different.

Sometimes building sites are such that neither abandoning the site nor inexpensive corrective measures are available options. By way of an example, the author recalls a site for a proposed high school in western Pennsylvania. The site was located over coal mine voids, with a history of some nearby subsidence. The improvement via filling the mine voids with a sand slurry was considered but deemed rather inappropriate for the site; the cost was high and the probable success of totally filling the voids in a reliable manner was doubtful. The adopted solution was to project the building loads down to the rock formation, via drilled piers (caissons) through the mine voids. Although a costly approach, this was deemed a viable alternative to abandoning the site; the school board felt strongly that the location of this facility was almost imperative to the functional needs of the community, thereby overcoming the added cost.

In this chapter we shall briefly evaluate some methods for site improvement.

5.2 COMPACTION

Of the number of methods used for improving the soil character of a site, compaction is usually the least expensive and by far the most widely used. It is a procedure employed frequently to densify in-situ soils, and is virtually the universal method for controlled fills (also referred to as engineered fills). Benefits from compaction include:

1. Increase in soil strength and improved bearing capacity.
2. Reduction in the voids (reduced void ratio and increased density); reduction of settlement and permeability.
3. Reduced shrinkage.

The increased soil strength and improved bearing capacity is attained by the increase in the values for ϕ, c, and γ (i.e., angle of internal friction, cohesion, and unit weight). Reduction in the void ratio may be achieved by particle reorientation subsequent to (1) alteration of the soil structure, (2) crushing and changes in the geometry of the soil grains, and (3) distortion of the grains. The reduced shrinkage benefits come as a result of a higher resistance to deformation and of a smaller void ratio (for all practical purposes, consolidation and shrinkage are solely due to a reduction in the void ratio).

Typically, the fill to be compacted is placed in somewhat uniform layers, which vary in thickness from about 0.2 m to perhaps as much as 0.5 m. In practice, the new fill is brought to the site, perhaps via a truck, dumped, and subsequently spread in a somewhat uniform thick layer by a bulldozer. The layers are then compacted via mechanical compactors, with each layer compacted to a specified density. Generally, the degree of compaction is designated as a percentage of optimum, based on a specified test (e.g., Standard Proctor, ASTM D-698-78 or AASHTO T-99; Modified Proctor, ASTM D-1557-78 or AASHTO T-180). The compaction is accomplished by rolling each layer. Each layer is tested and approved by a qualified technician prior to the placement of subsequent layers. If the degree of compaction is not obtained, it is likely that (1) the material is too dry or too wet or (2) additional passes with the roller must be made. It is the technician in the field who must make a relatively fast assessment of this information and thereby be able to guide the contractor. He tells the contractor that (1) the material passes and to proceed with the next lift, or (2) if the material does not pass, what should be done (e.g., provide more compaction effort, or add water for a dry soil, or decrease the moisture content via evaporation, change the borrow source). Generally, the technician provides but a rough estimate of the degree of compaction in the field since a more accurate determination of the moisture in the soil is normally determined via drying of the soil.

It is also common to designate the quality of the material. For example, for maximum density, a well-graded mixture of some gravel, sand, silt, and clay is specified, such that a specified density (unit weight) is attained at optimum (e.g., common values may range between 128 to 133 pcf). Occasionally, the moisture content is also specified. For example, in order to minimize the potential of shrinkage in clays, it is sometimes desirable to compact a clay at a moisture content of perhaps 3 or 4% wet of optimum.

Formations may be compacted to varying degrees for some depths with surface-type vibrators. However, the increase in density becomes smaller with depth, with rather minimal improvement beyond approximately 1 m in depth, particularly for the more cohesive-type soils. Some noticeable increase in density in sands has been realized in projects with which the author was involved to the depth of approximately 1.5 m, although some density improvement to depths in excess of 2 m were reported by others (8, 26). Interestingly, the maximum density from surface compaction occurs not at the surface, but at approximately 0.5 m below. Compaction of granular fills has been accomplished via repeated droppings of large weights (21, 22), with densification realized to depths exceeding 6 m.

With proper control, the engineered fill is frequently better than the soil on which it is placed. However, control means maintaining the quality of the fill, the thickness of the lifts, the moisture content, proper selection of the type and weight of compactor, and the number of passes the compactor must make.

Density–Water Content Test

The degree of compaction in the field is commonly measured relative to a *soil density–water content test*, generally run in the laboratory. The basis for such a test is usually attributed to R. Proctor, who pioneered this effort in connection with his construction of dams in the late 1920s, and to subsequent applications of several

Figure 5.1 Mold and hammer used to conduct density–moisture content tests. The upper collar is removed subsequent to the compaction of the last layer (3 layers for Standard and 5 for Modified), which is placed such that it is about even with the top of the lower section. Any excess is removed once the collar is removed and, subsequently, the weight is measured. See Table 5.1 for properties of mold and hammers. (Courtesy of Soiltest, Inc.)

Figure 5.2 Mechanical compactor and mold assembly for performing density–moisture tests.

articles in this regard (29). Subsequent efforts induced a standardization for some widely used compaction standards. Proctor's work was influential in showing that subjected to a given compaction effort, the soil increases in density with an increase in moisture content up to a certain point; beyond this point there is a decrease in density with further increase of moisture content. Figure 5.3 is a graphical depiction of this statement. Although various modifications of methods and procedures have been developed for determining the density–water content relationship, the basic test essentially described by and named after Proctor, remains by far the most widely used test for this purpose.

The details for the performance of density–water content tests can be obtained from a number of laboratory manuals dealing with soil testing (3, 17, ASTM or AASHTO specifications, etc.). Table 5.1 gives the adopted standards for some of the more common density–water content tests. Briefly, the test consists of compacting a soil sample, which is placed into a cylindrical mold of specified size in a designated number of layers. Each layer is tamped by a freely falling hammer of designated mass, imparting a specific number of evenly distributed loads over the surface of each layer. Figure 5.1 shows the mold and hammer used when the Proctor test is run manually; Figure 5.2 shows the mechanical version of the apparatus used to run this test. When all the layers are compacted, the soil in the mold is weighed and the water content measured. This is repeated several times for the same soil, but at a slight increase in moisture content with each complete test. Each test result is plotted as a point on a unit weight–water content test curve, as shown in Fig. 5.3.

Figure 5.3 depicts the results from the standard as well as the modified Proctor tests for the same soil. They are frequently referred to as *compaction-control charts*. Typical of such test results, the modified Proctor test (1) yields a maximum density at a somewhat lower optimum water content and (2) yields a maximum den-

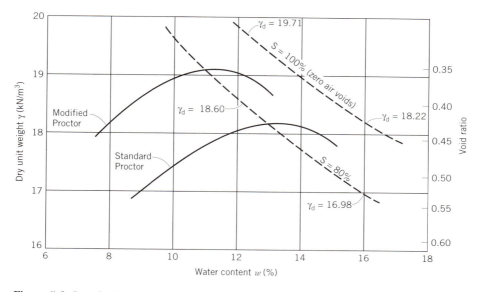

Figure 5.3 Standard and modified Proctor compaction tests, run-of-bank mixture (gravelly, silty sand, $G = 2.65$).

Table 5.1 Adopted Standards Related to Optimum Density–Water Content Test

Test	Reference		Hammer		Mold Volume	Layers	Blows per Layer	Energy* per Test
	ASTM	AASHTO	Weight	Fall				
Standard Proctor	D-698-78	T-99	2.49 kg (5.5 lb)	305 mm (12 in)	9.44×10^{-6} m^3 ($\frac{1}{30}$ ft^3)	3	25	592 kJ/m^3 (12,375 ft-lb/ft^3)
Modified Proctor	D-1557-78	T-180	4.54 kg (10 lb)	457 mm (18 in)	9.44×10^{-6} m^3 ($\frac{1}{30}$ ft^3)	5	25	2695 kJ/m^3 (56,250 ft-lb/ft^3)

* 1 kJ = 1 kN-m.

sity which, as might be expected, is larger than that from the standard Proctor test.

One notes that the energy expended via the modified Proctor test is approximately four times that for the standard test. The increase in maximum density, however, is not commensurate with the much larger proportion of expended energy. In fact, the increase in density from the modified Proctor test is generally only a few percent, varying somewhat with different types of soil and gradation, but it seldom exceeds 10% of the standard tests.

The dry density* could be expressed as a function of the void ratio, as given by Eq. 5-1:

$$\gamma_{dry} = \frac{G\gamma_w}{1 + e} \qquad (5\text{-}1)$$

Figure 5.3 relates the dry density–water content for the particular soil tested. Similarly, the dry density could be related to the water content and degree of saturation via Eq. 5-2:

$$\gamma_{dry} = \frac{G\gamma_w}{1 + wG/S} \qquad (5\text{-}2)$$

Figure 5.3 shows dry density calculations at a water content of 12 and 16% for the corresponding values of $S = 100\%$ and $S = 80\%$. Furthermore, the dry density and wet density could be expressed by Eq. 5-3:

$$\gamma_{dry} = \frac{\gamma_{wet}}{1 + w} \qquad (5\text{-}3)$$

The curve representing 100% saturation is a theoretical limit, something never reached in practice since this implies zero air voids; that is, some air voids will always remain within the mass.

Degree of Compaction

Typical specification requirements for compacted fills (frequently referred to as controlled fills) will specify the percentage of compaction based on maximum density. Sometimes both the density and the optimum moisture content are designated. For example, a specification of 95% of maximum density is frequently designated for the interior of buildings. Correspondingly, a range of water content from optimum could be tolerated without negative effects. That is, referring to Fig. 5.4, a range on the dry or wet side of the optimum can be tolerated without undue energy requirements. On the other hand, if the material is too dry or too wet, optimum compaction is difficult, if not impossible, to obtain. A soil that is too dry is likely to form lumps, which must be crushed with additional energy for an increase in density. On the other hand, a slight increase in the water content may not only reduce the strength of the lumps, but it may also lubricate the particles during the compaction shifting. Of course if excessive water is permitted within the mass, pore pressures are developed from the applied energy, while the mass is merely shifted

---

* It is common to discuss compaction in terms of dry density as a substitute for dry unit weight.

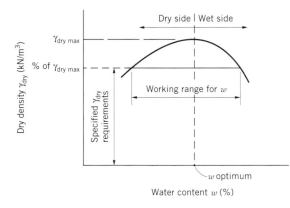

Figure 5.4 Variations in γ_{dry} with w near peak density.

and not packed. Needless to say, the range of water content for maximum benefit could be obtained by drying (e.g., scarifying) the soil if possible, if too wet; or by adding additional moisture (e.g., via spray-bar trucks) in order to maintain a most favorable moisture condition, if too dry.

Soil Types

Granular material will behave differently than a cohesive material under a given compaction effort and a given moisture content. Furthermore, the selection of equipment and the subsequent results from the use of this equipment will yield different results for the two types of material. For example, clays compacted dry of optimum display a particle arrangement somewhat independent of the type of compaction (31). On the other hand, when subjected to compaction under moisture conditions higher than the optimum (i.e., wet of optimum), the particle orientation is significantly affected, as are the strength, permeability, and compressibility of the clay. This is further discussed in Section 5.3.

In general, while the shear strength varies with soil types, samples compacted dry of optimum appear stronger and more stable than those compacted wet of optimum. Similarly, while increasing the compactive effort reduces the permeability via a reduction of voids, the permeability is also increased with an increase in the water content within the range on the dry side of optimum. However, a slight decrease in permeability is experienced if the water content is increased on the wet side of optimum.

Clay behaves more uniquely. The compressibility of two saturated clay samples at the same density is affected by both stress and the water content at the time of compaction. That is, the sample compacted on the wet side is more compressible than the one compacted on the dry side at low stresses, while the sample compacted on the dry side is more compressible than that compacted on the wet side at high stresses. Likewise, clays compacted on the dry side tend to shrink less upon drying and swell more when subjected to moisture than those compacted on the wet side. Hence, it is quite apparent that the engineer must weigh the effect of loading or stresses (e.g., from the addition of building loads, surcharge) and changes in water content and pore-water pressures as well as the subsequent effects that these changes in load or moisture conditions, or both, may have on the intended function of the compacted stratum.

The engineering properties of cohesionless soils are significantly affected by the relative density of the soil, and not as much by the many variables cited in connection with the compaction of cohesive soils. Generally, an increase in density increases the shear strength of the soil and reduces its compressibility. On the other hand, for a given compacting effort, the density also increases with an increase in water content, up to a point; then it decreases with a further increase in water content. Hence, density is usually the only specified criterion for the compaction of cohesionless soils; the degree of moisture is not a specified criterion, as may be frequently the case for cohesive soils. For optimum density, the material selected is usually a well-graded material, which may range in size from as much as 15 cm (6 in.) in diameter to clay (e.g., run of bank is a common selection for this type of fill). Many specifications will dictate either a relative density or a percentage of maximum density as the criterion for the degree of compaction.

Equipment

A variety of equipment for compaction purposes is available, with the choice for the proper equipment usually left to the engineer. Usually it is advisable to use vibration-type equipment (e.g., vibratory rollers, shown in Figs. 5.5 and 5.6; tampers shown in Fig. 5.7a) for granular soils, and equipment that penetrates the stratum (e.g., sheep-foot roller shown in Figs. 5.7b, 5.8, 5.9) for the more cohesive material such as silt or clay. Many of these units, such as sheep-foot rollers, vibratory rollers, and tampers, are common inventory of most of the larger construction firms or equipment dealers, who are also good sources for information regarding other types of equipment as well.

Figure 5.5 Bulldozer, on the left, spreading soil into a reasonably uniform layer of 6–10 in. A vibratory, smooth-face roller, on the right, is specifically effective in compacting granular soils.

Figure 5.6 Vibratory roller compacting a granular layer. Dump trucks discharge borrow material, which is to be spread by the bulldozer and subsequently compacted by the vibratory roller.

Performance Control

Compaction control and inspection are done by a qualified technician under the auspices of an engineer. Within the scope of duties, the technician performs field density tests, observes the placement of the fill (e.g., layer thickness, consistency of material), makes water content tests, and generally guides the contractor. Observations and density readings are made, usually a designated number for every lift (layer) or for a specified volume of fill placed. It is common to take such readings for each layer, for every 500 to 1000 m^2 depending on the importance of the site. Within the scope of inspection, when working with clay, compaction may result in a relatively smooth interface between lifts, and thereby create a potential seepage conduit—indeed undesirable in the case of a dam. This is less likely to be the case if sheep-foot rollers are used in the process.

In-Place Field Tests

A number of methods for determining field densities are available. Among these are the nuclear method (ASTM D-2922-70), the rubber balloon method (ASTM D-2167-66-1977, Fig. 5.11a), and the sand-cone method (ASTM D-1556-66-1974, Fig. 5.11b). The author has developed a method more expedient and more accurate than others, which has been used extensively under his supervision during the past 20 years (6). Briefly, the method consists of digging a hole in the compacted stratum, say 8 to 12 cm in diameter and for a similar depth. All the soil is carefully extracted from the hole and weighed via a portable field scale. A representative sample of the soil is preserved and dried, and subsequently the moisture content is determined. From this information both the dry and the wet states of the soil can be determined. Up to this point the procedure is similar to that followed with the rubber balloon or the sand-cone methods. For the determination of volume,

a

b

Figure 5.7 Hand-operated compactors are typically used in trenches or for backfilling behind walls and other areas with limited access. (a) Mechanical vibratory tamper, sometimes referred to as a jumping-jack, used to compact the bottom of a narrow excavation. (b) This hand-operated compactor is a modified version of a sheep-foot roller.

a glass or transparent plastic graduate is filled with a uniform-size dry sand to a point that will ensure an adequate volume of sand to fill the hole. With the initial level of the sand noted, the sand is poured into the hole until the hole is filled to the level judged by the technician to be that of the original ground surface. The new level of the sand in the graduate, shown in Fig. 5.12, is now observed, with the difference being the volume of the hole. The unit weight of the soil may thus be determined by merely dividing the weight of the soil from the hole by the volume of the hole.

Figure 5.8 Bulldozer pulling a sheep-foot roller (compactor), compacting a typical 6–10-in. thick layer of usually cohesive soil. The thickness of the layer is developed by the pans as they discharge, or by the bulldozer if its dumped in a pile by a dump truck (see Figs. 5.10 and 5.22).

Figure 5.9 Close-up view of a sheep-foot roller.

5.3 COMPACTED CLAYS

Compacted silts and clays are much less likely to fit into a well-defined category regarding their behavior and characteristics than granular soils. For example, while granular soils show gain in strength from compaction and subsequent increase in density, some silt and clays may, under certain conditions (e.g., method of compaction, molding water) display a decrease in strength past a certain increase in density.

Figure 5.10 Self-loading and unloading earthmovers, also referred to as pans or scrapers, are commonly used for moving earth (cut or fill) within building sites. The larger units are usually equipped with front and back engines (push–pull) and are controlled by one operator in the front. Auxiliary push is sometimes provided during the cutting phase by a bulldozer, as shown in (b).

For this reason it is advisable to determine the characteristics of a given fine-grained soil via laboratory tests of that soil and not depend on general data available in the literature. Nevertheless, much may be gained from the findings of others in this regard. Generally, considerable experimental data show that the compaction of clay at various water contents results in

1. Change in particle structure or grain arrangement.
2. Change in engineering properties.

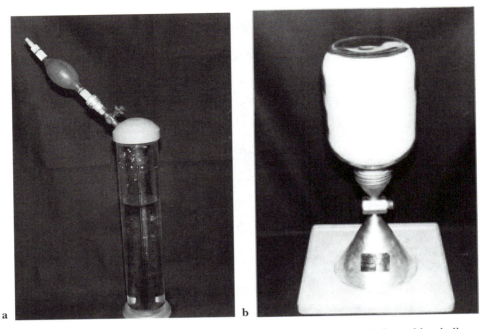

a   b

Figure 5.11 Apparatus used in performing field density tests via (a) the rubber-balloon method and (b) sand-cone method. (Courtesy of Soiltest, Inc.)

Figure 5.12 Apparatus used in performing field density tests via the method developed by the author.

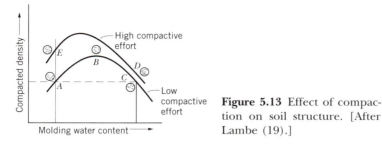

Figure 5.13 Effect of compaction on soil structure. [After Lambe (19).]

Figure 5.13 shows the effects of compaction on soil structure as presented by Lambe (19). A possible explanation for the change in structure is tied to the change in electrolyte concentration. At small water contents, such as points A or E located at dry of optimum, the concentration of electrolytes is relatively high; this impedes the diffuse double layer of ions surrounding each clay particle from full development. The result is low interparticle repulsion and subsequent flocculation of the colloids and, thereby, a lack of significant particle orientation of the compacted clay. On the other hand, as the water content is increased, say to point C or D located wet of optimum, the electrolyte concentration is reduced, and there is an increase in repulsion between clay particles, a reduction of flocculation, and thus an increase in particle orientation. Typical data illustrating this relationship between particle orientation and water content are given in Figs. 5.14 and 5.15.

A pronounced change in engineering properties of a compacted clay results from a change in the structure of the soil. The shrinkage of samples compacted dry of optimum is less than for those compacted wet of optimum, as shown in Fig. 5.16. Conversely, samples compacted dry of optimum swell more than those

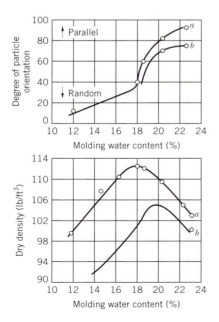

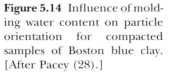

Figure 5.14 Influence of molding water content on particle orientation for compacted samples of Boston blue clay. [After Pacey (28).]

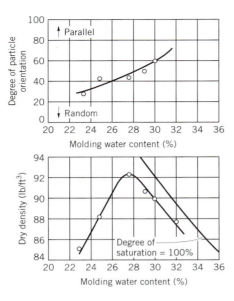

Figure 5.15. Influence of molding water content on particle orientation and axial shrinkage for compacted samples of kaolinite. [After Seed and Chan (31).]

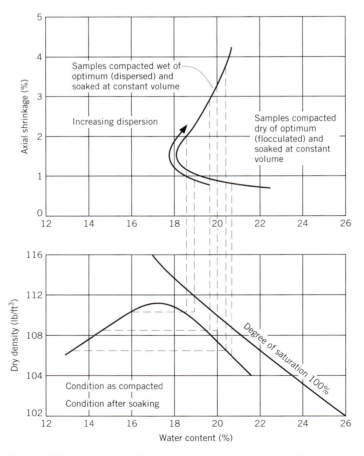

Figure 5.16 Influence of soil structure on shrinkage of silty clay. [After Seed and Chan (31).]

compacted wet of optimum. Likewise, as indicated by Fig. 5.17 for a kaolinite clay, molding water and the corresponding particle orientation exert a significant influence on the stress-strain relationships of compacted clay.

In general, the more flocculated samples develop their maximum strength at low strains and exhibit much deeper stress–strain slopes, whereas the dispersed samples display much flatter stress–strain curves, reaching maximum strength at much higher strains.

Although not applicable to all clays, and depending on the particle content and grain structure or arrangement, Fig. 5.17 depicts a general relationship between water content and the density, particle orientation, and strength of samples of silty clay subjected to kneading compaction. One notes that a significant reduction in strength results when the strength is determined at relatively low shear strain (5% in this case) beyond certain densities and certain water contents (e.g., compaction at wet of optimum). Conversely, strength appears to increase somewhat with density

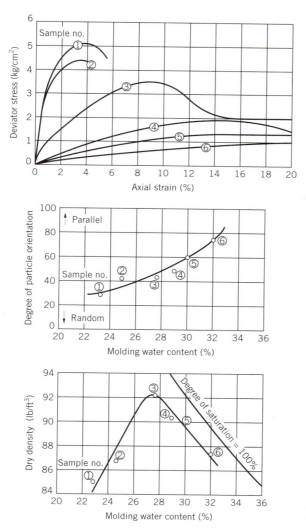

Figure 5.17 Influence of molding water content on structure and stress vs. strain relationship for compacted samples of kaolinite. [After Seed and Chan (31).]

when it is determined at high strains, for various conditions of water content, and irrespective of flocculated or dispersed soil structure. Also, clays compacted dry of optimum display much higher permeability than those compacted wet of optimum.

Generally, the soil characteristics seem to be related to the method of compaction (e.g., kneading, impact, laboratory, static). They are similar for samples compacted dry of optimum but are different when the samples are compacted wet of optimum. Furthermore, the strength determined at low strains tends to increase in the following order of compaction: kneading, impact, vibratory, static (31).

5.4 VIBRATORY COMPACTION

As mentioned in Section 5.2, conventional surface-type compactors, both the static and the vibrating types, exert a compacting effect to only a relatively shallow depth, generally less than 2 m. For greater depths other procedures are sometimes used; some of these are briefly described in the following paragraphs.

Vibroflotation

Vibroflotation is a compaction technique employed mostly for cohesionless soils. The basic equipment used in conjunction with this method is shown in Fig. 5.18 (5). Supplemented by some auxiliary units, the essence of the equipment is the vibroflot probe, which is freely suspended from a crane and equipped with water jets at the upper and lower ends. It consists of a cylindrical tubular section that houses eccentrically rotating weights that induce horizontal vibratory motion.

While vibrating and with water jetting, as shown in Fig. 5.19a, the probe is lowered under its own weight to the required penetration depth. When this depth is reached, the water flow is decreased and reversed from the lower to the upper interior jets, as shown in Fig. 5.19c. While still vibrating, the vibroflot unit is raised in incremental steps as a suitable granular material (e.g., river gravel) is shoved into the hole, which, by now, is enlarged by the horizontal vibratory motion. This material is then compacted by vibrating the probe at each incremental position for a certain time. The final result is essentially a column of granular material, as shown in Fig. 5.19d. Simultaneously, the adjoining soil formation is likewise denser than originally.

Vibroflotation is most effective for loose sands, particularly those below the water table. According to Brown (5), the range of the suitable size for most effective composition falls within zone B of Fig. 5.20, with zones A and C less desirable. Zone C appears appreciably more difficult to compact, whereas zone A, the zone containing coarse material such as gravels, may pose installation obstacles and subsequent economic problems if the penetration rate of the vibroflot is too low.

Terra Probe

The basic equipment employed for this method consists of a vibrodriver, a probe suspended from a crane, and an electric generator. The probe is an open-ended pipe that is vibrated down into the stratum with the aid of water jets. The weight, the vibration, and the water jets facilitate the penetration process. Once the depth of penetration has been reached, the still-vibrating probe is extracted at a slow

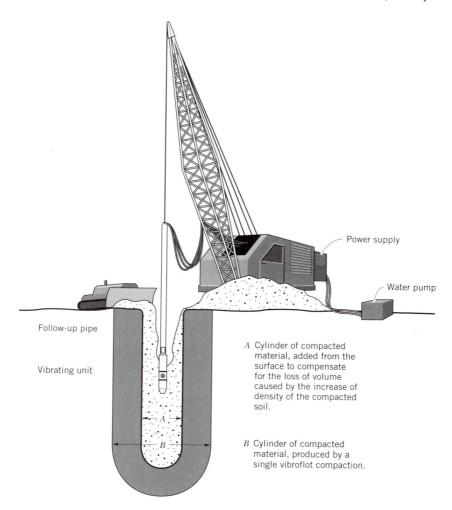

Power supply

Water pump

Follow-up pipe

Vibrating unit

A Cylinder of compacted
material, added from the
surface to compensate
for the loss of volume
caused by the increase of
density of the compacted
soil.

A

B

B Cylinder of compacted
material, produced by a
single vibroflot compaction.

Figure 5.18 Vibroflotation equipment. [After Brown (5).]

uniform rate and with continuous water jetting. This is repeated over the site to
be compacted.

   Although the terra probe and vibroflotation methods are somewhat similar, a
relative comparison of the results obtained via the two methods may be more
problematical than unique. Soil types, mechanical installation procedures, and
equipment are influential factors in this regard. Generally, however, (1) the extraction rate for the terra probe is higher than that for the vibroflot, and (2) more
probes may be needed for the terra probe than for vibroflotation to achieve equivalent results (4).

Pounding

This process consists of dropping a heavy weight from designated heights to achieve
densification of loose cohesionless soils or fills. Successful applications of this

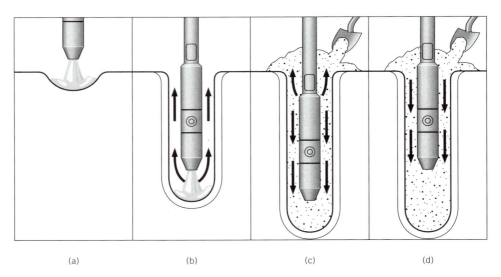

(a)   (b)   (c)   (d)

Figure 5.19 Vibroflotation compaction process. [After Brown (5).]

method have been reported for various site conditions, including a former garbage dump (21, 23). In some cases the weight ranged up to 40 tons, and the dropping height exceeded 30 m. Cohesive soils may be liquefied by this process.

Menard and Broise (23) propose a formula relating the required energy to the required depth of densification:

$$W \cdot H = D^2 \tag{5-4}$$

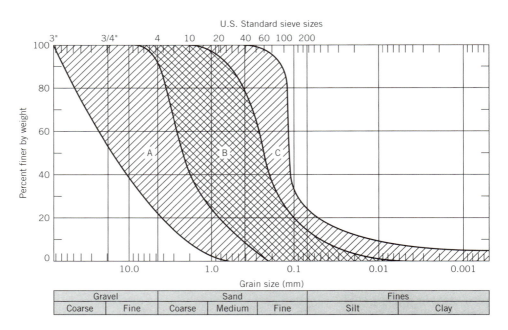

Figure 5.20 Soils suitable for vibroflotation. [After Brown (5).]

where W = weight (metric tons)
        H = height of drop (m)
        D = densification (m)

As might be expected, the rather high intensity of vibration induced by the dropping weight may affect adjacent structures. The effects are much greater than those from the vibroflotation or terra-probe methods, and they approach those from blasting—another method sometimes used for deep-soil densification. It should be used with caution in the proximity of adjacent buildings in order to minimize the potential dangers of damage to such buildings.

5.5 PRELOADING

The method consists of subjecting a building site to an artificial load, generally in the form of added fill or surcharge, prior to the building loads. The duration of the load may be several months or perhaps years, commensurate with the desired results. It is governed by an acceptable designated deformation or rate of settlement of the preloading. Sometimes a system of vertical sand or gravel drains is employed in conjunction with the preloading, in order to decrease the time of consolidation and, therefore, permit an earlier use of the site.

Preloading can be employed for virtually all sites with some beneficial results. However, it is limited to compressible strata, soft silt and clays, or sanitary landfills. It is not usually associated with granular formations where consolidation is an insignificant factor. Furthermore, economic considerations may indeed be a governing factor. In this regard the availability of soil or some other material to be used for preloading may have a decisive bearing on the economic feasibility of the method.

5.6 DEWATERING

As we have noted in various sections of this text, water can have a pronounced influence on the stability or behavior of a soil formation. In other words, the change in effective stress and, subsequently, the strength and stability of the soil mass may be directly tied to the change in pore-water pressure. For example, an increase in pore-water pressure results in a decrease in the effective stress, a reduction of soil strength, and perhaps a reduction in stability. Conversely, a reduction in water levels may have just the opposite effect—an increase in intergranular pressure, strength, and stability. Dewatering is a method of improving the soil properties by reducing the water content and the pore-water pressure.

Vertical sand or gravel drains are commonly used to dewater and, thereby, increase the consolidation and settlement of soft, saturated, and compressible soils, such as under embankments or for stabilizing liquefiable soil deposits (32). Horizontal drains are common means of dewatering and subsequently stabilizing natural slopes against seepage and erosion. Ditches along highways may serve the purpose of not only channeling the surface runoff from the road but also dewatering the base and perhaps the subbase of the road. Wellpoints are sometimes used to dewater a soil formation to greater depths.

Electroosmosis is sometimes employed during dewatering to facilitate the water movement. Electroosmosis develops a time-dependent imbalance in the pore-water pressure in a fine-grained soil, thereby expediting the water movement through the soil into an installed drain or wellpoint; see for example, References 1, 25, 35.

Routinely tied to considerations regarding the use of dewatering as a means of improving or stabilizing the site are factors such as cost versus benefit, feasibility, and effectiveness, and the time schedule to achieve a designated objective. In addition, one should evaluate the effect (e.g., settlement) dewatering may have on adjacent structures.

5.7 CHEMICAL STABILIZATION

Various chemicals added to a soil may yield one but more likely a number of changes in a soil formation: (1) reduce permeability of the soil (e.g., in dam construction, excavation infiltration); (2) increase soil strength; (3) increase bearing capacity; (4) decrease settlement; and (5) produce a stiffening of loose sand formation and thus minimize undesirable effects, such as from vibrations.

The designated objective of the engineer is the central consideration point for the selection of the method and chemicals to be used. Numerous cases of successful chemical stabilization attempts have been described in the literature. Indeed, an attempt to reflect even briefly on the very large number of methods and techniques and materials used in connection with this method of soil stabilization is perhaps unwarranted and may be even counterproductive. Instead, the discussion will be limited to a general overall view of the more common methods and chemicals.

Chemical stabilization may be employed for surface soils, for subsurface formations, or for both (36). Surface treatment, common in connection with subgrades or bases for pavement construction, generally consists of mechanically mixing the soil with a chemical (cement, bitumen, lime, bentonite, or some other chemicals) in place or by a batch process. Lime is an effective agent to be mixed with fine-grained soils with high plasticity. Subsurface treatment generally entails the impregnation of the subsurface formation with the chemical. Some of the more common additives are cement grout, sodium silicate (7), and calcium chloride solutions.

Grouting generally refers to the process of injecting a stabilizing substance into the soil stratum under pressure. The substance might be a Portland cement, a cement–sand mixture, a bentonite or a solution of sodium silicate or a number of other chemicals (2, 16, 27, 34, 37). Some of these solutions will penetrate the subsurface formation more easily than others, depending on the viscosity and the rate of chemical reaction or hardening. Some of these form a gel within a matter of seconds (a potential advantage when unstable conditions are critical), whereas others take perhaps hours to set or harden. The setting can be and often is controlled. Furthermore, the distance of penetration is not only related to the applied pressure, but it is also closely tied to the size of the soil particles and the permeability of the stratum. The coarser grain soils will normally permit much easier penetration for longer distances than will silt or clay formations. Grouting with either cement or chemicals is a relatively expensive form of soil stabilization. Thus, except for special instances where the desired effect cannot be obtained by more conventional stabilization techniques, its use is closely scrutinized from an economical point of view. In many instances an alternate method may be more economical. In addition, the degree of consistency and effectiveness over a layered system may require close monitoring of the installation effort as well as of the final results, particularly when subsurface conditions are not uniform or well defined (e.g., erratic layer formation, crevices in rock). In fact, a detailed evaluation of the soil formation, a careful evaluation of the method of injection, a close scrutiny of the type and desired characteristics of the stabilizing agent, and an economic assessment of the benefit are recommended prerequisites to the use of this method.

5.8 GEOSYNTHETICS

Geosynthetics is a name given to a family of man-made, sheet or net-like products derived from plastics or fiberglass compounds. Belonging to this group are *geotextiles, geogrids, geonets,* and *geomembranes.* Their use has increased dramatically in the last two decades to enhance soil properties or serve a special need.

The uses and design of geosynthetics vary so widely that it is virtually impossible to enumerate or describe them all. Indeed, these are beyond the scope of this text. For those interested, Koerner (16) provides an extensive and detailed coverage of use and design of geosynthetics material. Here the coverage will be limited to a general overview of and introduction to the uses of geosynthetics.

The following are but a few of many dozens of applications of geosynthetic materials in the Civil Engineering field. Figures 5.21 through 5.24 are examples of such applications.

- Reinforce soft soil; increase bearing capacity
- Strata separation

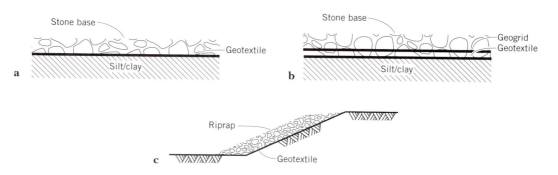

Figure 5.21 Examples of uses of geosynthetic products in construction. (a) The geotextile prevents fine soil particles from permeating into the coarse aggregate base (say, from traffic vibrations) and thereby reduce the shear strength (bridging) of the stone base. Water can move through the geotextile fabric. (b) The geotextile restricts soil but not water movement through it. The geogrid is added, as shown, to improve the bearing capacity of the base when pumping is a problem; geotextiles alone are not as effective. (c) Geotextiles and riprap as used for erosion control on slopes.

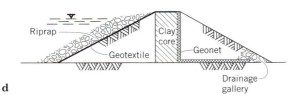

d

Figure 5.21 (*continued*) (d) Geotextile and riprap used on the upstream face of a dam to minimize erosion, and particularly at the higher elevation of the dam where wave action effects are more pronounced. The geonet directs any water permeating through the clay core into the drainage gallery, thereby reducing the hydraulic gradient and subsequent erosion of the downstream face.

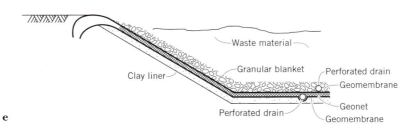

e

Figure 5.21 (*continued*) (e) In this landfill liner, the upper geomembrane restricts the downward flow and directs it to a perforated pipe drain. If any water should penetrate this member (e.g., via rips, lap joints), the geonet will conduct the water to a drain, placed over the second geotextile. The clay liner acts as a filter (further insurance) for any fluid that may find its way through this composite blanket.

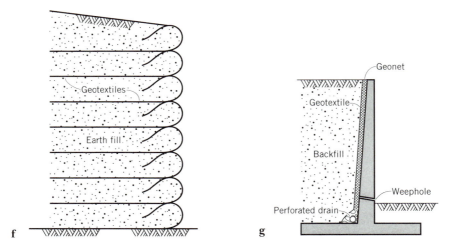

f   **g**

Figure 5.21 (*continued*) (f) Geotextiles permit the water to escape but retains the soil in this version of a reinforced-earth wall. (g) In this drainage arrangement behind a reinforced concrete wall, the geotextile prevents the soil particles from plugging up the geonet, which directs the water to the drain and weepholes.

Figure 5.22 Dump truck discharging granulated slag over a geofabric (geotextile) sheet placed over a soft (pumping) silt and clay. The geofabric permits moisture to permeate through it (perforated texture), but will inhibit the soil to permeate into the slag base and thus destroy the shear strength of the base. Also, it develops a more stable base for compacting the fill.

- Filtration
- Drainage
- Moisture barriers
- Retention walls, embankments, and slope stability
- Erosion control

Figure 5.23 A layer of slag is spread over the geofabric awaiting compaction for a truck driveway.

Geotextiles

Sometimes also referred to as geofabrics because some of the geotextiles resemble a woven fabric, geotextiles constitute the largest segment of the geosynthetics group. They are usually shipped to the construction site in rolls and are used in various situations for various needs as follows:

Strata Separation   In this function, the geotextile is placed between two different materials so as to preserve their individual characteristics and function. A common example is the use of a geotextile fabric to separate a fine-grain soil from the base (generally crushed stone) of a pavement (as shown in Fig. 5.21a). The fabric is

a

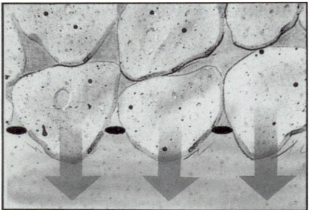

b

Figure 5.24 Geogrid sheets are used to stabilize a soft base (e.g., for roads, parking lots). (a) The aggregate penetrates the grid and locks into the base material (b) The grid sheets spread over the base (note overlap of joints). (Courtesy of Huesker, Inc.)

c

Figure 5.24 (*continued*) (c) A dozer spreads the new fill over the geogrid, usually 8–10 in. layers. (Courtesy of Huesker, Inc.)

placed between the fine-grain and the coarse material base in order to keep the fine-grain soil from intruding into the stone base, thereby destroying the shear or "bridging" strength of the base. The typically small perforations (slits) in the geotextile are large enough to allow water to permeate up or down but small enough to prevent the fine-grain soil from migrating up into the coarse base.

Reinforcement   Because of their tensile strength, geotextiles are sometimes placed over weak (poor bearing capacities) soils to form a reinforcement, somewhat analogus to reinforcing bars in concrete construction. Generally, a layer of controlled fill is placed over the geotextile, thereby creating a form of composite that spans over the weak soil. Indeed, inherent in the separation function described above is the added benefit of some reinforcement of the soft subsoil.

Filtration   Because water is permitted to move through the openings of the geotextiles without permitting the soil particles to do likewise, the effect is one of some filtration. Examples of such an application include the use of geotextiles as filters in hydraulic fills, to protect drains against soil infiltrations, to separate stone bases from fine-grain soils in various types of paved and unpaved construction, and for riprap construction and slope stability; see Figs. 5.21c, 5.21d, 5.21f, 5.21g.

Drainage Examples of geotextiles used for drainage include the construction of drainage galleries in earth dams (Fig. 5.21d); as a drainage blanket between two different formations (Figs. 5.21a, 5.21b); and under railroad ballast.

Geogrids

Geogrids resemble nets, to some extent, since they have relatively large apertures, which vary in size from 1 in. to as much as 4 in. Their primary purpose is to reinforce a soil or stone formation (Fig. 5.21b). Like the geotextiles, they are generally shipped to the job site in rolls. Examples of their uses are as follows:

Pavements (roads and parking lots) Typically, they are used in situations where the soil is relatively weak. Usually geogrids are used in conjunction with geotextiles; the geotextiles are placed first over the fine-grain soil, followed by a geogrid and stone blanket. This significantly increases the stiffness and bearing capacity of the strata. A common situation for their use is to reinforce soft and compressible soils for both paved and unpaved surfaces subjected to various types of loads, including moving loads.

Slopes and Embankments The general purpose is to reinforce, as part of the construction of new slopes, the slope or embankment against potential slip failure.

Reinforced Earth Walls Here the benefit is derived using the geogrid as a reinforcement to construct a fairly steep wall, somewhat similar to a reinforced earth construction (Fig. 5.21f).

Bearing Capacity Geogrids increase the bearing capacity of weak soils. Typically, these soils are fine grain, depict poor shear strength, and generally are of poor bearing capacity. The geogrid is placed over such soils followed by a controlled fill (usually large, crushed stone), which interlocks with the geogrid to form a composite blanket of stone and geogrid.

Geonets

Somewhat similar to geogrids, geonets have a net-like appearance. They are generally extruded with intersecting ribs. They find frequent use as a drainage medium (small series of pipes) under roads, solid waste sites, and behind retaining walls (Figs. 5.21d, 5.21e, 5.21g). For roads, the typical construction consists of placing the geotextile over the virgin soil, followed by the geonet, followed by another layer of geotextile. On top of the upper geotextile layer is the road base. The geonet is usually sloped toward a perforated drain or ditch, thereby permitting the water to move into the drain or ditch. For some landfills, where surface drainage is to be prevented from permeating down, the usual construction consists of an impervious plastic layer (geomembrane) over the virgin soil, followed by the geonet and, on top, the geotextile. This system permits vertical drainage through the geotextile into the geonet, which subsequently carries that water to a perforated drain. For retaining walls, the geonet is placed directly behind the wall followed by the geotextile. The geotextile separates the geogrid from the backfill. As such, it permits the water to be drained from the backfill into the geonet, which subsequently directs it into a perforated drain or weepholes. Geonets, too, are delivered to the job site in rolls.

Geomembranes

These are impervious (not perforated), continuous, thin plastic sheets, typically shipped in rolls for large projects, but folded for small projects. They are used primarily for lining and cover for liquid or solid waste storage facilities, with the typical function of limiting or eliminating the liquid or vapor transmission (Fig. 5.21e). Frequent uses of such material are in landfill construction, in limiting moisture infiltration to potentially unstable soils or expansive shales, and as vapor barriers under floor slabs.

As one might expect, geosynthetics may be subject to deterioration from some chemicals, ultraviolet light, biological hazards, animal and bacteria attack, and temperatures, especially heat. Thus, it is typical that these materials are placed and covered within a reasonably short time with suitable material to protect them from attack.

5.9 GEOENVIRONMENTAL—LANDFILLS

Estimates from surveys conducted by the U.S. Environmental Protection Agency, the Congressional Budget Office, the Chemical Manufacturers Association, and others indicate that hazardous waste generated in this country may exceed 300 million tons per year. It may cause a wide range of problems, including an increase in mortality and a variety of illnesses to humans and damage to our environment.

Hazardous waste may be found in the air, in bodies of water, and in the soil. The objective here is merely to provide an overview of the problem for treatment or management of ground-stored materials. Chemical treatment procedures of in-ground contaminants are beyond the scope of this book. A detailed approach to such treatments may be found in many (and a growing) number of sources; C. A. Wenz (38) presents particularly good coverage of this aspect.

In dealing with in-ground waste-disposal sites, we deal with two types: (1) sites that are engineered (designed) and constructed for the purpose of safe storage of waste, and (2) sites that have been used as dumps without proper regard, and in many instances without any regard, to the environment. For the discussion here, we shall refer to the former as "engineered" and to the latter as "unprotected" landfill sites.

Engineering a Landfill Site

When designing a landfill site, the engineer's goal is to develop a facility that will provide a reasonable degree of safety to humans and to the environment during its operational as well as its post-closure life. While costs are always a consideration, a totally fail-free facility for an indefinite time span is an unrealistic and perhaps unattainable goal, even without economic constraints; general degeneration of construction material with time, and man- or nature-induced destructive forces on the facility are factors that may adversely affect the life of the landfill facility.

The process of designing a hazardous waste site should include the following:

Site Selection   The topography and hydrology of the area are considered essential pieces of information. For example, the proximity and quality of the ground water and nearby users and geological faults are key ingredients when assessing the risks from the effects by a possible leak of the landfill facility during its operational as well as its post-closure life.

Planning and Design   The following data are considered relevant in the planning and design of a landfill facility:

- Anticipated amount and type of waste stored, the projected quantity and quality of material to be treated, and the selection of the treatment system
- Life expectancy of the landfill facility during its operating life
- The topography and soil characteristics near the site, and the climate conditions
- Surface and ground water data in the proximity of the site, and the monitoring of ground and surface water during the operating and post-closure phases of the facility
- Provisions for venting gaseous products
- Selection of durable (strong and immune to attacks by acids, alkaline substances), impermeable liners for the landfill
- Provisions for closure and post-closure of the landfill
- Compliance with governing regulations (local, state, federal)

Construction of a Landfill Site

Figure 5.25 depicts the basic components of an engineered landfill facility. The final "cover" is installed after the facility's working life. The construction sequence is usually as follows:

1. The site is excavated, generally bowl-shaped, with the bottom and sides reasonably smooth. Side slopes steeper than $30°$ may pose excavation problems, may introduce "anchoring" and stress problems in the geomembrane, and may sometimes create difficulty in placing any granular material (e.g., sand) over the inner geomembrane.

2. A clay liner, normally about 12-in. thick is placed over the excavated surface. This provides a degree of safety against puncturing of the outer geomembrane (e.g., from rock wedges that may protrude in the excavation surface). Also, it provides an added degree of safety regarding filtration, should a contaminated liquid escape the geomembrane container. A clay liner may be unnecessary if the subsurface material is cohesive and relatively free of broken rock.

3. The outer (bottom) membrane is placed over the clay, glued or fused at joints, and "anchored" at the top. A rather loose liner is advisable to reduce the potential for tearing during construction or due to waste-load shifting, and the like.

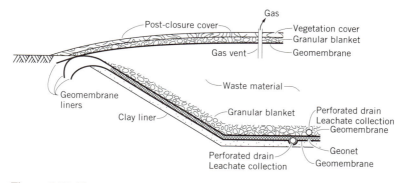

Figure 5.25 Elements of an engineered landfill facility. The post-closure cover follows the functional life of the facility.

4. A perforated pipe is placed at the bottom of the bowl over the outer liner to collect and discharge any liquid that may escape the inner liner.

5. The next layer may be a granular (e.g., sand), highly permeable material or a geonet whose primary function is to facilitate drainage of any hazardous liquid that may inadvertently escape the inner geomembrane liner, to be collected by a second perforated pipe. A geonet also provides an added degree of reinforcement to the geomembrane liner against tearing.

6. The inner (top) geomembrane is then placed and anchored to the top of the bowl. A perforated pipe is placed at the bottom of the bowl to collect and discharge the hazardous liquid from the facility for subsequent treatment.

7. A granular blanket about 8–12-in. thick is placed over the inner liner to facilitate flow to the perforated pipe.

8. The facility is capped following the functional life of the landfill to seal it from surface water. In order to direct surface runoff away, a small slope is constructed. Some additional soil cover, including some surface vegetation, may be placed over the geomembrane cap to keep it from external damage and to provide an acceptably esthetic (and possibly functional) site. A vent is sometimes installed to permit the escape of gases that may be generated in the landfill.

Unprotected Disposal Sites

Not all disposal sites are hazardous. Refuse, trash, and household garbage are biodegradable and, thus, do not pose severe, long-term problems to human health or the environment. On the other hand, many synthetic chemicals and hydrocarbons, if they should enter the water supply, may pose long-range problems to humans and the environment. Hazardous substances may induce alterations in the genetic/hereditary makeup (*mutagenicity*), abnormality in various forms (*teratogenicity*), various cancers (*carcinogenicity*), and other effects. Examples of toxic substances include lead, mercury, silver, cadmium, chromium, barium, arsenic—substances commonly associated with many industrial processes and wastes.

Many hazardous substances were and are still found in many disposal sites. Until about the late 1970s governmental regulations were lax if not nonexistent. Correspondingly, the hazardous waste problem grew almost unimpeded. Now, we are faced with not only the waste but also the increased evidence of potential harm these substances are capable of inflicting on us.

Some hazardous sites are known, and efforts are being taken to clean them up. Others are still undiscovered, and harmful substances remain untreated, posing the danger of contaminating our ground water. This presents enormous implications when one realizes that as much as 95% of the world's total fresh water resources are found in ground water, stored predominantly in aquifers. Once contaminated, ground water is difficult (sometimes impossible) to restore to a purified state.

Drilled wells are common means for monitoring ground water. Samples are extracted for testing, evaluation (volume, quantity, and type), and subsequently for developing methodologies for treatment. Figure 5.26 shows the basic features of a typical monitoring well. If the hole remains open (does not collapse), the steel casing is not needed. If a steel casing is needed, it is advanced (pounded) downward as the drilling auger also advances downward. Once the auger is extracted (the soil is removed by the auger flights), the pipe (usually plastic) and screen are lowered into the hole. A filter material (e.g., sand or gravel) is placed between the pipe and the casing, as the casing is extracted to just above the ground water level. The infiltrating fluid may then be extracted via pumping.

Another means of exploration and sampling is via excavating with a backhoe or excavator. The extraction of any liquid is by pumping from the open excavation. This approach has some disadvantages: limited depth, length of time the excavation will stay open without installed bracing, likely infiltration of surface runoff/rain water, and so on.

Frequently the most effective, but also quite possibly the most costly, remedial approach is the removal of the ingredients from a contaminated site via excavation. The excavated contaminated material is then transported for subsequent treatment at a designated treatment plant. Sometimes treatment at the contaminated site is possible and is usually less expensive.

The method of treatment of on-site pollutants may encompass three categories: *chemical, biological,* and *physical.* Here we shall merely scan over some aspects of

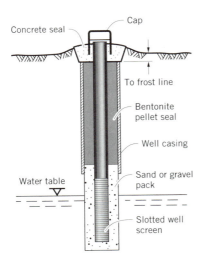

Figure 5.26 A typical ground water–monitoring well.

these categories; specifics regarding the processes is beyond the scope of this text. Briefly, the categories encompass some of the following treatments:

Chemical Methods

- Neutralize acids and alkaline contaminants
- Precipitate (e.g., of heavy metals) via coagulation and flocculation
- Convert toxic pollutants to less harmful materials via oxidation or reduction
- Destroy harmful organisms via disinfection
- Reduce the release of hazardous constituents via stabilization processes

Biological Methods

- Develop microbes that render organic compounds harmless via aerobic methods
- Introduce oxygen into the waste via aeration, to neutralize the harmful effects
- Convert harmful sludge to harmless via anaerobic methods

Physical Methods

- Separate large solids via screening
- Settle suspended solids via gravity (sedimentation)
- Separate low-density and hydrocarbon solids via flotation
- Filter suspended solids through a porous media such as sand
- Incinerate, consisting of heating the contaminants to very high temperatures (1600 to 1800°F) and ''burning off'' their harmful properties
- Others: evaporation, distillation, stripping, degasification

BIBLIOGRAPHY

1. Banarjee, S., and J. K. Mitchel, ''In-Situ Volume-Change Properties by Electro-Osmosis—Theory,'' *ASCE J. Geotech. Eng. Div.*, Apr. 1980.

2. Benzekri, M., and R. J. Marchand, ''Foundation Grouting of Moulay Youssef Dam,'' *ASCE J. Geotech. Eng. Div.*, Sept. 1978.

3. Bowles, J. E., *Engineering Properties of Soils and Their Measurements*, McGraw-Hill, New York, 1970.

4. Brown, R. E., and A. J. Glenn, ''Vibroflotation and Terra-Probe Comparison,'' *ASCE J. Geotech. Eng. Div.*, Oct. 1976.

5. Brown, R. E., ''Vibroflotation Compaction of Cohesionless Soils,'' *ASCE J. Geotech. Eng. Div.*, Dec. 1977.

6. Cernica, J. N., ''Proposed New Method for the Determination of Density of Soil in Place—Proposed New Technique,'' *ASTM Geotech. Test. J.*, Sept. 1980.

7. Clough, G. W., W. M. Kuck, and G. Kasali, ''Silicate-Stabilized Sands,'' *ASCE J. Geotech. Eng. Div.*, Jan. 1979.

8. D'Appolonia, D. J., R. V. Whitman, and E. D'Appolonia, ''Sand Compaction with Vibratory Rollers,'' *ASCE Specialty Conf. Placement and Improvement of Soil to Support Structure*, 1968.

9. D'Appolonia, D. J., ''Soil–Bentonite Slurry Trench Cutoffs,'' *ASCE J. Geotech. Eng. Div.*, Apr. 1980.

10. Foster, C. R., "Reduction in Soil Strength with Increase in Density." *Trans. ASCE,* vol. 120, 1955.

11. Giroud, J. P., and L. Noiray, "Geotextile-Reinforced Unpaved Road Design," *ASCE J. Geotech. Eng. Div.,* vol. 107, Sept. 1981.

12. Gray, D. H., and H. Ohashi, "Mechanics of Fiber Reinforcement in Sand," *ASCE J. Geotech. Eng. Div.,* vol. 109, Mar. 1983.

13. Hoare, D. J., "Synthetic Fabrics as Soil Filters: A Review," *ASCE J. Geotech. Eng. Div.,* vol. 108, Oct. 1982.

14. Holtz, W. G., and H. J. Gibbs, "Engineering Characteristics of Expansive Clays," *Trans. ASCE,* 1956.

15. Houlsby, A. C., "Engineering of Grout Curtains to Standards," *ASCE J. Geotech. Eng. Div.,* Sept. 1977.

16. Koerner, R. M., *Design with Geosynthetics,* 2nd ed., Prentice Hall, Englewood Cliffs, N.J., 1990.

17. Krizek, R. J., and T. Perez, "Chemical Grouting in Soils Permeated by Water," *ASCE J. Geotech Eng. Div.,* vol. 111, July 1985.

18. Lambe, T. W., *Soil Testing for Engineers,* Wiley, New York, 1951.

19. Lambe, T. W., "The Structure of Compacted Clay," *ASCE Soil Mech. Found. Div.,* vol. 84, no. SM2, May 1958.

20. Leshchininsky, D., and R. H. Boedeker, "Geosynthetic Reinforced Soil Structures," *ASCE J. Geotech. Eng. Div.,* vol. 115, Oct. 1989.

21. Lukas, R. G., "Densification of Loose Deposits by Pounding," *ASCE J. Geotech. Eng. Div.,* Apr. 1980.

22. Majumdar, D. K., "Effect of Catalyst in Soil–Resin Stabilization," *ASCE J. Geotech. Eng. Div.,* June 1975.

23. Menard, L., and Y. Broise, "Theoretical and Practical Aspects of Dynamic Consolidation," *Geotechnique,* vol. 25, 1975.

24. Mitchell, J. K., "The Fabric of Natural Clays and Its Relation to Engineering Properties," *Proc. Highw. Res. Board,* vol. 35, 1956.

25. Mitchell, J. K., and S. Banarjee, "In-Situ Volume-Change Properties by Electro-Osmosis—Evaluation," *ASCE J. Geotech. Eng. Div.,* Apr. 1980.

26. Moorhouse, D. C., and G. I. Baker, "Sand Densification by Heavy Vibratory Compactor," *3rd Proc. Soil Mech. Found.Div.,* ASCE 1968.

27. O'Rourke, J. E., "Structural Grouting of Oroville Dam Coreblock," *ASCE J. Geotech. Eng. Div.,* May 1977.

28. Pacey, J. G., Jr., "The Structure of Compacted Soils," M.S. thesis, Mass. Inst. Technol., Cambridge, Mass., 1956.

29. Proctor, R. R., "Fundamental Principles of Soil Compaction," *Eng. News Rec.,* Aug. 31, Sept. 7, 21, 28, 1933.

30. Seed, H. B., R. L. McNeill, and J. DeGuenin, "Increased Resistance to Deformation of Clay Caused by Repeated Loading," *ASCE J. Soil Mech. Found. Div.,* vol. 84, no. SM2, May 1958.

31. Seed, H. B., and C. K. Chan, "Structure and Strength Characteristics of Compact Clays," *ASCE J. Soil Mech. Found. Div.,* Oct. 1959.

32. Seed, H. B., and J. R. Booker, "Stabilization of Potentially Liquefiable Sand Deposits Using Gravel Drains," *ASCE J. Geotech. Eng. Div.,* July 1971.

33. Solymar, Z., V. Samsudin, J. Osellame, and B. J. Purnomo, "Ground Improvement by Compaction Piling," *ASCE J. Geotech. Eng. Div.,* vol. 112, Dec. 1986.

34. Vinson, T. S., and J. K. Mitchell, "Polyurethane Foamed Plastics in Soil Grouting," *ASCE J. Geotech. Eng. Div.,* June 1972.

35. Wan, T. Y., and J. K. Mitchell, "Electro-Osmotic Consolidation of Soils," *ASCE J. Geotech. Eng. Div.,* May 1976.

36.   Warner, J., "Strength Properties of Chemically Solidified Soils," *ASCE J. Geotech. Eng. Div.,* Nov. 1972.

37.   Warner, J., and D. R. Brown, "Planning and Performing Compaction Grouting," *ASCE J. Geotech. Eng. Div.,* June 1974.

38.   Wentz, C. A., *Hazardous Waste Management,* McGraw Hill, New York, 1989.

39.   Wilson, S. D., "Small Soil Compaction Apparatus Duplicates Field Results Closely," *Eng. News Rec.,* vol. 145, no. 18, Nov. 1950.

40.   Yamanouchi, T., N. Miura, N. Matsubayashi, and N. Fukuda, "Soil Improvement with Quicklime and Filter Fabric," *ASCE J. Geotech. Eng. Div.,* vol. 108, July 1982.

C H A P T E R 6

Spread-Footing Design

6.1 INTRODUCTION

Footings are foundation components that transmit the load from the structure to soil or rock. Included in the category of footings are those that support a single column, referred to as *isolated,* or *spread* footings. If the footing supports two or more columns it is classified as a *combined* footing. *Trapezoidal* and *strap* footings are special versions of combined footings, patterned to meet certain conditions or restrictions; so are *mat* foundations. *Pile caps* are special footings needed to transmit the column load to a group or cluster of piles. *Wall* footings support wall loads. Figure 6.1 depicts the general characteristics of these footings. Spread footings and wall footings are covered in this chapter; the others will be discussed in Chapter 7.

Concrete is almost always the material used in footings. It is strong and durable, is reasonably resistant to ordinary (sometimes even abnormal) acidic or alkaline conditions of soils, and is a convenient and economical construction material, workable and adaptable to field construction and requirements (4, 6, 7, 8).

Concrete footings may be plain or reinforced, with reinforcement running in one or two (or more) directions, depending on the direction(s) of flexure. For example, spread footings are usually reinforced in two directions, whereas the flexure steel in wall footings runs in the transverse direction to the wall.

Footing shapes usually vary with specific requirements and design needs. For spread or isolated footings, square shapes are common and usually most economical, but rectangular shapes are used if space is limited in one direction, or when loads are eccentric in one direction. The typically desired objective is to select the footing

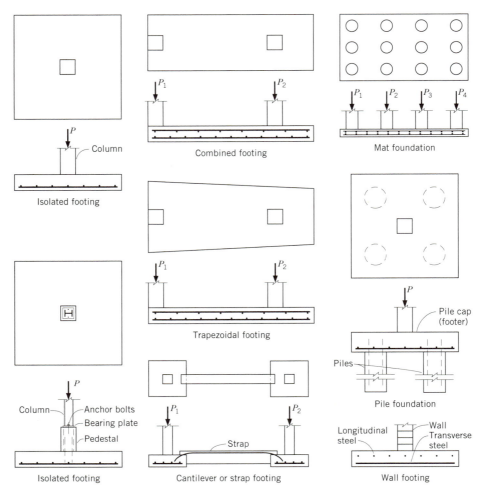

Figure 6.1 Typical configurations for various types of footings.

shape that makes the soil pressure (bearing pressure) as uniform as possible. Furthermore, footings may be of uniform thickness or may be sloped or stepped.

A steel *base plate* is used to spread the load from the steel column to the concrete, and thus ensure against crushing of the concrete. *Pedestals* are short concrete columns used to transmit loads from steel columns to footings at some depth in the soil strata. Embedment of steel columns, or contact with soil, should be avoided in order to protect the steel from possible hostile elements such as acids or alkaline substances that may be present in soils, and that subsequently cause corrosion of the steel.

Spread footings are designed to satisfy a combination of flexure, shear, and bearing. Much of the footing design procedure is adapted to yield a design that conforms with practical considerations and experimental findings. For example, the ϕ factors, discussed in the following section, are introduced so as to make the theoretical evaluations of members fit more consistently with experimental findings.

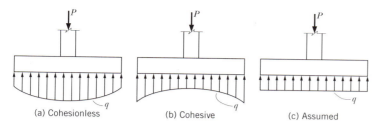

Figure 6.2 Soil pressure distributions under rigid footings.

Research by F. E. Richart (11) and J. Moe (9) provided early and useful guidelines for the analysis of spread footings. Their work, and documents such as the ACI Code and AASHTO specifications form the basis for the currently used guidelines and methods for spread-footing design. The ACI Code (ACI 318-86) (1) will be the code most adhered to in this text.

In Chapter 2, methodologies were presented for estimating pressures in subsurface formations induced by various types of loads (e.g., concentrated, line loads, rectangular and other surface shapes). Subsequently, in Chapter 4, we integrated many considerations, criteria, and procedures in an attempt to formulate some fundamental concepts toward a primary goal: determine the allowable bearing capacity, q_a, of the soil. In this chapter we will be using an allowable bearing pressure (q_a), with the implied assumption that the numerous data, soil-structure interaction, and various other considerations have been accounted for.

The pressure distribution under a footing depends, among other things, on footing rigidity, shape, footing depth, and soil properties. Generally, for ordinary spread footings resting on cohesionless formations, the pressure distribution is as shown in Fig. 6.2a. For combined and larger footings, the distribution may vary toward a more uniform shape near the middle two-thirds of the footing. On the other hand, for cohesive soils, the distribution appears to be opposite that for cohesionless; for this condition, the shape may approach that shown in Fig. 6.2b.

It is seldom that the engineer deals with a soil stratum that is totally cohesive or totally cohesionless; the more likely case is a mixture of cohesive and cohesionless material. Furthermore, it is rather difficult to place a numerical value for the pressure distributions for the two cases mentioned above. Thus, it is a widely accepted practice, and certainly convenient if perhaps not a totally accurate assumption, to use a uniform pressure distribution rather than a variable one. Hence, it will be the uniform pressure distribution (Fig. 6.2c) that will be assumed and discussed in this and subsequent chapters.

6.2 ELEMENTS OF REINFORCED CONCRETE DESIGN

In basic mechanics courses it is common practice when analyzing structural members subjected to bending to assume that stress is proportional to strain. The triangular stress distribution (e.g., depicted by Fig. 6.3a for the compression face) serves as the basis for what is known as the *Working-Stress Design* (*WSD*). Indeed, as early as about 1900 investigators detected appreciable variations from the "triangu-

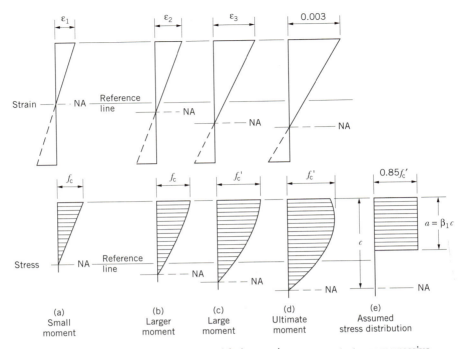

Figure 6.3 Stress–strain distribution with increasing moments to compressive failure. (a) The well-known straight-line distribution based on WSD. (b) through (e) Stress distributions associated with the Ultimate-Strength Design (USD).

lar'' stress distribution for concrete members. Yet the WSD approach was the predominant method until 1963, when the ACI Code gave both the WSD and *Ultimate-Strength Design* (*USD*) equal status.

The ACI Code (or simply Code) places almost total emphasis on the Ultimate-Strength Design (or just the Strength Design). Thus, the guide related to the design and analysis of concrete foundations herein will reflect on the latest Code (ACI 318-86), with virtually total emphasis on and conformity to the Strength Design. Furthermore, it is assumed that the student is familiar with the Strength Design. Nevertheless, but a brief summary of some relevant points will be presented here. A more comprehensive guide will be the Code itself, and references will be made to relevant sections of the Code as deemed necessary.

Load Factors (Code Sec. 9.2)

The Code specifies that the service loads be converted to ultimate loads by the following expressions:

When Dead and Live Load only
$$P_u = 1.4D + 1.7L \tag{a}$$

When Wind Load is included
$$P_u = 0.75 \, (1.4D + 1.7L + 1.7W) \tag{b}$$

When Dead and Wind Load only (L is absent)

$$P_u = 0.9D + 1.3W \tag{c}$$

Where D, L, and W are the service Dead, Live, and Wind loads, respectively. However, for any combination of D, L, and W, required P_u shall be not be less than Eq. (a).

ϕ Factors (Code Sec. 9.3)

The nominal design strength is multiplied by a ϕ factor, or capacity reduction factor, in order to account for uncertainties related to materials strengths, variations from drawings, workmanship, accuracy of calculations, exactness of equation, and other possible deviations from ideal or assumed conditions. The Code designates reduction ϕ factors as follows:

Flexure, without axial load	0.9
Axial tension and axial tension with flexure	0.9
Axial compression and axial compression with flexure:	
Members with spiral reinforcement conforming to Code Sec. 10.9.3	0.75
Other reinforced members	0.7
Shear and torsion	0.85
Bearing on concrete (see also Code Sec. 18.13)	0.70
Unreinforced footings	0.65

USD; Flexure

Strains in concrete members subjected to bending vary almost linearly with the distance from the neutral axis. Stress, on the other hand, varies in a linear fashion, as shown in Fig. 6.3a, only until f_c equals approximately $0.5 f_c'$. As the bending moment is increased, the linearity in the stress distribution is lost; in the process, the neutral axis is lowered.

In a paper presented in 1942, C. S. Whitney (12) showed that a simplified rectangle stress block of intensity equal to $0.85 f_c'$ and a depth of $\beta_1 c$, as depicted in Fig. 6.4c, gave acceptable results. Although the Code permits the use of any stress distribution if the resulting equations yield results that compare favorably with test data, the rectangular stress block has found virtually universal adoption; it will be the one focused on herein.

Reference is made to Fig. 6.4c.

$$\text{From } \Sigma F_x = 0, \qquad C = T.$$

Thus,

$$0.85 f_c' \, ab = A_s f_y$$

or

$$a = A_s f_y / 0.85 f_c' \, b \tag{6-1}$$

The Code limits the tensile reinforcement to a maximum of $0.75 p_b$ (Sec. 10.3.3). This is designed for the steel to yield prior to sudden concrete failure. Thus, the ultimate strength will be governed by the tensile steel.

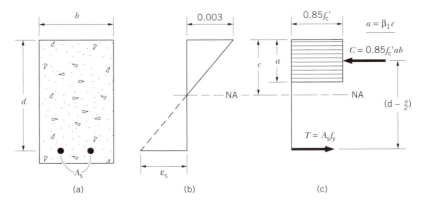

Figure 6.4 Relevant criteria used for the development of USD equations. (a) Beam section, (b) strain, (c) equivalent stresses and resultant couple.

Hence, from ΣM_c:

$$M_n = T\left(d - \frac{a}{2}\right) = A_s f_y \left(d - \frac{a}{2}\right)$$

The ultimate moment $M_u = \phi M_n$ or, for $\phi = 0.9$

$$M_u = \phi M_s = 0.9 A_s f_y \left(d - \frac{a}{2}\right) \qquad (6\text{-}2)$$

The expression for a from Eq. 6-1 can be written as:

$$a = \frac{A_s f_y}{0.85 f'_c b} = \frac{f_y}{0.85 f'_c}\left(\frac{A_s}{bd}\right) d = \frac{f_y d}{0.85 f'_c}\,(p)$$

where $p = A_s/bd$. Hence, substituting for a in Eq. 6-2,

$$M_u = 0.9 A_s f_y \left(d - \frac{d}{2}\,\frac{f_y}{0.85 f'_c}\,p\right) = 0.9 A_s f_y d \left(1 - p\frac{f_y}{1.7 f'_c}\right)$$

or

$$M_u = 0.9 A_s f_y d (1 - 0.59 p f_y / f'_c) \qquad (6\text{-}2a)$$

Also,

$$p_{max} = 0.75 p_b; \qquad p_{min} = 200/f_y$$

where $p_b = A_s/bd$, %, at balanced design.

The neutral axis may be located by using the linear strain relationship depicted in Fig. 6.4b:

$$c/d = 0.003/(0.003 + f_y/E_s)$$

Using a value of $E_s = 29 \times 10^6$ psi, and solving for c

$$c = 87,000(d)/(87,000 + f_y) \qquad (6\text{-}3)$$

From Fig. 6.4c, and for balanced design (i.e., $p_b = A_s/bd$)

$$c = a/\beta_1 = p_b f_y d/0.85 f'_c \beta_1 \qquad (6\text{-}4)$$

Equating the two expressions, Eqs. 6-3 and 6-4, we get

$$p_b f_y d / 0.85 \beta_1 f'_c = 87,000(d) / (87,000 + f_y)$$

$$p_b = (0.85 \beta_1 f'_c / f_y) [87,000 / (87,000 + f_y)] \tag{6-5}$$

In SI units

$$p_b = (0.85 \beta_1 f'_c / f_y) [600 / (f_y + 600)] \tag{6-5a}$$

where $\beta_1 = 0.85$ for $f'_c \leq 4000$ psi, and reduced at a rate of 0.05 for each 1000 psi in excess of 4000 psi, but β_1 shall not be less than 0.65. Table 6.1 provides some values for p_{max} for some commonly used values for f'_c and f_y.

Shear

The Code stipulates that $V_u = \phi V_n$, and $V_n = V_c + V_s$; V_u = ultimate shear; V_c = shear strength provided by the concrete; V_s = shear strength provided by the shear reinforcement; V_s is commonly absent in footings. The Code provides the following expressions for V_u when evaluating shear forces in footings.

Section Evaluated	Expression		Code Section
	Fps	SI	
Wide beam	$V_u = \left(2\phi \cdot \sqrt{f'_c}\right) b_w d$	$V_u = \left(\dfrac{\phi}{6} \cdot \sqrt{f'_c}\right) b_w d$	11.3
Diagonal tension or punching shear	$V_u = \left(4\phi \sqrt{f'_c}\right) b_w d$	$V_u = \left(\dfrac{\phi}{3} \sqrt{f'_c}\right) b_w d$	

where (see Fig. 6.5)

$\begin{aligned}
V_u &= \text{shear force (lb; Newtons, respectively)} = v b_w d \\
v &= \text{allowable shear stress (psi; MPa, respectively)} \\
b_w &= \text{width of footing (in.; mm, respectively)} \\
d &= \text{effective depth of footing (in.; mm, respectively)} \\
\phi &= 0.85 \\
f'_c &= \text{28 day compressive strength (psi; MP}_a\text{, respectively)}
\end{aligned}$

Table 6.1 $p_{max} = 0.75 p_b$, (%)

f'_c (ksi)	β_1	f_y (ksi)			
		40	50	60	75
2.5	0.85	0.023	0.017	0.013	0.010
3	0.85	0.028	0.021	0.016	0.012
3.5	0.85	0.032	0.024	0.019	0.014
4	0.85	0.037	0.028	0.021	0.016
5	0.80	0.044	0.032	0.025	0.018
6	0.75	0.049	0.036	0.028	0.021

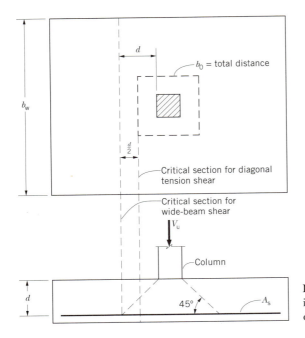

Figure 6.5 Critical section for investigating wide beam and diagonal tension stress.

Development (Bond) and Splices

The Code specifies expressions for determining basic development lengths for reinforcement bars (Sec. 12.2) as summarized in Table 6.2.

The basis development length shall be multiplied by the applicable factor or factors for

Top reinforcement	1.4
Reinforcement with $f_y > 60,000$ psi	$(2 - 60,000/f_y)$
Lightweight aggregate concrete	1.33
When bars spacing > 6 in. and bar cover is 3 in. or more	0.8
Reinforcement is in excess of required $l_d \geq 12$ in.	(A_s req'd/A_s provided)

Table 6.2 Development Lengths for Bars in Tension

Bar Size	l_d	C Values	
		Fps	S.I.
#11 and smaller bars	$C_1 A_b f_y / \sqrt{f'_c}$	0.04	0.02
but not less than	$C_2 d_b f_y$	0.0004	0.06
#14 bars	$C_3 f_y / \sqrt{f'_c}$	0.085	25
#18 bars	$C_4 f_y / \sqrt{f'_c}$	0.125	35
Deformed wire	$C_5 d_b f_y / \sqrt{f'_c}$	0.03	0.38

The development length for compression bars (see Code Sec. 12.3) shall not be less than

Fps $\qquad l_d = 0.02 f_y d_b / \sqrt{f'_c}$, or $0.0003 f_y d_b$, or 8 in.

SI $\qquad l_d = 0.24 f_y d_b / \sqrt{f'_c}$, or $0.044 f_y d_b$, or 200 mm

Critical Sections for Footings

The Code designates specific critical sections; Sec. 15.4 designates the critical locations for moment, while Sec. 15.5 covers the critical sections for shear. The critical sections differ, depending on column type or wall (e.g., concrete, steel, or masonry) as depicted by Fig. 6.6. For masonry walls, the author recommends the critical section for shear at the face of the wall—a somewhat conservative approach.

Bearing on Concrete (Sec. 10.15)

$$q_{brg} \leqq 0.850 f'_c A_1$$

This may be multiplied by $\sqrt{A_2 / A_1}$, but not more than 2 when the supporting surface is wider on all sides than the loaded area. (See Example 6.1 for determining $\sqrt{A_2 / A_1}$.)

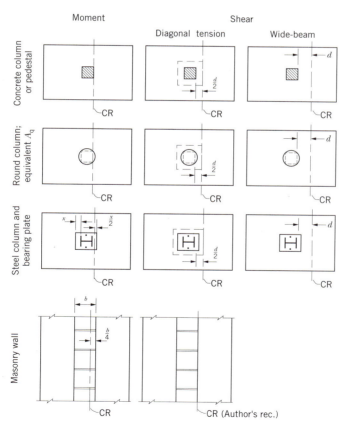

Figure 6.6 Critical footing sections for moment, and shear.

Minimum Footing Depth (Sec. 15.7)

6 in. above bottom reinforcement

12 in. on piles

Transfer of Forces at Base of Column or Pedestal

As related to footings, the Code (Sec. 15.8) requires that reinforcement dowels (or mechanical connectors) be used to transfer all compressive force that exceeds concrete bearing of either footing or column/pedestal, with $A_s = 0.005A_g$. For short pedestals (say up to 5 or 6 ft), the pedestal reinforcement may be cast into the footing. However, for longer pedestals or columns, it is more practical to use dowels, four or more for stability.

Miscellaneous Requirements

Spacing of bars (see Secs. 7.6 and 3.3): clear dist. $\geqq d_b$ or 1 in.; greater than $1.33 \times$ max. size of aggregate and not more than $3 \times$ depth of footing, or 18 in.

Lap Splices   The Code provides requirements for splice lengths as follows (Secs. 12.15 and 12.16):

Tension
$$\begin{cases} l_s = l_d \text{ when } f_s < f_y/2 \text{ and not over 75\% of the tension bars are spliced within} \\ \quad \text{one lap length} \\ l_s = 1.3 l_d \text{ when } f_s < f_y/2 \text{ but over 75\% of the tension bars are spliced within} \\ \quad \text{one lap length} \\ l_s = 1.3 l_d \text{ when } f_s > f_y/2 \text{ and not over 50\% of the tension bars are within one} \\ \quad \text{lap length} \\ l_s = 1.7 l_d \text{ when } f_s > f_y/2 \text{ and more than 50\% of the tension bars are within} \\ \quad \text{one lap length.} \end{cases}$$

Compression
$$\begin{cases} \text{Splice lengths of deformed bars in compression shall be (Sec. 12.16) } l_s = l_d, \\ \text{but not less than } 0.0005 f_y d_b, \text{ nor } (0.0009 f_y - 24) d_b \text{ for } f_y > 60,000 \text{ psi, nor 12 in.} \end{cases}$$

Modulus of Elasticity $E_c = 57,000 \sqrt{f_c'}$ (Sec. 8.5).

Reinforcement Cover   Minimum of 3 in. when exposed to earth (Sec. 7.7). Limits for reinforcement of compression members $0.01A_g \leqq A_s < 0.08A_g$ (Sec. 10.9).

6.3 SPREAD-FOOTING DESIGN: GENERAL CRITERIA

The name spread, or isolated, footing encompasses both square and rectangular footings that support a single column. The load may be concentric, in which case the footing is likely to have a square shape. If the load is eccentric (e.g., bending moment is present) or if restricted by space, the footing may assume a rectangular shape. In all instances, however, the objective is toward a uniform soil pressure distribution as much as practical.

Shear stresses usually govern the thickness of reinforced footings. Flexure may control the thickness of plain footings. Diagonal tension stress (or punching shear) always controls the thickness of reinforced square footings subjected to concentric loads; this is frequently true for combined footings as well. Wide-beam shear may govern the footing thickness for long footings, generally when the length to width ratio exceeds about $2:1$. Thus, if the depth of rectangular footings is calculated based on diagonal shear governing, a check should be made for wide-beam stress. The larger of the two effective depths is to be used.

Shear failures never occur on vertical planes. Instead, the diagonal tension cracks develop on planes sloped at approximately 45°, as shown in Figs. 6.7b and 6.8. Thus, for footings, a truncated pyramid is "punched out" from the footing. Subsequent to a detailed 1962 report (2), the Joint ACI-ASCE Committee recommended that diagonal tension shear stress be calculated on planes at a distance of $(d/2)$ from the face of the column, as shown in Fig. 6.7a; Fig. 6.7c shows the forces acting on the cut-out section based on such an assumption. For rectangular footings, the shear stress is to be checked across the width of the footer at a distance d from the face of the column, as shown in Fig. 6.7a.

From Fig. 6.7c $\Sigma F_y = 0$,

$$P_u = 2v_c d(x + d) + 2v_c d(y + d) + (x + d)(y + d)q \tag{a}$$

but

$$P_u = BLq$$

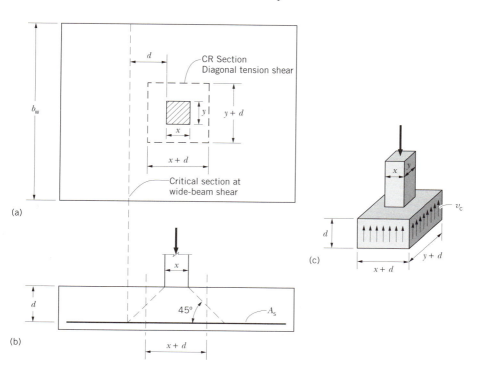

Figure 6.7 Diagonal tension, or punching shear, in reinforced footings.

Figure 6.8 Diagonal tension failure in beams. Typical diagonal tension reinforcement to prevent such failures in beams are "stirrups," which are usually placed vertically for convenience but at a 45° angle, perpendicular to the cracks, for maximum efficiency. In footings, the common practice is to increase the footing thickness instead of using stirrups.

Thus, expanding and rearranging terms, Eq. (a) becomes

$$BLq = d^2(4v_c + q) + d(2v_c + q) + xyq$$

or, for rectangular columns

$$(BL - xy)q = d^2(4v_c + q) + d(2v_c + q)(x + y) \qquad (6\text{-}6)$$

For square columns, $x = y$. Thus, Eq. 6-6 becomes

$$(BL - x^2)(q/4) = d^2(v_c + q/4) + dx(v_c + q/2) \qquad (6\text{-}6a)$$

For round columns of diameter $2r$, Eq. 6-6a becomes

$$(BL - A_{col})q/\pi = d^2(v_c + q/4) + 2dr(v_c + q/2) \qquad (6\text{-}6b)$$

In the above equations, q represents the ultimate soil pressure. The allowable soil pressure q_a is part of the given data; so are x, y (or r), and f'_c; v_c is the allowable shear stress. Hence, one proceeds to solve for d, then to check for wide-beam shear stress, and subsequently to complete the design.

At this point the general procedure for design of isolated footings subjected to concentric load is as follows:

Given: Column dimensions and reinforcement; column loads (L, D); f_c' for footing and column; f_y for footing and column; q_a.

Solution:

1. Find $P_u = 1.4D + 1.7L$ (W usually absent)
2. Determine B and L; $B = (D + L)/q_a$. For a unique solution, B or L is fixed.
3. Find $q_u = P_u/BL$
4. Determine v_c for diagonal tension
5. Determine d via applicable Eqs. 6-6, 6-6a, or 6-6b. For a rectangular footing, check wide-beam shear; the larger d governs.
6. Compute the required flexure steel, development length, etc., in each direction. The moment is computed at the critical sections as specified by Code (Sec. 15.4).
7. Compute column bearing stress and select the appropriate dowels.
8. Complete a design drawing showing all details (footing dimensions, reinforcement size, spacing, cover, etc.).

6.4 SPREAD-FOOTING DESIGN: CONCENTRIC LOADS

In the majority of cases, square footings are associated with concentric column loads. However, if eccentrically loaded, a square footing is generally not the most efficient shape even if the column location is eccentrically shifted with respect to the centroid of the footing in order to ensure uniform bearing pressure. In that case a rectangular shape is generally more desirable; it provides a more favorable moment of inertia about the axis of rotation, among other things.

Square Footings

A concentrically loaded square footing is usually reinforced with equal steel in both directions. In the case of square columns (or equivalent circular ones) the effective d is commonly designated at the interface of the two steel grids, as shown in Fig. 6.9. The selected A_s over that required is generally sufficient to compensate for a slight loss in d for the upper level of bars; conversely, this yields a slight gain in d for the lower row of bars, thus somewhat balancing the effects.

For rectangular concrete columns, the moment is to be calculated at the wider face, line *a-a* in Fig. 6.9. This yields a slightly larger moment than with respect to the narrow face, and correspondingly a larger A_s. In using the same A_s in both directions, the bottom layer of steel should be placed perpendicular to line *a-a*.

The diagonal shear stress always controls for reinforced square footings supporting concentric-load columns. Wide-beam shear should be checked, however, for a square footing where the column is located eccentrically with the centroid of the footing. For an unreinforced footing, d is generally controlled by concrete flexure stresses. Again, the observations are made predicated on a uniform soil pressure—the desired objective as much as practical.

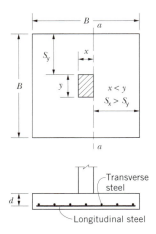

Figure 6.9 Section for calculating bending moment for footing supporting a rectangular concrete column.

Transverse steel

Longitudinal steel

Example 6.1

Given   (The quantities in parentheses are SI equivalents.)

Column	Footing
$D = 275$ kips (1223 kN)	$f'_c = 3$ ksi (20.7 MPa)
$L = 200$ kips (890 kN)	$f_y = 60$ ksi (414 MPa)
18 in. × 18 in. (0.46 m × 0.46 m)	$q_a = 3$ ksf (144 kPa)
8 #8 bars	
$f_y = 60$ ksi (414 MPa)	
$f'_c = 4$ ksi (27.6 MPa)	

Find   Suitable square footing design.

Solution   (Previously outlined steps will be noted again)

1. $P_u = 1.4(275) + 1.7(200) = 725$ kips
2. $B = L = \sqrt{(275 + 200)/3} = 12.58$ ft, say 12.75 ft
3. $q_u = P_u/B^2 = 725/(12.75)^2 = 4.46$ ksf
4. $v_c = 4\phi\sqrt{f'_c} = 4(0.85)\sqrt{3000} = 186.2$ psi $= 26.8$ ksf
5. Use Eq. 6-6a to determine d:

$$(B^2 - x^2)\, q/4 = d^2(v_c + q/4) + d(v_c + q/2)x$$

In our case $x = 18$ in. $= 1.5$ ft. Hence,

$$(12.75^2 - 1.5^2)(4.46/4) = d^2(26.8 + 4.46/4) + d(26.8 + 4.46/2)(1.5)$$

Solving, $d = 1.86$ ft $= 22.32$ in.; use $d = 22.5$ in.

(Checking for wide-beam shear is not necessary; for square footing diagonal shear stress governs.)

6. The critical section for bending for a concrete column is at its face: Using a 1-ft wide strip, Fig. 6.10, we get,

$$M = 4.46(5.625)^2/2 = 70.56 \text{ k-ft} = 846.70 \text{ k-in.}$$
$$a = A_s f_y/0.85 f_c' b = 60A_s/0.85 \times 3 \times 12 = 1.96A_s$$
$$M_u = \phi A_s f_y (d - a/2) = 0.9A_s(60)(22.5 - 1.96A_s/2) = 846.7 \text{ k-in.}$$

Solving, $A_s = 0.72 \text{ in.}^2/\text{ft}$

$$A_{tot} = 0.72 \times 12.75 = 9.18 \text{ in.}^2$$

Check p: $p = A_s/bd = 0.72/(12)(22.5) = 0.0027$

$$0.75p_b = 0.016 \text{ (See Table 6.1)}; \qquad p_{min} = 0.002$$
$$0.016 > 0.0027 \gg 0.002 \qquad \text{O.K.}$$

Use #8 bars; $A_s/\text{bar} = 0.79 \text{ in.}^2$

$$\text{Spacing} = (A_s \text{bar}/A_{\text{req'd}})(12) = (0.79/0.72)(12)$$
$$= 13.08 \text{ in.; use 13 in.}$$

Using 12 #8 bars @ 13 in. c.c., $A_s = 9.42 \text{ in.}^2 > 9.18 \text{ in.}^2$ O.K.
Development length (Sec. 12.2)

$$l_d = 0.04A_s f_y/\sqrt{f_c'} = (0.04)(0.79)(60,000)/\sqrt{3000} = 34.4 \text{ in., say 35 in.}$$

or

$$l_d = 0.0004d_b f_y = (0.0004)(1)(60,000) = 24 \text{ in.}$$

Thus, use 35 in.

7. Bearing (Sec. 10.16)

$f_c = 0.85\phi f_c' \sqrt{A_2/A_1}$ (see Fig. 6.11).   In our case, $\sqrt{A_2/A_1} = 6$;   use 2

Thus $f_c = 0.85(0.7)(3000)2 = 3570 \text{ psi} = 3.57 \text{ ksi}$

$$f_{c \text{ actual}} = 475/(18)^2 = 1.466 \text{ ksi} < 3.57 \text{ ksi} \qquad \text{O.K.}$$

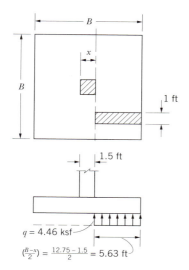

Figure 6.10 Critical section for calculating bending in a footing supporting a concrete column.

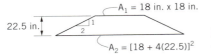

Figure 6.11 Frustrum used when determining A_2.

Dowels
Since $f_{c\ actual} < f_{c\ allow}$ for bearing,

$$A_{dow} = 0.005A_{g\ col}; \text{min. of 4 bars}$$
$$A_{dow} = 0.005(18)^2 = 1.62 \text{ in.}^2$$

Use 4 #6 bars, $A_s = 1.76$ in.^{2}
Dowel diameter must not exceed longitudinal bar diameter by more than 0.15 in. In our case dowel bar diameter = 0.75 < 1 in. O.K.
Dowel development length (Sec. 12.3.2)

$$l_d = 0.02d_b f_y/\sqrt{f_c'} = 0.02(0.75)60,000/\sqrt{3000} = 16.43 \text{ in.} > 8 \text{ in.}$$

or

$$l_d = 0.003d_b f_y = 0.003(0.75)(60,000) = 13.5 \text{ in.} > 8 \text{ in.}$$
$$l_d = 16.43 \text{ in.} < d = 18 \text{ in.}$$

If needed, the l_d could be modified (Sec. 12.3.3.1), as

$$l_d' = l_d(A_{req'd}/A_{prov}) = 16.43(1.62/1.76) = 15 \text{ in.}$$

but, practically, $l_d \approx$ say 23 in. or 24 in., would be advisable into the footing "secured" to main reinforcement, and say 16.5 in. into column.

8. Details of the design are shown in Fig. 6.12. Reinforcement cover (Sec. 7.71a) is 3 in. Thus, $D = 22 + 1 + 3 = 26$ in. If we were to use $d = 22.5$ in. to center of top steel (instead of to the interface of two layers of reinforcement), $D = 22.5 + 0.5 + 1 + 3 = 27$ in. Since A_s furnished = 9.42 in.^{2} > 9.18 in.^{2},

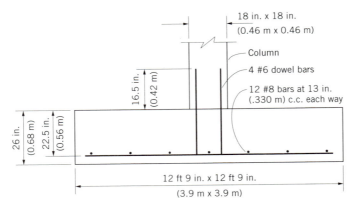

Figure 6.12 Dimensions and details depicting the final design.

Figure 6.13 Reinforcement in both directions. For square footings, the amount of reinforcement is equal steel in each direction. For rectangular or combined footings, the steel varies as dictated by the bending moments in each direction. Note the concrete blocks used to support the steel, thereby providing for reinforcement cover. Although such blocks are common, steel "seats" are frequently specified for this purpose.

using a $d = 22$ in. versus $d = 22.5$ in. will generally not alter $A_{s\,\text{req'd}}$. However, a check:

$$A_s(22.0 - 1.96A_s/2) = 846.7/(0.9)(60)$$

Solving $A_s = 0.735$ in.^{2}/ft $= 9.36$ in.^{2}/12.75 ft

9.36 in.^{2} < 9.42 in.^{2} available

O.K. to use 26 in. for overall depth, as shown.

Rectangular Footings

A concentrically loaded rectangular footing requires different reinforcement in the two directions, as depicted in Fig. 6.13, since the moments in the "long" and "short" directions are different (see Example 6.2).

The effective depth, d, is calculated based on diagonal shear stress, but is checked for wide-beam stress. As with square footings, it is common to assign d to the interface of the two reinforcement layers. Which of the two layers is placed on the bottom may depend on the A_s selected versus that required. For deeper footings (say $d = 20$ in. or larger), placement usually makes but a slight difference in the overall scope for design (see Example 6.1).

Example 6.2 _____

Given   The same data used for Example 6.1 (the quantities in parentheses are SI equivalents).

<table>
<tr><td colspan="1">Column</td><td>Footing</td></tr>
</table>

Column	**Footing**
$D = 275$ kips (1223 kN)	$f_c' = 3$ ksi (20.7 MPa)
$L = 200$ kips (890 kN)	$f_y = 60$ ksi (414 MPa)
18 in. $\times$ 18 in. (0.46 m $\times$ 0.46 m)	$q_a = 3$ ksf (144 kPa)
8 #8 bars	
$f_y = 60$ ksi (414 MPa)	
$f_c' = 4$ ksi (27.6 MPa)	

Find   Suitable rectangular footing design for $L/B = 1.4$

Solution
1. $P_u = 1.4(275) + 1.7(200) = 725$ kips
2. $L = 1.4B$; $A = 1.4B^2 = (275 + 200)/3 = 158.33$ ft^2
   $B = 10.63$ (say 10 ft 8 in.); $L = 14.89$ ft, say 15 ft; $A = 160$ ft^2
3. $q_u = 725/160 = 4.53$ ksf
4. $v_c = 4\phi\sqrt{f_c'} = 4(0.85)\sqrt{3000} = 186.2$ psi $= 26.8$ ksf
5. Use Eq. 6-6a to determine d

$$(BL - x^2)(q/4) = d^2(v_c + q/4) + dx(v_c + q/2)$$
$$(160 - 1.5^2)(4.53/4) = d^2(26.8 + 4.53/4) + d(1.5)(26.8 + 4.53/2)$$
$$d^2 + 1.56d - 6.4 = 0$$
$$d = 1.86 \text{ ft} = 22.4 \text{ in., say } 22.5 \text{ in.}$$

Check wide-beam shear at Sec. 1-1, in Fig. 6.14.

$$V = (4.88)(4.53) = 22.11 \text{ kips} = 22,110 \text{ lbs.}$$
$$v_c = V/bd = 22,110/12 \times 22.5 = 81.87 \text{ psi} < 2\phi\sqrt{f_c'} = 93.1 \text{ psi} \qquad \text{O.K.}$$

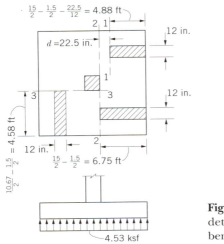

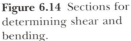

Figure 6.14 Sections for determining shear and bending.

6. The moment in the long direction is computed at Sec. 2-2. Using a 1-ft wide strip, Fig. 6.14

$$M = 4.53(6.75)^2/2 = 103.2 \text{ k-ft} = 1238.4 \text{ k-in.}$$
$$a = 1.96A_s$$
$$M_u = \phi A_s f_y (d - a/2) = 0.9A_s(60)(22.5 - 1.96A_s/2) = 1238.4$$

Solving, $A_s = 1.07$ in.²/ft (long direction)

$$A_{tot} = 1.07 \times 10.67 = 11.42 \text{ in.}^2 \qquad \text{(long direction)}$$

Check p: $p = A_s/bd = 1.07/(12)(22.5) = 0.00396$

$$0.75 p_b = 0.016 \text{ (Table 6.1)} > 0.00396 > 0.002 \qquad \text{O.K.}$$

Use 15 #8; $A_s = 11.85 > 11.42$ in.²   O.K.
Spacing say 8.5 in. c.c = 119 in., starting at approximately 4.5 in. from edge.
In short direction: The moment is determined at Sec. 3.3.

$$M = 4.53(4.58)^2/2 = 47.5 \text{ k-ft} = 570.1 \text{ k-in.}$$

Thus, $a = 1.96A_s$

$$M_u = \phi A_s f_y (d - a/2) = 0.9A_s(60)(22.5 - 1.96A_s/2) = 570.1$$

Solving for A_s in short direction,

$$A_s = 0.42 \text{ in.}^2/\text{ft}$$
$$p = 0.42/(12 \times 22.5) = 0.00156 < 0.002; \qquad \text{use } 0.002$$

Thus total steel in short direction $= 0.002(12 \times 22.5) = 0.54$ in.²/ft.

$$A_{s\,tot} = 15 \text{ ft} \times 0.54 = 8.3 \text{ in.}^2$$

If we use 11 #8 bars, $A_s = 8.69$ in.²   O.K.
Spacing say 1 ft 5 in. = 170 in. starting at approximately 5 in. from edge (Code Sec. 15.4.4.2)

$$A_s = \left(\frac{2}{\dfrac{L}{B} + 1} \right) A_{s\,tot} = \left(\frac{2}{(1.4 + 1)} \right) 8.69$$

7. Bearing

$$f_c = 0.85\phi f_c' \sqrt{A_2/A_1}; \qquad \sqrt{A_2/A_1} = 6.0; \qquad \text{use } 2$$

Thus, $f_c = 0.85(0.7)(3000)(2) = 3570 \text{ psi} = 3.57 \text{ ksi.}$

$$f_{c\,act} = 475/(18)^2 = 1.466 \text{ ksi} < 3.57 \text{ ksi.} \qquad \text{O.K.}$$

Dowels: See Example 6.1 for dowel design; same dowel criteria apply here.

8. Details for the design are shown in Fig. 6.15.

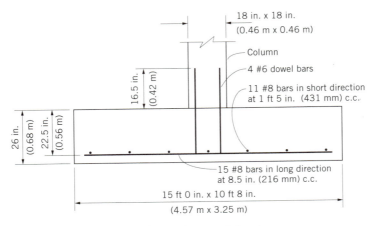

Figure 6.15 Final design dimensions and details.

6.5 ECCENTRICALLY LOADED SPREAD FOOTINGS

Figure 6.16 is a schematic depiction of a column that supports beams from several spans and a crane rail. The ensuing "load" on the column, and subsequently on the footing, may be a combination of a vertical load and moment(s), as shown by the lower figure.

The vertical dead load transmitted to a column may be fairly constant unless the building function changes significantly with time. The live-load and/or wind-load

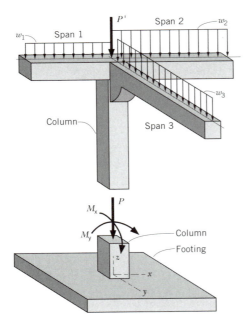

Figure 6.16 Example of a loading condition that may induce eccentric loading in two directions.

effects, however, may vary. For example, the support in Fig. 6.16 may experience loads (both vertical and horizontal) that change in both magnitude and duration during the life of the structure. Similarly, the wind loads may vary not only in intensity and duration but also in direction. Thus, it is perhaps apparent that some discretion and engineering judgment must be used in establishing reasonable load parameters to be used in designing the various support components, including the footing.

The source for the effects of eccentricity on the footing may be either a concentric vertical load and moment combination (Fig. 6.17a) or a column located eccentrically to the centroid of the footing (Fig. 6.17b).

The corresponding soil pressures for these load conditions are shown in Figs. 6.17a and 6.17b. It is prudent to proportion the footing size such that $q_{min} \geq 0$ (q is never negative); indeed, in order not to overstress the soil under some points of the footing, and to eliminate tilting of column and footing, a uniform soil distribution is desirable (Fig. 6.17c).

The difficulty in establishing a fixed location of the load centerline relative to the footing centroid is inherent in the problem of assigning reliable values and orientations of the direct loads, P, and of moments, particularly the live and wind loads. Thus, our design is perhaps somewhat hypothetical in a strict sense—perhaps only as good as our assumptions. Once the soil distribution is formulated, however, the general design procedure remains as presented in the two preceding sections.

In a general sense, the net soil pressure distribution associated with eccentric loads is obtained by superimposing the pressures resulting from the direct vertical load to those from moment. That is, from Fig. 6.18,

$$q = (P/A) \pm (M_x c_1/I_x) \pm (M_y c_2/I_y) \tag{6-7}$$

The moments can be expressed as equivalent to the load P shifted an eccentricity e: that is,

$$M_x = Pe_y \quad \text{and} \quad M_y = Pe_x$$

Also,

$$I_x = LB^3/12 \quad \text{and} \quad I_y = BL^3/12$$

Hence, Eq. 6-7 becomes

$$q = (P/BL)[1 \pm (6e_x/L) \pm (6e_y/B)] \tag{6-8}$$

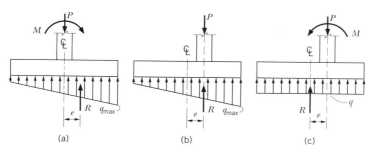

(a)   (b)   (c)

Figure 6.17 Soil pressures resulting from eccentric loading.

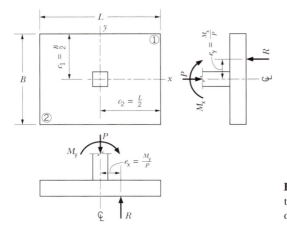

Figure 6.18 Footing subjected to eccentric loading in two directions.

It may be apparent that locating the column at the centroid of the footing is creating an undesirable soil distribution; point 1 in Fig. 6.18 may be much "over-stressed" while the reverse is true for point 2. For an existing installation one may have to make the best of the situation. On the other hand, for a new design, a remedy would be to locate the column toward point 2; this is a somewhat cumbersome, try-and-miss procedure if one's objective is to get a relatively uniform soil distribution. The procedure is rather straightforward for only one moment (see Example 6.3).

For a vertical load and a two-moment condition, one alternative is to resolve, vectorially, the two moments into one. Then, the footing might be oriented (rotated) in the *x-y* plane, space permitting, such that the length is in the direction of the resultant moment. In essence, this converts the two-moment case into a single-moment condition.

Prior to assigning a fixed value for M_x or M_y onto the footing, one should closely evaluate the rigidity of attachment or anchorage at the column–footing interface (e.g., dowels for concrete columns and anchor bolts for steel columns). Although concrete columns and dowels can be designed to transmit significant moments into the base (footing), steel columns and anchor bolts require appreciably more scrutiny; crushing of concrete under heavy eccentric loading, bearing stress induced by hooks (anchors) on the footing concrete, plate stiffness, and tolerances are frequent points to evaluate.

Example 6.3 _____

Given	*Column*	*Footing*
	$D = 180$ kips (800 kN)	$f'_c = 4$ ksi (27.6 MPa)
	$L = 150$ kips (668 kN)	$f_y = 60$ ksi (414 MPa)
	20 in × 20 in. (500 mm × 500 mm)	$q_a = 4$ ksf (190 kPa)
	8 #8 bars	
	$f_y = 60$ ksi (414 MPa)	
	$f'_c = 4$ ksi (27.6 MPa)	
	$M_y = 250$ k-ft (327 kN-m)	

Find   A suitable design such that $L/B = 1.4$, assuming a reasonably uniform q.

Solution   The depicted orientation (Fig. 6.19), relative to the applied moment, provides a favorable resistance (i.e., I_{yy} is larger than I_{xx}). Hence, $e = 250/(180 + 150) = 0.75$ ft $= 9$ in. This location of P yields a uniform q. Thus,

$$A = (180 + 150)/4 = 82.5 \text{ ft}^2$$
$$1.4B^2 = 82.5$$
$$B = 7.67 \text{ ft} = 7 \text{ ft } 8 \text{ in.}$$
$$1.4B = 10.75 \text{ ft} = 10 \text{ ft } 9 \text{ in.}$$
$$P_u = 1.4 (180) + 1.7(150) = 507 \text{ kips}$$

and

$$q_u = 507/82.5 = 6.15 \text{ ksf}$$

For punching shear, (square column: $x = y = 20$ in. $= 1.67$ ft)

$$(BL - x^2)(q/4) = d^2(v_c + q/4) + d(v_c + q/2)x$$
$$v_c = 4\phi\sqrt{4000} = 214 \text{ psi} = 31 \text{ ksf}$$
$$(82.5 - 2.78)(6.15/4) = d^2(31 + 6.15/4) + d(31 + 6.15/2)1.67$$

Simplifying,

$$d^2 + 1.75d - 3.77 = 0$$

and

$$d = 1.25 \text{ ft} = 15 \text{ in.}$$

Check wide beam Sec. 2-2, 1-ft strip (Fig. 6.20).

$$V = (4.04)(6.15) = 24.85 \text{ kips}$$

Actual $v_c = v/bd = 24.85/(12)(15) = 0.138$ ksi. Allowable v_c is

$$v_c = 2\phi\sqrt{f'_c} = 0.1075 < 0.138 \qquad \text{N.G.}$$

Thus, we increase d. To find the required d, use allow v_c; that is,

$$0.1075 = \frac{4.04 \times 6.15}{(12)d}$$

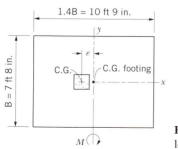

Figure 6.19 Eccentrically loaded footing.

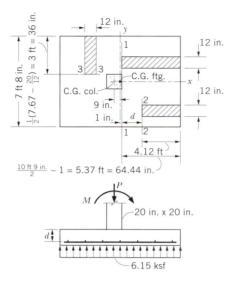

Figure 6.20 Sections for calculating bending and shear.

Hence, $d_{req'd} = 19.3$ in., say 19.5 in.
Critical section for longitudinal bending is 2-2.

$$M = 6.15(5.37)^2/2$$
$$= 88.67 \text{ k-ft} = 1064 \text{ k-in.}$$
$$M_u = \phi A_s f_y (d - a/2)$$

where,

$$a = A_s f_y / 0.85 f'_c b = A_s(60) / 0.85(4)(12) = 1.47 A_s$$

Thus,

$$M_u = 0.9 A_s(60) \left(19.5 - \frac{1.47 A_s}{2} \right) = 1050 A_s - 39.69 A_s^2 = 1064 \text{ in.-lb}$$

Simplifying,

$$A_s^2 - 26.41 A_s + 26.81 = 0$$

Solving,

$$A_s = 1.24 \text{ in.}^2/\text{ft}; \qquad A_{s\,tot} = (1.24)(7.67) = 9.5 \text{ in.}^2$$

Check $p = 1.24/(12)(15) = 0.0069 > 0.02 < 0.021$ (Table 6.1); O.K.
Using 10 #8 bars, $A_s = 9.48$ in.^{2} < 9.5 in.^{2} req'd ($\approx$ close); O.K. Place these longitudinal bars in bottom grid, thus more than negating the slight deficiency in A_s via a larger d. Spacing 7.5 in. c.c. with approx. 5 in. on each side.

$$l_d = 04 A_b f_y / \sqrt{f'_c} = 0.04(0.79)(60,000) / \sqrt{4000} = 30 \text{ in.}$$

or

$$l_d = 0.0004 d_b f_y = 0.0004(1)(60,000) = 24 \text{ in.}$$

Thus, required $l_d = 30$ in. < 64.44 in. available; O.K. In the short direction, (see Fig. 6.20)

$$M = (6.15)(3)^2/2 = 27.675 \text{ k-ft} = 332 \text{ k-in.}$$

Thus,

$$1050 A_s - 39.69 A_s^2 = 332$$

or, rearranging,

$$A_s^2 - 26.41 A_s + 8.36 = 0$$

solving for A_s,

$$A_s = 0.312 \text{ in.}^2/\text{ft; based on } p = \left(\frac{200}{60,000}\right) bd = 0.80 \text{ in.}^2$$

$$A_{s\,\text{tot}} = 0.80 \times 10.75 = 8.60 \text{ in.}^2$$

Code Sec. 15.4.4.2 directs: For reinforcement in short direction, a portion of the total reinforcement shall be distributed uniformly over a band width (centered on centerline of column or pedestal) equal to the length of short side of footing, as $A_{s\,\text{band}} = [2/(\beta + 1)]A_s$ in short direction; β = ratio of long side to short side of footing. Remainder of reinforcement required in short direction shall be distributed uniformly outside center band width of footing.

$$\beta = 10.75/7.67 = 1.40$$

Thus

$$A_{s\,\text{band}} = [2/(1.4 + 1)]A_s = (2/2.4)(6.45) = 5.375 \text{ in.}^2$$

Using 7 #8 bars, $A_s = 5.53$ in.^{2}. Add 2 #8 bars on each side of strip. Place bars above the longitudinal mat.

$$A_{s\,\text{tot}} = 7.11 > 6.45 \text{ in.}^2; \qquad \text{O.K.}$$
$$l_{d\,\text{req'd}} = 30 \text{ in.} < 36 \text{ in. avail.;} \qquad \text{O.K.}$$

Bearing   Check is not deemed necessary since both concretes (footing and column) have equal strength.

Dowels   The size of dowels and development length must be adequate to transmit the column moment into the footing (rigid connection). Rather than calculating dowel size and placement on the basis of moment transmitted, is it advisable to use the same bar size and number of dowels as column steel. In our case, use 8 #8 dowels, with 30 in. into column and 30 in. into footing, bent as required.

Details   Fig. 6.21 shows the completed design. Figure 6.22 illustrates dowels placed prior to pouring concrete for a pit bottom.

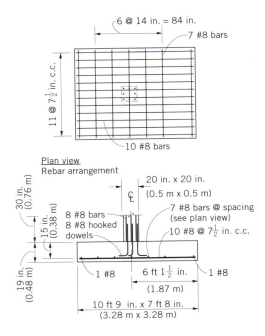

Figure 6.21 Final design dimension and details.

6.6 BEARING PLATES; ANCHOR BOLTS

A metal column does not bear directly on a concrete footing for perhaps obvious reasons: (1) the typical allowable stresses in the steel would far exceed the allowable stresses in the concrete (Code Sec. 15.8), with possible crushing failure, and (2) the "connection" details of the steel column to footing would be impractical.

Metal bearing plates are used to transfer (spread) a column load over a sufficient area on the concrete footing in order to reduce the bearing stresses on the concrete. The use of embedded anchor bolts provides a practical and expedient technique to attach steel columns (via the base plate) to concrete footings. Thus, metal plates and anchor bolts are typical base connections of steel columns to footings.

The bearing plate is cut to size and anchor bolt holes are drilled or punched in the fabricating shop. The plate is then welded to the column, usually in the shop but, occasionally, in the field. The anchor bolts are anchored into the concrete (for new construction they are usually set prior to pouring of the concrete) and, subsequently, they secure or bolt down the plate to the footing.

Bearing Plates

Steel base plates must be sufficiently large and sufficiently stiff to distribute the steel column load to the concrete, such that the allowable concrete strength is not exceeded. That is, the area of the plate, A_p, is

$$A_p = P/f_b \tag{6-9}$$

Figure 6.22 Reinforcement placement for machinery pit. Note the dowels (vertical bars) being secured to the bottom and cantilevered up for later lapping with the vertical wall reinforcement. The bottom is poured first, with the dowels remaining exposed above the slab. Since less steel is needed for the upper section, where the moment is less than at the bottom, the dowel lengths are sometimes calculated to have the dowels also serve as part of vertical reinforcement.

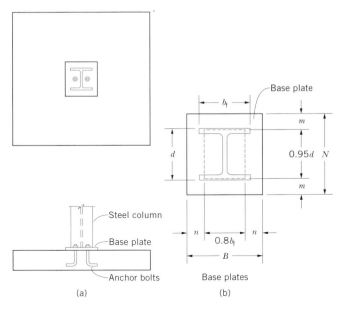

Figure 6.23 Aspects of base plate design.

where P = column load
A_p = area of plate (Fig. 6.23b) = BN
f_b = actual bearing stress or pressure $\leqq f_c$; f_c = allowable bearing stress

Allowable bearing stress, f_c, is given by two specifications, as follows:

| | Support Area Occupied by Base Plate | |
Specification*	Full	Less Than Full
AISC (3)	$0.35f_c'$	$0.35f_c'\sqrt{A_2/A_1} \leqq 0.7f_c'$
(Sec. 1.5.5)		
ACI	$0.85\phi f_c'$	$0.85\phi f_c'\sqrt{A_2/A_1} \leqq 1.7\phi f_c'$
(Sec. 10.15.1)	$= 0.6f_c'$	$= 0.6f_c'\sqrt{A_2/A_1} \leqq 1.2f_c'$

where A_1 = loaded area
A_2 = the area of the lower base of the largest frustrum of a pyramid having for its upper base the loaded area A_1 and having side slopes of 1 vertical to 2 horizontal (see Example 6.1).

The plate thickness is determined by treating the plate as a cantilever beam, fixed at the edges of a rectangle whose sides are $0.80b_f$, and $0.95d$, as shown in Fig. 6.23b. The column load P is assumed to be uniformly distributed over the area

* The AISC specifications are the more conservative of the two and are the more widely used.

$B \times N$. The allowable bending stress is $= 0.75 f_y$. The cantilever bending moment for a unit plate width is

$$M = f_b(m)^2/2, \quad \text{or} \quad M = f_b(n)^2/2; \quad I = 1 \cdot (t)^3/12$$

Thus, from

$$\sigma = \frac{Mc}{I} = \frac{M(t/2)}{I}$$

we have

$$0.75 f_y = f_b(m^2/2)(t/2)/(t^3/12) = 3 f_b m^2/t^2$$

Thus,

$$t = m\sqrt{4 f_b/f_y} \tag{6-9a}$$

Similarly,

$$t = n\sqrt{4 f_b/f_y} \tag{6-9b}$$

The larger t obtained from these two equations governs; thus, only the larger of the two values for m or n need to be used to determine the thickness.

Anchor Bolts

As the name implies, anchor bolts are used to secure the base plate to the footing. Figure 6.24 shows a variety of shapes. Other shapes may be available through the manufacturer's designation.

If reinforcing bars are used as anchors, they are usually welded to steel plates and anchored to the concrete via bond, with the development length calculated the same as for grade 60 reinforcing bars (10). For best results, the bars are cast in the concrete, rather than imbedded via grout-type mixtures in drilled holes.

Inserts are seldom used in newly poured footings. When anchors are used in "existing" concrete, the recommended procedure is to use expansion-type inserts, as depicted in Fig. 6.25. All expansion inserts are proprietary, and design values may be found in the manufacturer's catalog; however, the tensile strength may be

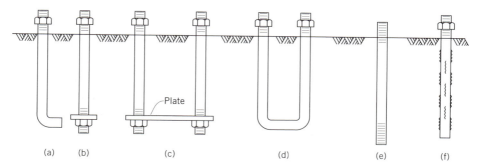

Plate

(a)   (b)   (c)   (d)   (e)   (f)

Figure 6.24 Various types of anchor bolts. (e) Typifies a bolt or stud that may be used in an expanding sleeve.

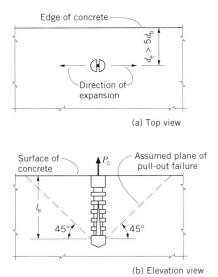

Edge of concrete

$d_e > 5d_b$

Direction of expansion

(a) Top view

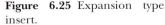

Surface of concrete

P_c

Assumed plane of pull-out failure

l_e

45° 45°

(b) Elevation view

Figure 6.25 Expansion type insert.

estimated from Eq. 6-10 (10, p. 6–6, eq. 6.5.2b)

$$\phi P_c = 10.7 l_e^2 \lambda \sqrt{f_c'} \qquad (6\text{-}10)$$

where (Fig. 6.25), P_c = nominal tensile strength of concrete element; l_e = embedment length; ϕ = 0.85; λ = 1 for normal weight concrete (typical for footings), and 0.75 for all lightweight concrete.

Because of slip of the anchor in the hole, deeper embedment does not proportionately increase the capacity of the anchor. Edge distances for expansion inserts, however, are more critical than for cast-in-place anchors. Expanding the inserts in the direction of the edge should be avoided. The anchorage strength depends entirely on the lateral force (wedge action) on the concrete. Therefore, it is advisable to limit their use to connections with more shear than tension applied to them.

High-strength bolts are seldom used since pull-out or bond generally controls the design. The A-307 or A-36 material is the most frequently used in footings. Table 6.3 provides recommended design guidelines for such bolts. Bolts of larger than 4-in. diameter are seldom used for column anchorages.

The typical cast-in-place installation consists of setting the anchor bolts into the wet concrete. The spacing and location of the bolts are typically held in place by a plywood template with the bolt holes drilled to the size and configuration of the steel base plate. The bolts are then suspended and aligned in the appropriate place where the column base plate is to be located. Adequate threaded length above the concrete must be provided in order to permit an adjustment of the base plate during construction. Once the concrete has hardened, the plywood template is removed and, subsequently, the base plate of the column can be placed in this location. Typically, the top of the footing is neither perfectly level nor smooth. Hence, wedges are generally used to level the base plate. In turn, there is a void created by such wedges. This void must then be filled with an expansion-type grout, in order to provide uniform distribution of the column load onto the footing.

Table 6.3 Anchor Bolt Properties; Recommended Loads, A-307 & A-36 Steel

Diameter (in.)		Area (in.²)			Threads per in.**	Recommended Allowable Tensile Loads (kips)			
						A-307		A-36	
Basic Major D	Root K	Gross A_D	Root A_K	Tensile Stress*	n	Design Strength	Service Load	Design Strength	Service Load
$\frac{5}{8}$	.514	.307	.202	.226	11	7.46	4.52	8.14	4.97
$\frac{3}{4}$	.627	.442	.302	.334	10	11.02	6.69	12.02	7.36
$\frac{7}{8}$	.739	.601	.419	.462	9	15.25	9.23	16.63	10.16
1	.847	.785	.551	.606	8	20.00	12.11	21.82	13.33
$1\frac{1}{8}$	.950	.994	.693	.763	7	25.18	15.26	27.47	16.78
$1\frac{1}{4}$	1.075	1.227	.890	.969	7	31.98	19.38	34.88	21.32
$1\frac{3}{8}$	1.171	1.485	1.05	1.16	6	38.28	23.20	41.76	25.52
$1\frac{1}{2}$	1.296	1.767	1.29	1.41	6	46.53	28.20	50.76	31.02
$1\frac{3}{4}$	1.505	2.405	1.74	1.90	5	62.71	38.00	68.40	41.79
2	1.727	3.142	2.30	2.50	$4\frac{1}{2}$	82.51	50.00	90.00	54.99
$2\frac{1}{4}$	1.977	3.976	3.02	3.25	$4\frac{1}{2}$	107.26	65.00	117.00	71.49
$2\frac{1}{2}$	2.193	4.909	3.72	4.00	4	132.01	80.00	144.00	87.97
$2\frac{3}{4}$	2.443	5.940	4.62	4.93	4	162.71	98.60	177.48	108.44
3	2.693	7.069	5.62	5.97	4	197.03	119.00	214.92	131.32
$3\frac{1}{4}$	2.943	8.296	6.72	7.10	4	234.32	142.00	255.60	156.18
$3\frac{1}{2}$	3.193	9.621	7.92	8.33	4	274.92	166.60	299.88	183.23
$3\frac{3}{4}$	3.443	11.045	9.21	9.66	4	318.81	193.20	347.76	212.49
4	3.693	12.566	10.6	11.1	4	366.34	222.00	399.60	244.16

* Tensile stress area = $0.7854(D - 0.9743/n)^2$

** For basic major diameters of $\frac{1}{4}$ to 4 in. inclusive thread series is UNC (coarse);

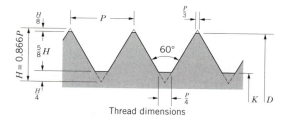

Thread dimensions

for $4\frac{1}{4}$-in. diameter and larger, thread series is 4UBN.

Normally, contractors prefer to place, atop the footing during the column erection, a cement grout soft and thick enough to provide an adjustable base; the excess amount of grout would be squeezed out as dictated by elevation requirements. However, this may not be a solution if the column has to be readjusted back and forth during the erection, thereby recreating an uneven surface. Remember, the objective is to obtain a uniform distribution of the column load onto the footing.

Example 6.4 _____

| *Given* | *Column* | *Footing* |

Given

Column	**Footing**
$D = 180$ kips (800 kN)	$f'_c = 4$ ksi (27.6 MPa)
$L = 150$ kips (668 kN)	$f_y = 60$ ksi (414 MPa)
10 W 60 Steel section	$q_a = 4$ ksf (190 kPa)
$f_y = 36$ ksi (248 MPa)	

Find   A suitable square footing design.

Solution   **Preface**   At a first glance, one may be inclined to proceed first with the design of the plate. This would appear as an obvious choice since the plate needs are designed anyway, and since its size figures into the determination of critical sections for punching and diagonal shear evaluations, and so on. However, for the plate design one needs $\sqrt{A_2/A_1}$, (AISC or ACI specifications), which is a function of the effective depth d. Thus, we have an interesting paradox: We need d to find the plate size and we need the plate size to find d.

One possible "out" from this apparent dilemma is to calculate d via Eq. 6-6 by first estimating x and y dimensions, perhaps 1 to 2 in. larger than the column dimensions. Another approach is to merely guess at a reasonable d, an expedient but not a crucial step since an "exact" d will subsequently be determined once the plate size has been finalized. In most instances $\sqrt{A_2/A_1} > 2$; thus, the value of 2 is used in most instances. This will become more apparent as the calculations develop.

Plate Design   A conservative value of d may be estimated by relating a reasonable thickness to the footing required; experience and judgment are helpful but not indispensable ingredients. In our case, the footing area $= A = (180 + 150)/4 = 82.5$ ft^2. Thus, $B = \sqrt{82.5} = 9.083$ ft. or 9 ft 1 in. Hence, a $d = 12$ in. appears quite minimal at this juncture; the final d will be determined based on punching shear later.

For a 10 W $\times$ 60 column section, $d = 10.22$ in.; $b_f = 10.08$ in. Thus, as a start, let us assume a 12 in. $\times$ 12 in. plate; the dimensions may be altered later to satisfy bearing requirements and physical needs for welding of column to plate, and so on. Also, say we assume a $d = 12$ in. (conservative in view of the footing size of 9 ft 1 in. $\times$ 9 ft 1 in., etc.). Hence,

$$A_1 = 12 \times 12 = 144 \text{ in.}^2$$
$$A_2 = [12 + (4 \times 12)]^2 = 3600 \text{ in}^2.$$

Thus,

$$\sqrt{A_2/A_1} = \sqrt{3600/144} = 5; \qquad \text{use 2}$$

The allowable $f_c = 0.35(4000)(2) = 2800$ psi $= 2.8$ ksi (AISC). Hence, the required plate area is

$$A_p = P/f_c = (180 + 150)/2.8 = 117.86 \text{ in.}^2$$

The 12 in. $\times$ 12 in. that is originally estimated is adequate since $A_p = 144 > 117.86$ in.^{2} required. This also provides nearly 1 in. beyond the column dimensions for welding the column to plate, both inside and outside of flanges. The actual bearing stress on the concrete, f_b, is

$$f_b = (180 + 150)/144 = 2.29 \text{ ksi} < 2.8 \text{ ksi}; \quad \text{OK}$$

Thus from Fig. 6.23b,

$$m = [12 - 0.95(10.22)]/2 = 1.15 \text{ in.}$$
$$n = [12 - 0.8(10.08)]/2 = 1.97 \text{ in.}$$

From Eq. 6-9b, the plate thickness is

$$t = n\sqrt{4f_b/f_y} = 1.97\sqrt{4(2.29)/36} = 0.994 \text{ in., say 1 in.}$$

Thus, plate is 12 in. $\times$ 12 in. $\times$ 1 in. thick.

Effective Depth The effective depth, d, may be determined based on diagonal shear. From Fig. 6.26 (also see Fig. 6.6), equivalent column dimensions may be estimated as

$$y = 12 - 1.97 = 10.03 \text{ in.} = 0.836 \text{ ft}$$
$$x = 12 - 1.15 = 10.85 \text{ in.} = 0.904 \text{ ft}$$

Also, for $f'_c = 4$ ksi, $v_c = 31$ ksi. Thus, from Eq. 6-6

$$(BL - xy)(q) = d^2(4v_c + q) + d(2v_c + q)(x + y)$$

where $B = L = 9$ ft 1 in. $= 9.083$ ft

$$P_u = 1.4(180) + 1.7(150) = 507 \text{ kips}$$

Hence, $q_u = 507/82.5 = 6.14$ ksf. Therefore,

$$[(9.083)^2 - (0.836)(0.904)](6.14) = d^2[4(31) + 6.14]$$
$$+ d[2(31) + 6.14](0.836 + 0.904)$$

Simplifying,

$$(82.5 - 0.755)(6.14) = d^2(130.14) + d(118.56)$$

Solving,

$$d = 1.56 \text{ ft} = 18.74 \text{ in.}; \quad \text{use } d = 19 \text{ in.}$$

Checking for wide-beam shear is not necessary; for square footings diagonal shear controls.

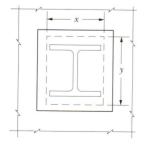

Figure 6.26 Base plate size transformed into equivalent column dimensions.

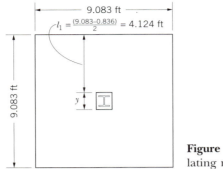

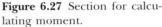

Figure 6.27 Section for calculating moment.

The critical section for bending is at midpoint between column and edge of plate. From Fig. 6.27 (also see Fig. 6.26)

$$l_1 = (9.083 - 0.836)/2 = 4.124 \text{ ft}$$
$$M = (6.14)(4.124)^2/2$$
$$M = 52.2 \text{ k-ft} = 626.4 \text{ k-in.}$$
$$a = A_s f_y/0.85 f_c' b = 1.96 A_s$$

Thus,

$$A_s(19 - 0.98 A_s) = 626.4/(0.9)(60) = 11.6$$

Solving,

$$A_s = 0.63 \text{ in.}^2/\text{ft}; \qquad \text{or } A_{s\,\text{tot}} = (0.63 \text{ in.}^2/\text{ft})(9.083 \text{ ft})$$

Thus,

$$p = A_s/bd = 0.63/(12)(19) = 0.00276 > p_{\min} = 0.002$$
$$< P_{\max} = 0.021 \quad \text{(Table 6.1)} \qquad \text{O.K.}$$

Use 10 #7 bars each way, $A_s = 6$ in.^{2} > 5.72 in.^{2} req'd; O.K.
Spacing: 11 in. c.c. at approximately 5 in. from each edge.

Anchor Bolts   Since no bending moment is to be projected into the footing via the base connection, we may assume a pin-type design. Hence, let us assume two, 1-in. diameter A-36 bolts of type shown in Fig. 6.28 (hook-type). The design strength (Table 6.3) is 21.82 kips.

(ACI 12.2.2) $l_d = 0.04 A_s f_y/\sqrt{f_c'} = 0.04(0.79)36,000/\sqrt{4000} = 18$ in.

but not less than $l_d = 0.0004 d_b f_y = 0.0004(1)(36,000) = 14.4$ in.

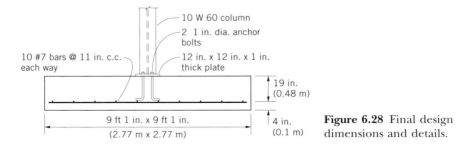

Figure 6.28 Final design dimensions and details.

Use $l_d = 19$ in. into concrete, with a $8d_b = 8$ in. hook (ACI 12.5). The two bars can be tied to the main footing reinforcement to hold them in place for the concrete pour. A temporary template (say of plywood) may hold the upper part of the bolts in proper alignment and suspension. The upper, threaded portion of the bolts should extend above the concrete an amount sufficient to account for the plate, the leveling, and two nuts; in our case approximately 4 in. would appear sufficient.

6.7 PEDESTALS

As mentioned in Section 6.1, steel columns should not be embedded into or have contact with the soil; this is to avoid corrosion of the steel. Thus, when the footing is at some depth in the ground, **pedestals** are used as the intermediary short concrete column links between steel columns and footings.

The Code (Art. 2.1) defines pedestal as "upright compression member with a ratio of unsupported height to average least lateral dimension of less than 3 in." If we let H = the unsupported height, and B = the least lateral dimension of the pedestal, the above definition can be abbreviated as

$$H/B < 3$$

The Code Sec. 318-86 version governs only reinforced concrete; previous codes made provisions for plain concrete design. Furthermore, the Code (Sec. 10.9.1) stipulates that the area of longitudinal reinforcement for compression (column) members shall be not less than 0.01 nor more than 0.08 times the gross area A_g of the section. From a practical point of view, however, when A_s approaches $0.08A_g$, there is "crowding of steel" reinforcement, particularly if accompanied by lap splicing, anchor bolts, and related ties required for the longitudinal steel; this makes the placement of the concrete quite difficult.

All longitudinal reinforcement shall be encircled by lateral ties. The size of the tie bars shall be #3, for longitudinal (in the pedestal) bars up to #10 size. The tie-bar size shall be #4 for longitudinal bar sizes of #11 or larger and for bundled longitudinal bars (Sec. 7.10.5.1). Their vertical spacing shall not exceed 16 longitudinal bar diameters, 48 tie bar diameters, or least column dimension of the compression member (Sec. 7.10.5.2). Also, ties shall be arranged such that every corner and alternate longitudinal bar shall have lateral support provided by the corner of a tie with an included angle of not more than 135° and no bar shall be farther than 6 in. clear on each side along the tie from such a laterally supported bar (Sec. 7.10.5.3). Ties shall be located vertically not more than half a tie spacing above the top of the footing (Sec. 10.5.4). The author recommends such minimum spacing for the upper part of the pedestal, as well, in order to avoid spalling of the shell of concrete outside the ties or spirals. Similarly, ties should enclose the anchor bolts in order to minimize the effects of spalling and to increase the pull-out resistance of the bolts.

Typically, the pedestals are designed as short columns. For axially loaded columns (i.e., no bending moment), the factored axial load, P_u, may be determined as follows.

For *tied reinforcement* (Sec. 10.3.5.2)

$$P_u = 0.80\phi[0.85 f_c'(A_g - A_s) + f_y A_s] \qquad (6\text{-}11a)$$

For *spiral reinforcement* (Sec. 10.3.5.1)

$$P_u = 0.85\phi[0.85 f_c'(A_g - A_s) + f_y A_s] \qquad (6\text{-}11b)$$

where ϕ = 0.7 for tied and 0.75 for spiral reinforcement
A_g = gross area of section, in.^{2}
A_s = total area of longitudinal reinforcement, in.^{2}
f_y = specified yield strength of longitudinal steel, psi

Example 6.5

Given

Column	Pedestal
D = 180 kips (800 kN)	f_c' = 4 ksi (27.6 MPa)
L = 150 kips (668 kN)	f_y = 60 ksi (414 MPa)
10 W 60 steel section	H = 3 ft (0.91 m)
Base plate 12 in. × 12 in. × 1 in.	below a 6 in. floor slab

Find   A suitable pedestal design.

Solution   Typically, the minimum pedestal cross section is equal to the area of the base plate. In fact, quite frequently, the pedestal dimensions are enlarged, step-wise, as shown in Fig. 6.29 to provide bearing area for adjoining slabs.

$$P_u = 1.4(180) + 1.7(150) = 507 \text{ kips}$$

Column Dimensions   Assume $A_s = 0.02\, A_g$, and tie reinforcement. Thus,

$$P_u = \phi(0.80)[0.85 f_c'(A_g - A_s) + f_y A_s]$$

or

$$507 = 0.7(0.8)[0.85(4)(A_g - 0.02 A_g) + (60)(0.02\, A_g)]$$
$$507 = (1.860 + 0.672)A_g$$

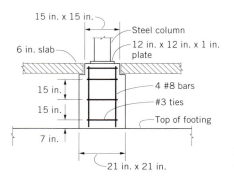

15 in. x 15 in.
Steel column
12 in. x 12 in. x 1 in. plate
6 in. slab
15 in.
15 in.
4 #8 bars
#3 ties
Top of footing
7 in.
21 in. x 21 in.

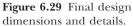

Figure 6.29 Final design dimensions and details.

or

$$A_g = 200.23 \text{ in.}^2; \qquad \text{use } 15 \text{ in.} \times 15 \text{ in.} = 225 \text{ in.}^2$$

A 21 in. $\times$ 21 in. size is arbitrarily used to provide a support area for the slab, as shown in Fig. 6.29.

Steel (longitudinal)   To find A_s: using $A_g = 225$ in the equation above,

$$507 = (0.7)(0.8)[0.85(4)(225 - A_s) + 60A_s]$$

Solving,

$$A_s = 2.48 \text{ in.}^2$$

Use 4 #8 bars (Sec. 10.9.2), $A_s = 3.16 \text{ in.}^2$
Ties   Use #3 bars.
Spacing

$$16 \text{ in.} \times 1 \text{ in.} = 16 \text{ in.}$$
$$48 \text{ in.} \times \tfrac{3}{8} \text{ in.} = 18 \text{ in.}$$
$$\text{Least dim.} = 15 \text{ in.}; \qquad \text{Use } 15 \text{ in.}$$

Details   are shown in Fig. 6.29.

6.8 WALL FOOTINGS

As the name implies, a wall footing supports a wall. In most cases, the load that the wall transmits to the footing may be considered uniform over the length of the wall. It follows, therefore, that the entire footing is given a uniform width, except in those cases where the soil properties vary appreciably over the length of the wall, or when columns become an integral load-bearing part with the wall, thereby necessitating an appropriate widening of the footing commensurate with an increase in load from the column. Footings may be "formed," as shown in Fig. 6.30, or when the excavation walls are stable against collapse, they may be poured without forms, as shown in Fig. 6.31.

For residential construction, footing depths (thickness) and widths are usually governed by local building codes or practical considerations. For example, the minimum footing sizes for ordinary homes specified by many local building codes is 8 in. thick by 16 in. wide, sufficient to accommodate common masonry sizes of 8–12-in. thickness widely used in home construction, and sufficient to overcome stresses. The concrete stresses are generally very low; hence, the vast majority of such footings are poured using a rather low-strength concrete (3000 psi is typical) and no steel reinforcement.

Normally, wall footings are analyzed for bending in only one direction, in a plane perpendicular to the face of the wall. The rigidity of the wall–footing composite eliminates virtually all bending stress in the longitudinal direction of the footing. Thus, the reinforcing steel in the longitudinal direction is only that needed to satisfy shrinkage requirements. Figure 6.32 depicts the typical steel reinforcement for a wall footing.

Design of a wall footing consists of considering a one-foot strip cut from the footing, as shown in Fig. 6.33. The projection is treated as a cantilever beam loaded

Figure 6.30 Forming for residential wall footing, typically without steel reinforcement.

Figure 6.31 Wall footings poured directly into excavation, without forming, at a significant cost savings in labor and materials.

Figure 6.32 Reinforcing steel for a commercial project. Frequently, the footing is poured directly into the excavation (without forming) when the excavation walls are stable (see Fig. 6.31), and may prove more economical; the savings from the additional concrete costs frequently more than offset labor and material cost for forming.

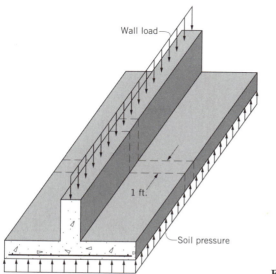

Figure 6.33 Wall footing.

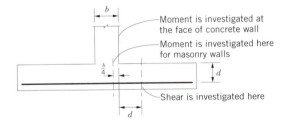

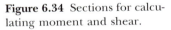

Figure 6.34 Sections for calculating moment and shear.

with the net or effective upward soil pressure, q_u. The maximum factored moment is computed at (ACI 15.4.2):

1. The face of the wall for footings supporting a reinforced concrete wall, or
2. Halfway between middle and edge of wall, for a footing supporting a masonry wall.

The shear is calculated at a section distance d from the face of the wall, with the allowable shear located at a stress based on the wide-beam criteria. Figure 6.34 shows the respective critical sections.

Example 6.6

Given

Wall	Footing
$D = 16$ k/ft; $L = 11$ k/ft	$f_y = 50$ ksi
13 in. thick	$f'_c = 3$ ksi
Reinforced concrete	$q_a = 3.75$ ksf
	Negligible overburden

Find A suitable wall footing.

Solution Assume a 12-in.-thick footing and 150 psf concrete unit weight (typical value for reinforced concrete). Thus, the net or effective soil pressure is:

$$q_n = 3750 - 150 = 3600 \text{ psf} = 3.6 \text{ ksf}$$

Thus, footing width B:

$$B = (16 + 11)/3.6 = 7.5 \text{ ft}$$

Bearing pressure:

$$q_u = [1.4(16) + 1.7(11)]/7.5 = 5.48 \text{ ksf}$$

Effective d: at a distance d from face

$$v_c = (0.85)(2)\sqrt{3000} = 93.11 \text{ psi}$$
$$V_u = [3.75 - (6.5 + 12)/12]5.48 = 12.10 \text{ kips}$$
$$d = 12.10/(93.11)(12) = 10.83 \text{ in., say 11 in.}$$
$$\text{total thickness} = 11 + 3.5 = 14.5 \text{ in.}$$

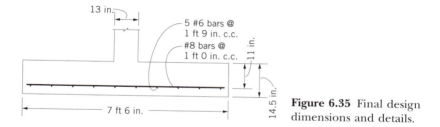

13 in.

5 #6 bars @
1 ft 9 in. c.c.
#8 bars @
1 ft 0 in. c.c.

11 in.

7 ft 6 in.

14.5 in.

Figure 6.35 Final design
dimensions and details.

A recheck is not necessary; a reevaluation for this d value would result in a
smaller V_u and subsequently a smaller d, and so on. A_s (in transverse direction):
cantilever length l' is

$$l' = 3.75 - 6.5/12 = 3.21 \text{ ft}$$
$$M_u = (5.48)(3.21)^2/2 = 28.22 \text{ k-ft} = 338.6 \text{ k-in.}$$

Thus, $M_u = \phi A_s (d - a/2)$, for $\phi = 0.9$.

$$a = A_s f_y/0.85 f'_c b = A_s(50)/(0.85)(3)(12) = 1.634 A_s$$

or

$$A_s(11 - 1.634 A_s/2) = 338.6/(0.9)(50) = 7.52$$

Solving,

$$A_s = 0.725 \text{ in.}^2/\text{ft of wall}$$
$$p = 0.725/(12)(11) = 0.0055 > 0.002 \qquad \text{O.K.}$$
$$< 0.021 = p_{max}$$

Use #8 bars, $A_s = 0.79$ @ 1 ft c.c.

$$l_d = 0.04 A_b f_y/\sqrt{f'_c} = (0.04)(1)(50,000)/\sqrt{3000} = 36.5 \text{ in.}$$
from face of wall, $l = 3.75 - 6.5/12 = 3.21 \text{ ft}$
$$= 38.5 \text{ in.} > 36.5 \text{ in.;} \qquad \text{O.K.}$$

Shrinkage A_s   (ACI 7.12.2.1)

$$A_s = (0.002)(12)(11) = 0.264 \text{ in.}^2/\text{ft of width}$$
$$\text{Tot } A_s = 7.5 \times 0.264 = 1.98 \text{ in.}^2$$

Use 5 #6 bars; spacing 1 ft 9 in. c.c. at approximately 3 in. from each side.
Details   Design details are given in Fig. 6.35.

Problems

6.1   Design square footings ($L = B$) for the following problems in Tables P6.1
and P6.1m. Note: all loads are concentric on column.

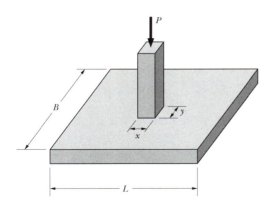

Figure P6.1 Dimensions to be used in conjunction with data for Problems 6.1, 6.2, 6.3, and 6.4.

Table P6.1

	Footing Data			(Concrete) Column Data					
Problem	f_y (ksi)	f'_c (ksi)	q_a (ksf)	$x; y$ (in.)	Reinforcement Bars	f_y (ksi)	f'_c (ksi)	D.L. (kips)	L.L. (kips)
(a)	60	3.0	4.0	18 × 18	10 - #10	60	4.0	275	200
(b)	50	3.0	3.0	15 × 15	8 - #8	60	4.0	220	120
(c)	50	3.5	2.0	12 × 12	6 - #6	50	3.5	90	130
(d)	60	4.0	5.0	21 × 21	10 - #10	60	3.5	350	300
(e)	60	3.0	2.0	18 × 15	8 - #7	50	4.0	170	170
(f)	50	3.0	3.5	21 × 18	10 - #8	50	3.5	300	220
(g)	60	4.0	3.0	15 × 12	6 - #8	60	4.0	130	80
(h)	60	3.5	2.5	16 × 12	10 - #7	60	3.0	150	100
(i)	60	3.0	4.0	18″ dia. (circular column)	10 - #10	60	4.0	200	170
(j)	50	3.0	3.0	15″ dia. (circular column)	8 - #8	60	4.0	220	100

Table P6.1m

	Footing Data			(Concrete) Column Data					
Problem	f_y (MPa)	f'_c (MPa)	q_a (KPa)	$x; y$ (mm)	Reinforcement Bars qty. @ dia. (mm)	f_y (MPa)	f'_c (MPa)	D.L. (kN)	L.L. (kN)
(a)	415	21	190	460 × 460	10 @ 32	414	28	1220	890
(b)	345	21	144	380 × 380	8 @ 25	414	28	980	530
(c)	345	24	100	310 × 310	7 @ 18	345	24	400	580
(d)	415	28	240	530 × 530	10 @ 32	414	24	1550	1330
(e)	415	21	100	460 × 380	8 @ 22	345	28	760	760
(f)	345	21	170	530 × 460	10 @ 25	345	24	1330	980
(g)	415	28	144	380 × 310	6 @ 25	414	28	580	360
(h)	415	24	120	410 × 310	10 @ 22	414	21	670	450
(i)	415	21	190	460 mm dia. (circular column)	10 @32	414	28	890	760
(j)	345	21	144	380 mm dia. (circular column)	8 @ 25	414	28	980	450

6.2 Design rectangular footings for the previous problems in Tables P6.1 and P6.1m, using the L/B ratios (ratio of footing length to footing width) given in Table P6.2.

Table P6.2

Problem	L/B
(a)	1.4
(b)	1.5
(c)	1.8
(d)	1.6
(e)	1.5
(f)	1.9
(g)	2.0
(h)	1.7
(i)	1.4
(j)	1.75

6.3 Design footings, bearing plates, anchor bolts, and pedestals (where applicable) for the following problems in Tables P6.3 and P6.3m: Note: All loads are concentric on column; $f_y = 36$ ksi (250 MPa), for columns and plates. Assume plate elevation is 0.0.

Table P6.3

	Footing Data					Column Data		
Problem	L/B	f_y (ksi)	f'_c (ksi)	q_a (ksf)	Elev. (ft) of Top of Footing	Column Section	D.L. (kips)	L.L. (kips)
(a)	1.0	60	3.0	4.0	-4.5	W14 × 120	275	200
(b)	1.0	50	3.0	3.0	-3.0	W12 × 79	220	120
(c)	1.8	50	3.5	2.0	0.0	W10 × 54	90	130
(d)	2.5	60	4.0	5.0	-5.5	W14 × 132	350	300
(e)	1.0	60	3.0	2.0	0.0	W12 × 72	170	170
(f)	2.0	50	3.0	3.5	-3.5	W14 × 99	300	220
(g)	1.0	60	4.0	3.0	-2.0	W 8 × 58	130	80
(h)	1.7	60	3.5	2.5	-4.5	W10 × 49	150	100
(i)	1.0	60	3.0	4.0	0.0	W12 × 79	200	170
(j)	1.8	50	3.0	3.0	-4.0	W10 × 60	220	100

Table P6.3m

Problem	L/B	f_y (MPa)	f'_c (MPa)	q_a (kPa)	Elev. (m) of Top of Footing	Column Section	D.L. (kN)	L.L. (kN)
(a)	1.0	414	21	190	−1.4	W360 × 180	1220	890
(b)	1.0	345	21	144	−1.0	W300 × 120	980	530
(c)	1.8	345	24	100	0.0	W250 × 80	400	580
(d)	1.2	414	28	240	−1.8	W360 × 200	1550	1330
(e)	1.0	414	21	100	0.0	W300 × 105	760	760
(f)	1.3	345	21	170	−1.2	W360 × 150	1330	980
(g)	1.0	414	28	144	−0.7	W200 × 85	580	360
(h)	1.7	414	24	120	−1.4	W250 × 75	670	450
(i)	1.0	414	21	190	0.0	W300 × 120	890	760
(j)	1.2	345	21	145	−1.3	W250 × 90	980	450

6.4   Design footings so that the soil pressure will be uniform for the following problems in Tables P6.4 and P6.4m.

Table P6.4

Problem	f_y (ksi)	f'_c (ksi)	q_a (ksf)	x; y (in.)	Reinforcement Bars	f_y (ksi)	f'_c (ksi)	D.L. (kips)	L.L. (kips)	M_x (kip-ft)	M_y (kip-ft)
(a)	60	3.0	4.0	18 × 18	12 #10	60	4.0	210	130	390	—
(b)	50	3.0	3.0	15 × 15	8 #8	60	4.0	200	90	260	—
(c)	50	3.5	2.0	12 × 12	6 #8	50	3.5	80	80	200	—
(d)	60	4.0	5.0	21 × 21	10 #11	60	3.5	290	260	—	400
(e)	60	3.0	2.0	18 × 15	8 #8	50	4.0	140	120	—	50
(f)	50	3.0	3.5	21 × 18	12 #8	50	3.5	250	180	—	180
(g)	60	4.0	3.0	15 × 12	6 #8	60	4.0	110	80	60	20
(h)	60	3.5	2.5	16 × 12	8 #8	60	3.0	100	90	50	50
(i)	60	3.0	4.0	18 in. dia.	10 #10	60	4.0	160	130	50	100
(j)	50	3.0	3.0	15 in. dia.	8 #7	60	4.0	200	60	100	50

Table P6.4m

Problem	f_y (MPa)	f'_c (MPa)	q_a (kPa)	x; y (mm)	Reinforcement Bars qty. @ dia. (mm)	f_y (MPa)	f'_c (MPa)	D.L. (kN)	L.L. (kN)	M_x (kN-m)	M_y (kN-m)
(a)	415	21	190	460 × 460	12 @ 32	414	28	930	580	530	—
(b)	345	21	144	380 × 380	8 @ 25	414	28	890	400	350	—
(c)	345	24	100	310 × 310	6 @ 25	345	24	360	360	270	—
(d)	415	28	240	530 × 530	11 @ 35	414	24	1290	1150	—	540
(e)	415	21	100	460 × 380	8 @ 25	345	28	620	530	—	70
(f)	345	21	170	530 × 460	13 @ 25	345	24	1110	800	—	240
(g)	415	28	144	380 × 310	6 @ 25	414	28	490	350	80	25
(h)	415	24	120	410 × 310	8 @ 25	414	21	440	400	70	70
(i)	415	21	190	460 mm dia.	10 @ 32	414	28	700	580	70	140
(j)	345	21	144	380 mm dia.	8 @ 22	414	28	890	260	140	70

6.5 For the problems from Tables P6.4 and P6.4m, design the footings so that the column is concentric with the footing.

6.6 Design wall footings for the following problems in Tables P6.6 and P6.6m.

Table P6.6

	Footing Data					Wall Data					
									D.L.	L.L.	
Problem	f_y (ksi)	f'_c (ksi)	q_a (ksf)	Thickness (in.)	Wall type	Reinforcement Bars/Spacing	f'_c (ksi)	f_y (ksi)	(kips/ft)		M (kip-ft/ft)
(a)	60	3.0	3.5	12	Block	—	—	—	4.0	1.8	—
(b)	60	3.5	2.5	12	Concrete	#7 @ 12 in.	4.0	60	6.5	3.3	—
(c)	60	4.0	2.5	12	Concrete	#6 @ 12 in.	3.5	50	3.8	2.1	—
(d)	50	3.0	3.5	8	Block	—	—	—	3.1	1.7	—
(e)	60	4.0	3.0	12	Concrete	#8 @ 6 in.	4.0	60	3.8	2.1	3.3
(f)	50	3.5	2.0	12	Concrete	#6 @ 8 in.	3.0	50	6.0	4.0	0.5
(g)	60	4.0	4.0	18	Concrete	#8 @ 6 in.	4.0	60	7.1	5.0	5.9
(h)	50	3.0	3.5	15	Concrete	#7 @ 6 in.	3.5	50	4.1	2.1	5.8

Table P6.6m

	Footing Data					Wall Data						
						Reinf. bars Bar/Spacing						
Problem	f_y (MPa)	f'_c (MPa)	q_a (ksf)	Thickness (mm)	Wall Type	(mm)	(mm)	f'_c (MPa)	f_y (MPa)	D.L. (kN/m)	L.L. (kN/m)	M (kN m/n)
(a)	415	21	170	300	Block	—		—	—	60	25	—
(b)	415	24	120	300	Concrete	22 @ 300		28	415	95	50	—
(c)	415	28	120	300	Concrete	20 @ 300		24	345	55	30	—
(d)	345	21	170	200	Block	—	—	—	—	45	25	—
(e)	415	28	144	300	Concrete	25 @ 150		28	415	55	30	15
(f)	345	24	100	300	Concrete	20 @ 200		21	345	88	58	2.5

6.7 A rectangular footing 5 ft × 10 ft supports a load P. Determine the value of P if the ultimate soil pressure 5 ft below the bottom of the footing is 1000 psf. Neglect the weight of the footing and soil; assume uniform pressure distribution.

6.8 Evaluate and compare the maximum actual load a W12 × 96 column can safely deliver to a 14 in. × 14 in. base plate on concrete if:
   (a) $t = \frac{3}{8}$ in.,
   (b) $t = \frac{1}{2}$ in. (Assume full bearing; $f'_c = 4000$ psi; $f_y = 36$ ksi.)

6.9 Referring to Problem 6.8, at what thickness of the 14 in. × 14 in. plate will bending of the plate and bearing stress on the concrete both simultaneously govern the maximum allowable load on the W12 × 96 column?

6.10 Determine the length of embedment needed to fully develop a $\frac{3}{4}$-in. A-36 anchor bolt, using the ACI code requirement for development of bars in tension. What is the minimum thickness of footing that this bolt could be placed in and still fully develop its strength?

6.11 Design the base plate and anchor bolts needed for a W12 × 120 column on a 9 ft × 6 ft × 18 in.-thick footing given the following data:

$$D = 65 \text{ kips (comp)}$$
$$L = 95 \text{ kips (comp)}$$

$$V = 20 \text{ kips (shear)}$$
$$f'_c = 4000 \text{ psi, GR 60 rebar, } f_y = 36 \text{ ksi (column \& plate)}$$

Use A36 bolts. Would this footing be stable?

6.12 Design the anchor bolts for a W8 × 40 column with a 10 in. × 10 in. × $\frac{5}{8}$ in. plate on a 6 ft × 6 ft × 16 in.-thick footing given the following data:

$$D = 45 \text{ kips (comp)}$$
$$L = 95 \text{ kips (comp)}$$
$$V = 16 \text{ kips (shear)}$$
$$f'_c = 3500 \text{ psi, GR 60 rebar, } f_y = 36 \text{ ksi (column \& plate)}$$

Use A36 bolts.

6.13 Given the following footing size and load data, find the maximum and minimum soil pressure:

$$7 \text{ ft} \times 7 \text{ ft footing}; \quad P = 100 \text{ kips}; \quad M_x = 80 \text{ k-ft}; \quad M_y = 60 \text{ k-ft}$$

6.14 Given the following footing size and corner soil pressures (Fig. P6.14), find the concentric column load and eccentricities e_x and e_y.

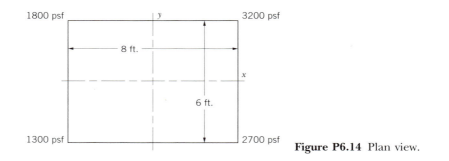

Figure P6.14 Plan view.

BIBLIOGRAPHY

1. American Concrete Institute, ''Building Code Requirements for Reinforced Concrete (ACI 318-83) (Revised 1986),'' Detroit, Mich.

2. ACI Committee 326, ''Shear and Diagonal Tension,'' *ACI J. Proc.,* vol. 59, Jan. 1962, p. 1; Feb. 1962, p. 277; Mar. 1962, p. 352.

3. American Institute of Steel Construction (AISC), New York.

4. Cernica, J. N., *Fundamentals of Reinforced Concrete,* Addison Wesley, Reading, Mass., 1963.

5. Concrete Reinforcing Steel Institute (CRSI), Chicago, Ill.

6. Ferguson, P. M., J. E. Breen, and J. O. Jirsa, *Reinforced Concrete Fundamentals,* Wiley, New York, 1988.

7. McCormac, J. C., *Design of Reinforced Concrete,* Harper and Rowe, Cambridge, Mass., 1986.

8. Mindness, S., and J. F. Young, *Concrete,* Prentice-Hall, Englewood Cliffs, N.J., 1981.

9. Moe, J., "Shearing Strength of Reinforced Concrete Slabs and Footings Under Concentrated Loads," Development Department Bulletin D47, Portland Cement Assn., Skokie, Ill., Apr. 1961.

10. Prestressed Concrete Institute (DCI), *Design Handbook,* 3rd ed., Parts 5 and 6, Chicago, Ill.

11. Richart, F. E., "Reinforced Concrete Wall and Column Footings," *ACI J.,* vol. 20, Oct. and Nov. 1948; *Proc.,* vol. 48, pp. 97, 237.

12. Whitney, C. S., "Plastic Theory of Reinforced Concrete Design," *Trans. ASCE,* vol. 107, p. 251.

CHAPTER 7

Special Foundations

7.1 INTRODUCTION

Isolated and wall footings, discussed in the previous chapter, when compared with other foundation systems, are usually economical and practical, but are generally limited to relatively light to moderate loads, and for building sites of good soil bearing. Special considerations and design features or schemes are sometimes adopted to overcome or accommodate imposed limitation by perhaps space, soil formations, loads, or functional concerns. Such obstacles or impositions are frequently resolved via the design of special foundations. The following are rather typical examples of such situations and associated possible solutions.

Conditions	*Possible Design Adaptations*
Property line restrictions	Combined footings
Closely spaced isolated footings	Combined footings
Weak or compressible strata	Piles or drilled piers or mat foundations
Poor soil or shallow rock	Piles or drilled piers
Very heavy loads	Piles or drilled piers
Bridging over pipes, trenches, etc.	Cantilever footings
Bridge piers or abutments	Footings–retaining wall combinations

Perhaps at this point in his or her career, the student realizes that foundation designs are not always unique; indeed options for other designs are often available.

Thus, typically, one searches for designs that are not only safe but are also economical. In this chapter we shall develop and delineate design procedures for some commonly used special foundations.

7.2 RECTANGULAR COMBINED FOOTINGS

If one were to locate a column very close to a property line, an isolated footing is likely to result in an uneven soil-pressure distribution (i.e., higher pressures near the property line). This is contrary to our typical design objective to achieve uniform soil pressure. In such a case, one alternative may be a rectangular-shaped, combined footing, as depicted by Fig. 7.1. The footing near the property line (P.L.) is connected with an adjacent one. The proportioning (B and L) of the combined footing is to be such that the resulting soil pressure is uniform.

As with many other reinforced concrete structures, the design of combined footings has not been totally standardized. Generally, it is assumed that the rectangular footing is a rigid member and, thus, the pressure is linear. This assumption is rather common, although we recognize that we deal with neither a truly rigid footing nor with an ideal, homogeneous soil. This approach yields a rather conservative design; the moments are somewhat larger than those obtained by treating the footing as a beam on an elastic foundation. However, its simplicity and "history of success" are perhaps sufficient reasons for regarding the method as a practical and viable one. The following is a summary of the procedure. Example 7.1 further illustrates the approach.

Given Typically included in the given part of the problem are column data (loads, sizes, reinforcement, location, and spacing), soil bearing, concrete strength (f_c'), and grade of reinforcement (f_y).

Objective The goal is to determine footing dimensions (width, length, thickness), steel reinforcement (bar sizes, spacing, placement, details, dowels), and relevant details for construction.

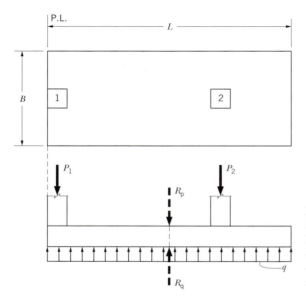

Figure 7.1 Rectangularly shaped combined footing. For uniform q, the resultant of the applied load is collinear with that of the soil pressure q.

Procedure The design is predicated on the assumption that the footing is rigid and that the soil pressure is uniform. The following explanation may illustrate the procedure:

Step 1 Convert the column loads to ultimate loads via $P_u = 1.4(\text{D.L.}) + 1.7(\text{L.L.})$. Then convert the allowable soil pressure to ultimate pressure via $q_u = (P_{1u} + P_{2u})q_a/(P_1 + P_2)$.

Step 2 Determine the footing length (L) and width (B). First determine the location of the load resultant distance ($\bar{x}$). This point coincides with the midpoint of L, thus yielding the value for L. B is then determined from $B = \Sigma P_u/Lq_u$.

Step 3 Draw shear and moment diagrams. The footing is treated as a beam, loaded with a uniform soil pressure (upward) and column loads (downward), which are treated as concentrated loads.

Step 4 Determine footing depth based on shear. Critical sections are at $d/2$ for diagonal tension (or punching shear) and at the d for a wide beam, the same as for spread footings. The critical section for wide-beam shear is investigated only at one point (max. shear) for punching shear. However, an investigation of a three- or four-sided zone for each column may have to be done; Example 7.1 will further illustrate this point.

Step 5 Determine the flexural reinforcing steel. The longitudinal (flexural) steel is designed using the critical moments (negative and positive) from the moment diagram. Thus, typically, combined footings will have longitudinal steel at both top and bottom of the footing, with a minimum percentage of $200/f_y$.

Step 6 Determine the steel in the short direction. The steel in the transverse direction is determined based on an equivalent soil pressure q' and subsequent moment, for each column. Even for stiff footings, it is widely accepted that the soil pressure in the proximity of the columns is larger than that in the zone between columns. Thus, for design, we account for this phenomenon by assuming an empirical effective column zone width of s. The soil pressure in this zone, q', is calculated as $q' = P_u/Bs$, where P_u is the ultimate column load, B the footing width and s an equivalent width of footer strip for the column in question. Commonly, the value of s is taken as the width of the column (in the longitudinal direction) plus about $0.75d$ on each side of that column.

Step 7 Evaluate dowel steel. The requirements are the same as for spread footings.

Step 8 Provide a drawing showing final design. This drawing is to show sufficient detail from which one may construct.

Example 7.1

Given $f'_c = 3$ ksi; $f_y = 60$ ksi; $q_a = 2.5$ ksf

	Column 1	**Column 2**
	12 in. × 12 in.	16 in. × 16 in.
	4 #7 bars	6 #8 bars
	D.L. = 80 kips	D.L. = 120 kips
	L.L. = 60 kips	L.L. = 110 kips

Edge of column 1 at P.L.; spacing: 20 ft c.c.

Find   A suitable rectangular combined footing.

Procedure   **Step 1**   Convert column loads to ultimate; then, convert q_a to q_u.

$$P_1 = 140 \qquad P_{1u} = 1.4(80) + 1.7(60) \qquad = 214$$
$$P_2 = \underline{230} \qquad P_{2u} = 1.4(120) + 1.7(110) = \underline{355}$$
$$\Sigma = \overline{370} \text{ kips} \qquad\qquad\qquad\qquad = \overline{569} \text{ kips}$$

$$\text{Ratio: ULT/ACT} = \text{U/A} = 569/370 = 1.54$$

$$q_u = (\text{U/A})\, q_a = (1.54)(2.5) = 3.85 \text{ ksf}$$

Step 2   Determine footing dimensions L and B (Fig. 7.2).

$$R\bar{x} = P_1 x_1 + P_2 x_2$$

or

$$569\bar{x} = 214(0.5) + 355(20.5)$$

$$\bar{x} = 12.98 \text{ ft, say 13 ft}$$

$$\text{Thus, } L = 2\bar{x} = 26 \text{ ft}$$

The "rounding" to 26 ft will prevent complete closure of the moment diagram, but negligibly.

$$B = (P_{1u} + P_{2u})/Lq_u = 569/26 \times 3.85 = 5.684 \text{ ft}$$

(Use 5 ft 9 in. for construction; see detail, Fig. 7.6.)

$$q' = 3.85 \times 5.684 = 21.885 \text{ k/ft, say } 21.89 \text{ k/ft}$$

Step 3   Draw shear (V) and moment (M) diagrams. The column loads are treated as concentrated loads acting at the centers of the columns. The shear and moment diagrams are shown in Fig. 7.3.

Step 4   Determine the footing depth based on wide-beam and diagonal-tension shear, as indicated in Fig. 7.4. Note that the diagonal-tension analysis reflects on a three-sided section for column 1 and a four-sided section for column 2. (If column 2 were close to the end of the footing, a three-sided analysis might be required; however, that's not the case here.)

We shall first determine d via a wide-beam analysis, then check for diagonal tension. From the shear diagram, the maximum shear is near column 2. At

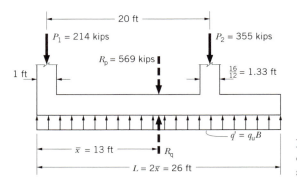

Figure 7.2 Pertinent aspects in determining footing dimension L and B for Example 7.1.

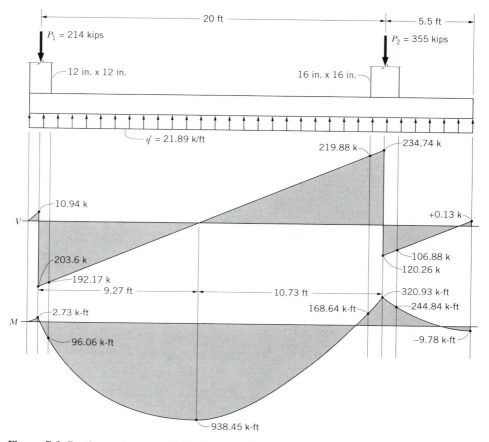

Figure 7.3 Pertinent features of the shear and moment diagrams for Example 7.1.

a distance d from the face of column 2,

$$V = 219.88 - 21.89d$$

Also, from $v_c = V/Bd$,

$$Bdv_c = 219.88 - 21.89d$$

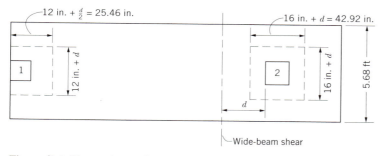

Figure 7.4 Illustrations of sections for investigating diagonal tension and wide beam shear.

In our case, $v_c = 2\phi\sqrt{f'_c} = 2(0.85)\sqrt{3000} = 93.1$ psi $= 13.4$ ksf
Hence,

$$(5.68)(13.4)d = 219.88 - 21.89d$$

Solving,

$$d = 2.24 \text{ ft} = 26.92 \text{ in.}$$

Using $d = 26.92$ in., we check diagonal tension at column 1.

$$V = \text{column load less upward soil pressure}$$

$$= 214 - Aq_u = 214 - A(3.85)$$

$$A = (12 + d/2)(12 + d) = (12 + 13.46)(12 + 26.92)$$

$$= 991 \text{ in.}^2 = 6.91 \text{ ft}^2$$

Hence,

$$V = 214 - (6.91)(3.85) = 187.5 \text{ kips}$$

$$\text{Perimeter} = 2(12 + d/2) + (12 + d) = 89.84 \text{ in.} = 7.5 \text{ ft}$$

$$v_c = 187.5/(7.5)(2.24) = 11.19 < 4\phi\sqrt{f'_c} \qquad \text{O.K.}$$

At column 2,

$$A = (42.92)^2/144 = 12.79 \text{ ft}^2$$

$$\text{Perimeter} = (42.92/12)4 = 14.31 \text{ ft}$$

$$V = 355 - 12.79(3.85) = 305.76 \text{ kips}$$

$$v_c = 305.76/(14.31)(2.24) = 9.54 \text{ ksi} < 4\phi\sqrt{f'_c} \qquad \text{O.K.}$$

Step 5 Determine the reinforcing steel, A_s.

$$A_s(d - a/2) = M/\phi f_y$$

for $f_y = 60$ ksi, $f'_c = 3$ ksi, $b = 12$ in., $a = A_s f_y/0.85 f'_c b = 1.96 A_s$
Hence,

$$A_s(26.92 - 0.98 A_s) = (938.45)(12)/(0.9)(60)(5.68)$$

Solving,

$$A_s = 1.44 \text{ in.}^2/\text{ft width}$$

$$p = 1.44/(26.92)(12) = 0.0044 > 200/f_y = 0.00333 \qquad \text{O.K.}$$

$$< p_{max} = 0.016 \quad (\text{see Table 7.1})$$

$$A_{s\,tot} = 1.44 \times B = 1.44 \times 5.68 = 8.18 \text{ in.}^2$$

Use 11 #8 bars ($A_s = 8.63$ in.^{2}) at 6 in. c.c. across top of footing (at approximately 4 in. from each side). Provided that $\frac{1}{3}$ of the bars extend the full length of the footing, the bars could be cut off as dictated by the moment requirements (moment diagram). However, the saving is not worth the effort (engineering, fabrication, placing, etc.). Thus, typically, all bars will run the full length of the footing.

Based on $200/f_y$, $A_{s\,min} = (0.0033)(26.92)(12) = 1.07$ in.^{2}. For the positive moment, $+A_s = 1.07 \times 5.68 = 6.08$ in.^{2} Use eight #8 bar; $A_s = 6.28$ in.^{2} > 6.08 in.^{2}; this A_s is larger than that required by positive $+M$. Place these bars at 8.5 in. c.c. (approximately 4 in. from sides) and say 4 in. from bottom (cover of 4 in. is advisable if excavation bottom is not closely controlled). Also, based on the moment diagram, running $\frac{1}{3}$ of $+A_s$ (three bars) the full length of the footing satisfies both Code and moment requirements. The other five bars could be cut off at, say, half-length and placed on the right half (column 2 side) of the footing.

Step 6 Determine A_s in the short direction. Reference is made to Fig. 7.5. The widths s_1 and s_2 are commonly assumed as equal to column width plus $0.75d$. Hence, $s_1 = 12 + 0.75(26.92) = 32.19$ in. $= 2.68$ ft and $s_2 = 16 + 1.5(26.92) = 56.38$ in. $= 4.70$ ft.

Column 1

$$q' = P/Bs_1 = 214/(5.68)(2.68) = 14.06 \text{ ksf}$$

$$l_1 = (5.68 - 1)/2 = 2.34 \text{ ft}$$

$$M_1 = (14.06)(2.34)^2(12)/2 = 461.92 \text{ k-in.}$$

Placing the transverse steel above the "positive" longitudinal steel,

$$d = (26.92 - 1) = 25.92 \text{ in.}$$

Thus,

$$A_s(25.92 - 0.98A_s) = 461.92/(0.9)(60)$$

Solving,

$$A_s = 0.334 \text{ in.}^2/\text{ft} = 0.895 \text{ in.}^2/s_1 \text{ width}$$

Based on p_{min},

$$A_s = 2.75 \text{ in.}^2$$

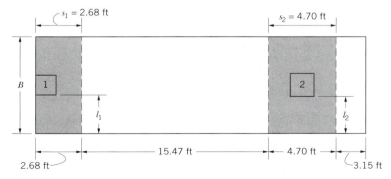

Figure 7.5 Sections for determining transverse steel reinforcement for the footing in Example 7.1.

Thus, use five #7, at 8 in. c.c.

$$A_s = 3.00 \text{ in.}^2 > 2.75 \text{ in.}^2 \qquad\qquad \text{O.K.}$$

$$l_d = 0.04(0.60)(60{,}000)/\sqrt{3000} = 26.3 \text{ in.}$$

l_d available $= 2.34 \times 12 = 28$ in.

For column 2, transverse $+A_s$

$$s_2 = 4.70 \text{ ft}; \quad q' = 355/(4.70)(5.68) = 13.30 \text{ psf}$$

$$l_2 = (5.68 - 1.33)/2 = 2.17 \text{ ft}$$

$$M_2 = (13.30)(2.17)^2(12)/2 = 375.71 \text{ k-in.}$$

Using $d = 25.92$ in. (transverse bars on top of longitudinal)

$$A_s(25.92 - 0.92A_s) = 375.71/(0.9)(60)$$

Based on p_{min},

$$A_s = 4.82 \text{ in.}^2/s_2 \text{ width}$$

Use nine #7 bars at 6.5 in. c.c.:

$$A_s = 5.40 \text{ in.}^2$$

Development length is same as above.
Based on p_{min},

$$A_s \text{ in } 15.47 \text{ ft-section} = 15.87 \text{ in.}^2$$

Thus, use 27 #7 bars at $6\frac{3}{4}$ in. c.c.

$$A_s = 16.2 \text{ in.}^2$$

For 3.15 ft section, use six #7 at 7 in. c.c.

Step 7   Evaluate dowel steel.

Column 1

$$\text{Allowable } f_c = 0.85(0.7)f_c' = 1.785 \text{ ksi} \qquad \text{(ACI 10.15)}$$

$$P = (12)(12)(1.785) = 257 > 214 \text{ kips}$$

Thus, dowels are not required for load transfer, but Code (Sec. 15.8) designates four bars such that $A_s \geqq 0.005\,A_g$.

$$A_s = 0.005(12)(12) = 0.72 \text{ in.}^2$$

Use four #6 bars; $A_s = 1.76$ in.^{2}.

Column 2

$$f_c = (0.85)(0.7)f_c'\sqrt{A_2/A_1}; \text{ use 2 if } \sqrt{A_2/A_1} > 2$$

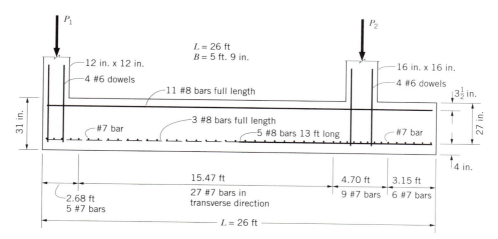

Figure 7.6 Design details.

In our case,

$$f_c = (0.85)(0.7)(3)(2) = 3.57 \text{ ksi}$$

$$P = (16)(16)(3.57) = 913.9 > 355 \text{ kips} \qquad \text{O.K.}$$

Use four #6 bars; $A_s = 1.76 > 16 \times 16 \times 0.005 = 1.28 \text{ in.}^2 \qquad$ O.K.

Step 8 Make a drawing showing design details (Fig. 7.6).

7.3 TRAPEZOID-SHAPED FOOTINGS

Let us visualize a foundation where two columns that support unequal loads are located at each end of a rectangular footing; such a condition is sometimes dictated by space restrictions and the like. In such a case, the resultant of the column loads would not fall over the center of a rectangular-shaped footing. Correspondingly, the soil presure would not be uniform (recall that our typical objective is uniform soil pressure). For very large column spacings (e.g., say greater than 25 ft), a cantilever footing (discussed in the next section) may be a somewhat more economical (i.e., less material) solution to such a problem. For smaller column spacings, a trapezoid-shaped footing, as shown in Fig. 7.7 for a two-column arrangement, is usually deemed suitable.

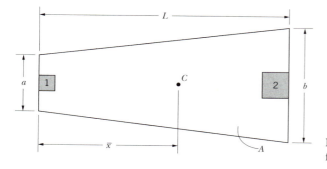

Figure 7.7 Trapezoid-shaped footing.

The following discussion and explanation reflect on the design approach for a two-column footing.

Referring to Figure 7.7, the area, A, is

$$A = (a + b)L/2 \qquad (7\text{-}1)$$

From $\bar{x} = \Sigma Ax / \Sigma A$, we get

$$\bar{x} = \frac{(aL)(L/2) + [(b-a)(L/2)]2L/3}{aL + (b-a)L/2}$$

$$\bar{x} = (L/3)(a + 2b)/(a + b) \qquad (7\text{-}2)$$

For the pressure to be uniform, the resultant of the column loads coincides (is collinear) with the resultant of pressure at the centroid (C) as shown.

The following is a summary of the procedure for the design of trapezoid-shaped footings:

Given Included in the given data are column information (loads, sizes, location, and spacing), length of footing (L), soil bearing values (q_a), concrete strength (f_c'), and grade of reinforcement (f_y).

Objective The goal is to determine footing dimensions (width, thickness), steel reinforcement (bar sizes, spacing, placement, details, dowels), and relevant details for construction.

Procedure The design is predicated on the assumption that the footing is rigid and that the soil pressure is uniform. The basic steps are:

Step 1 Convert the column loads to ultimate loads via $P_u = 1.4(\text{D.L.}) + 1.7(\text{L.L.})$; then convert the allowable soil pressure to ultimate; that is, $q_u = (P_{u1} + P_{u2}) q_a / (P_1 + P_2)$.

Step 2 Determine dimensions a and b via simultaneous solutions of two independent equations.

$$A = (a + b)L/2 \qquad (7\text{-}1)$$

$$\bar{x} = (L/3)(a + 2b)/(a + b) \qquad (7\text{-}2)$$

Thus, we solve for a and b.

Step 3 Draw the shear and moment diagrams. The footing is treated as a beam, loaded with a uniform soil pressure (upward) and column loads (downward), which are treated as concentrated loads. Note that while the pressure is uniform, the pressure force for unit length varies with the width [e.g., at the narrow end, the load is $a(q_u)$, and $b(q_u)$ at the wide end, etc.].

Step 4 Determine footing depth based on shear. Critical sections are usually checked for wide-beam shear at the narrow end and diagonal tension at the wide end.

Step 5 Determine the flexural reinforcing steel. Because the width varies, it is advisable to determine $-A_s$ at several points; the same is now required for $+A_s$ since it is typically governed by p_{min}.

Step 6 Determine the steel in the short direction. Assume an average length for the cantilever length; determine the equivalent lengths as for rectangular footings.

Step 7 Determine dowel steel, as for rectangular combined or spread footings.

Step 8 Provide a drawing with details for construction. Here some judgment is necessary to accommodate the steel arranagement in view of the variable width along the footing.

Example 7.2

Given

$$f'_c = 21 \text{ MPa}; \qquad f_y = 414 \text{ MPa}; \qquad q_a = 120 \text{ kPa}; \qquad L = 7 \text{ m}$$

Columns are at ends of footing, as shown in Fig. 7.8.

	Column 1	*Column 2*
	0.3 m × 0.3 m	0.4 m × 0.4 m
	4 25-mm bars	6 25-mm bars
	D.L. = 355 kN	D.L. = 535 kN
	L.L. = 265 kN	L.L. = 490 kN

Find A suitable trapezoid-shaped footing.

Procedure **Step 1** Convert P_1 and P_2 to P_{1u} and P_{2u}, and q_a to q_u:

$$P_1 = 620 \qquad P_{1u} = 1.4(355) + 1.7(265) = 947$$
$$P_2 = \underline{1025} \qquad P_{2u} = 1.4(535) + 1.7(490) = \underline{1582}$$
$$\Sigma \rightarrow \quad = 1645 \qquad\qquad\qquad\qquad\qquad\quad = 2529$$

$$\text{Ratio} = 2529/1645 = 1.5374$$

$$q_u = 184.5 \text{ kPa}$$

Step 2 Determine a and b.

$$\bar{x} = [(947)(0.15) + (1582)(6.8)]/2529 = 4.31 \text{ m}$$

For the trapezoid,

$$A = (a + b)L/2 = (a + b)(7)/2 = 3.5(a + b)$$

Also,

$$A = \Sigma P/q_u = 2529/184.5 = 13.71 \text{ m}^2$$

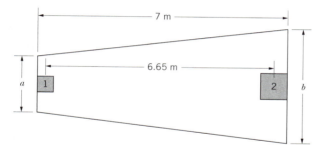

Figure 7.8 Trapezoid-shaped footing. For uniform q, the resultant of the column load s is collinear with that of the soil pressure q.

Thus,

$$13.71 = 3.5(a + b)$$

$$(a + b) = 3.92 \text{ m} \tag{1}$$

From $\bar{x} = (L/3)(a + 2b)/(a + b)$

$$4.31 = \frac{(7/3)(a + 2b)}{(a + b)}$$

Simplifying,

$$0.8471a - 0.1529b = 0 \tag{2}$$

Simultaneous solution of Eqs. 1 and 2 yields

$$b = 3.32 \text{ m}; \qquad a = 0.60 \text{ m}$$

Check via substituting in the expression for A or $\bar{x}$; for A,

$$3.5(3.32 + 0.6) = 13.71; \qquad \text{O.K.}$$

Step 3 Draw V and M diagrams, as shown in Fig. 7.9.

Step 4 Determine d. First, we'll evaluate wide-beam shear at the narrow end, and then check for diagonal tension at column 2. From the shear equation, Fig. 7.9,

$$V = 121(x - 0.15) + 35.8(x - 0.15)^2 - 930$$

at d from the inner face of column 1, $x = 0.3 + d$. Hence,

$$V = 121(d + 0.15) + 35.8(d + 0.15)^2 - 930$$

or

$$V = 35.8d^2 + 131.74d - 911 \tag{3}$$

Also,

$$v_c = 660 \text{ kPa} = V/bd$$

b at a distance d from inner face of column 1 is

$$b = 0.6 + (d + 0.3)(3.32 - 0.6)/7 = 0.3885d + 0.716$$

Thus,

$$660 = \frac{V}{(0.3885d + 0.716)d} = \frac{V}{(0.3885d^2 + 0.716d)}$$

or

$$V = 256.4d^2 + 472.56d \tag{4}$$

Equating Eq. 3 to Eq. 4,

$$256.4d^2 + 472.56d = (35.8d^2 + 131.74d - 911)$$

or

$$292.2d^2 + 604.3d - 911 = 0$$

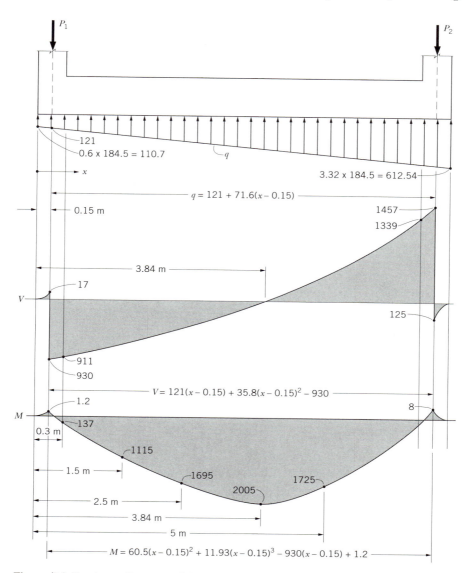

Figure 7.9 Pertinent features of the shear and moment diagrams for Example 7.2.

Solving,

$$d = 1.01 \text{ m}$$

Check v_c; at $x = 1.01 + 0.3 = 1.31$ m

$$b = 0.392 + 0.716 = 1.108$$
$$V = 121(1.16) + 35.8(1.16)^2 - 930 = -741.5$$
$$v_c = 741.5/(1.108)(1.01) = 662, \text{ close to } 660 \text{ kPa} \qquad \text{O.K.}$$

By inspection, required d at large end is less than 1.01 m.

Table 7.1

x, m	Width b, m	M kN-m	A_s cm^2 Per 1 m Width	For Width b	min.*	No. of 25-mm Bars
1.5	1.18	1115	36	42	42	8
2.5	1.57	1695	31	48.5	48.5	10
3.84	2.09	2005 (max)	27.3	57	57	12
5	2.54	1725	19	48.5	51	1

* A_s based on p_{min} is probably debatable, particularly near the wide end, since b varies so greatly. Thus, using 12–25-mm bars are deemd as satisfactory (no cutoffs) for the 4 m on "right" side.

Step 5   Determine flexural steel based on moments at several points (reference to Fig. 7.9).

$$\text{Use } A_s(d - a/2) = M/0.9f_y b$$
$$a = A_s f_y/0.85 f'_c b = 400\, A_s/0.85(21)(1) = 22.4 A_s$$

at $x = 1.5$ m, we have

$$A_s(1.01 - 22.4 A_s/2) = 1115/(0.9)(400)(1000)(1.18) = 2.65 \times 10^{-3}$$

or

$$A_s^2 - 0.09018 A_s + 0.000234 = 0$$

Solving,

$$A_s = 0.0036 \text{ m}^2 = 36 \text{ cm}^2/\text{m width}$$

Checking percent steel,

$$p = 0.0036/(1)(1.01) = 0.00356 > p_{min}; \qquad \text{O.K.}$$
$$< p_{max}; \qquad \text{(Table 6.1)}$$

Thus, total A_s at $x = 1.5$ m is $(A_s)(b) = (36)(1.18) = 42$ cm^2/m Similarly for points at $x = 2.5$ m; 3.84 m; 5 m. The results are summarized in Table 7.1.

Step 6 A_{s1} in short direction. The effective "transverse beam width" s_2 can be approximated as before. Referring to Fig. 7.10,

$$s_2 = 0.4 + 0.75(d) = 1.16 \text{ m}$$
$$L_2 = (3.095/2) - 0.4 = 1.147$$

$$M = qL^2/2 = \left(\frac{P}{A_{\text{strip}}}\right)(1.147)^2/2, \text{ etc.}$$

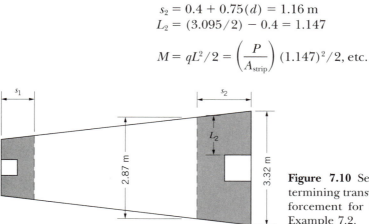

Figure 7.10 Sections for determining transverse steel reinforcement for the footing of Example 7.2.

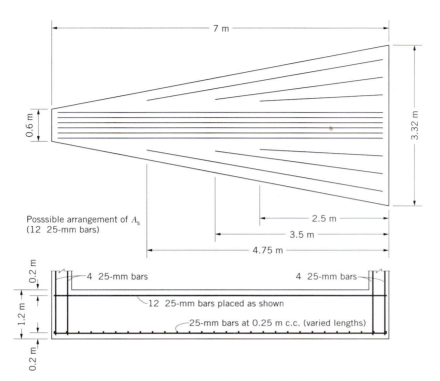

Figure 7.11 Illustration of final design details for footing in Example 7.2.

By inspection,

$$A_s' \leqq A_s \text{ based on } p_{min}$$

Thus, $A_{s\,min} = (5\text{-}25\text{-mm bars})$

$$A_{s\,min} = (0.002)(1.01) = 20 \text{ cm}^2/\text{m}$$

Practically, it's advisable to space these bars at about 0.25 m even spacing for the entire 7-m length, with varying length, for approximately 3 m near the wide end, to about 0.4 m near the narrow end. These bars would be tied to the $+A_s$ at about 0.2 m from the bottom of excavation. Other configurations of placement are, of course, possible.

Step 7 Four dowels, 25-mm bars are recommended for each column (see Example 7.1).

Step 8 Detail the drawing. Figure 7.11 shows a suitable layout. Because of the irregular shape, some practical improvisation is frequently a practical necessity.

7.4 STRAP FOOTING

Also referred to as a cantilever footing, a strap footing is a composite of two spread (isolated) footings connected by a rigid beam or strap, as shown in Fig. 7.12. The

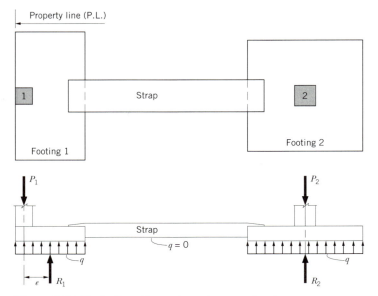

Figure 7.12 Typical configuration of a strap footing.

strap connects an eccentrically loaded footing (e.g., footing 1) with an interior footing, subsequently resulting in a uniform soil pressure and minimum differential settlement.

A strap footing may be somewhat more economical than a combined footing if distances between columns are large (say greater than 25 ft). It may also serve a special need of bridging over areas that cannot be loaded, such as pits, shallow culverts, tunnels, and the like. On the other hand, many practitioners shy away from its use because of the special effort and care needed for the construction detail in placement of reinforcement that such footings require.

The strap is designed as a rigid beam connected to the footings such that it overcomes rotational effects on eccentrically loaded footings; it is assumed to experience no soil pressure. This is accomplished by either forming the strap above the ground or by pouring the strap over a compressible formation, such as loose or spaded soil or semi-rigid styrofoam. Hence, the shear is a constant between the footings; the moment varies linearly.

The footings are treated as isolated footings. The interior footing (e.g., footing 2) is generally square-shaped and is designed as a spread footing, with appropriate negative (top) longitudinal steel provided to resist the negative moment transmitted via the strap. While this spread-footing approach also applies to footing 1, one carefully scrutinizes the zone near column 1 for some additional transverse steel requirements, as typically included for rectangular or trapezoid-shaped footings discussed in the preceding sections.

The following procedural summary and Example 7.3 illustrate the recommended approach for a strap-footing design.

Given Typically, included in the given part of the problem are column data (loads, sizes, reinforcement, location, and spacing), allowable soil bearing, q_a, concrete strength (f_c'), and grade of reinforcement (f_y).

Objective The goal is to (a) determine the footing dimensions (length, width, thickness) proportioned such that the soil pressure is reasonably uniform and differential settlement is minimal, (b) design the strap, (c) design the footings, and (d) show a drawing with pertinent details for construction purposes.

Procedure The design assumes no soil pressure under the strap (other than that necessary to support the weight of the strap; hence, the weight of the strap is negated). The footings are designed as isolated footings subjected to column loads and strap reactions.

Step 1 (a) Convert to P_u and q_u, as previously described. (b) Try a value for e. This establishes the position of R_1; subsequently, this influences the ratio of L_1 and B_1. An adjustment in e may be warranted if L_1/B_1 appears unreasonable. (c) From equilibrium (i.e., $\Sigma M = 0$ and $\Sigma F_y = 0$), determine the values for R_1 and R_2.

Step 2 Determine footing dimensions, L and B. Note that q will be uniform when R coincides with the centroid of that footing. Also, for minimum differential settlement, q should be the same for both footings.

Step 3 Draw the shear (V) and moment (M) diagrams.

Step 4 Design the strap as a beam. Use maximum V and M in the section between footings. Affix the strap to the footings to effectively prevent footing rotation.

Step 5 Design the footings as spread (isolated) footings with reinforcement in both directions including $-A_s$ steel to accommodate the negative moment. Some special assessment for the transverse steel near column 1 is recommended.

Step 6 Provide the final drawing showing details for construction.

Example 7.3

Given (Also see Example 7.1 for a rough comparison.)

	Column 1	*Column 2*
	12 in. $\times$ 12 in.	16 in. $\times$ 16 in.
	4 #7 bars	6 #8 bars
	D.L. = 80 kips	D.L. = 120 kips
	L.L. = 60 kips	L.L. = 110 kips

Edge of column 1 at P.L.; spacing 25 ft c.c. $q_a = 2.5$ ksf; $f_y = 60$ ksi; $f_c' = 3.5$ ksi for both footing and strap.

Find A suitable strap-footing design.

Procedure Reference is made to Fig. 7.13.

Step 1 Convert to P_u and q_u.

$$
\begin{array}{ll}
P_1 = 140 \text{ kips} & P_{1u} = 214 \text{ kips} \\
P_2 = \underline{230} \text{ kips} & P_{2u} = \underline{355} \text{ kips} \\
\Sigma \rightarrow \quad = 370 \text{ kips} & \rightarrow = 569 \text{ kips}
\end{array}
$$

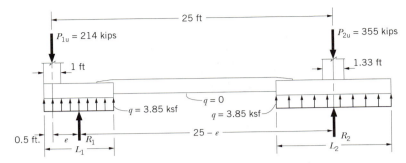

Figure 7.13 Pertinent given data for the design of a strap footing.

$$\text{Ratio} = 569/370 = 1.54$$

$$q_u = 3.85 \text{ ksf}$$

Try $e = 3$ ft. From $\Sigma M_2 = 0$, we have

$$(214)(25) - R_1(22) = 0$$
$$R_1 = 243.18 \text{ kips}$$

From $\Sigma M_1 = 0$

$$(355)(22) - (214)(3) - R_2(22) = 0$$
$$R_2 = 325.82 \text{ kips}$$

From $\Sigma F_y = 0$

$$243.18 + 325.82 = 569 \text{ kips}; \qquad \text{O.K.}$$

Step 2

$$L_1 = 2(0.5 + e) = 2(0.5 + 3) = 7 \text{ ft}$$
$$B_1 = R_1/(L_1)(q_u) = 243.18/(7)(3.85) = 9 \text{ ft}$$
$$L_1/B_1 = 0.775$$

The ratio appears reasonable. If specific clearance between footings is required, then adjustment of L_1 and L_2 would be in order. If we select to make footing 2 square

$$B = L = \sqrt{R_2/q_u} = \sqrt{325.82/3.85} = 9.2 \text{ ft (say 9 ft 3 in.)}$$

Step 3   Draw the shear (V) and moment (M) diagrams (Fig. 7.14)

Step 4   Design of strap. From Fig. 7.14, use $V = 28.56$ kips and $M = 546.83$ k-ft. *Observation:* While a deeper strap is more efficient than a wider but shallower one (i.e., larger I), too narrow a strap may pose some practical or functional concerns (e.g., limited space for reinforcement, lateral or torsional stiffness). The author recommends a range for strap widths b as $2 \text{ ft} \leq b \geq (L'/10)$ ft, and an effective depth, d, 50 to 100% larger than theoretically needed, in order to reduce the number of reinforcing bars (thus, accommodating spacing) and to minimize the effects of uneven excavation and perhaps eliminate the need for stirrups. $L' = $ distance between footings.

In our case $L'/10 = 1.4$ ft; hence, use $b = 2$ ft.

$$V = v_c bd; \qquad v_c = 2\phi\sqrt{f_c'} = 100.51 \text{ psi} = 14.44 \text{ ksf}$$

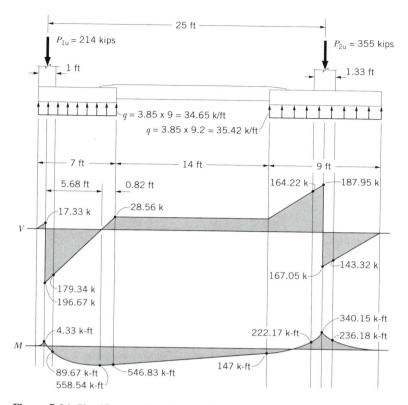

Figure 7.14 Significant values (e.g. peak, minimum, point of zero values) for the shears and moments of the strap footing in Example 7.3.

Thus, from $28,560 = (100.57)(24)d$, $d = 11.83$ in. Arbitrarily, use $d = 24$ in. Also, $M = 546.93$ k-ft $= 273.46$ k-ft/ft width $= 3281.58$ k-in./ft width. Hence,

$$3281.58 = \phi A_s f_y (d - a/2) = 0.9 A_s (60)(24 - a/2)$$

Where $a = A_s f_y / (0.85 f_c' b) = 1.68 A_s$. Thus,

$$3281.58 = 0.9(60)(24 - 0.84 A_s) A_s$$

Simplifying,

$$A_s^2 - 28.56 A_s + 72.33 = 0$$

Solving,

$$A_s = 2.74 \text{ in.}^2/\text{ft} = 5.48 \text{ in.}^2/2 \text{ ft width}$$

Use seven #8 bars

$$A_s = 5.5 \text{ in.}^2, \text{ at } 3.25 \text{ in. c.c.}$$
$$> p_{\min}; \quad \text{O.K.}$$

Step 5 Design of footings. Reference is made to Fig. 7.15.
Footing 1. For $b = 9$ ft and $v_c = 14.44$ ksf

$$(9)(14.44)d = 179.34 - 34.65d \quad \text{(see } V \text{ diagram)}$$

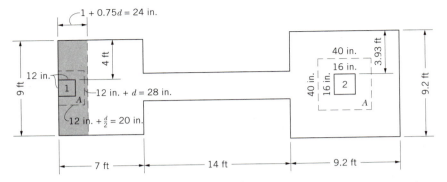

Figure 7.15 Sections for determining transverse steel reinforcement for the footing of Example 7.3.

Solving,

$$d = 1.09 \text{ ft} = 13 \text{ in.}$$

Check diagonal tension.

$$V = 214 - Aq_u = 214 - 3.85A$$

Where

$$A = (12 + d/2)(12 + d) = (12 + 6.5)(12 + 13) = 925 \text{ in.}^2 = 6.42 \text{ ft}^2$$
$$\text{Perimeter} = 2(12 + 6.5) + (12 + 13) = 62 \text{ in.} = 5.17 \text{ ft}$$

Hence,

$$V = 214 - 24.73 = 189.27 \text{ kips}$$

and

$$v_c = 189.27/(5.17)(1.09) = 33.587 > 4\phi\sqrt{f_c'} = 28.88 \text{ ksf}; \qquad \text{N.G.}$$

Thus, increase d to 16 in. Then, we have

$$A = 2(20)(28) = 1120 \text{ in.}^2 = 7.78 \text{ ft}^2$$
$$\text{Perimeter} = 2(12 + 8) + (12 + 16) = 68 \text{ in.} = 5.67 \text{ ft}$$
$$V = 214 - (7.78)(3.85) = 184 \text{ kips}$$

and

$$v_c = 184/(5.67)(1.33) = 24.4 < 28.88 \text{ ksf}; \qquad \text{O.K.}$$

Longituding steel (axial direction)

$$M_{\max} = 558.54 \text{ k-ft} = 6702.5 \text{ k-in.} = 744.7 \text{ k-in./ft width}$$
$$744.7 = 0.9(60)(16 - 0.84A_s)A_s$$

Simplifying,

$$A_s^2 - 19.04A_s + 16.41 = 0$$

Solving,

$$A_s = 0.91 \text{ in.}^2/\text{ft} = 8.19 \text{ in.}^2/9 \text{ ft width}$$

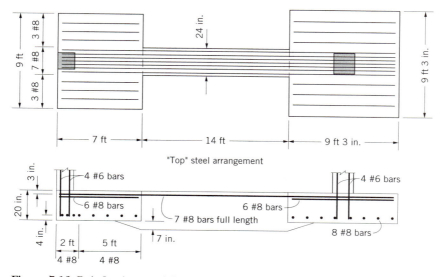

Figure 7.16 Reinforcing steel for the footing of Example 7.3.

Use 13 #8 bars (A_s = 10.20 in.). Extend the 7 #8 bars from the strap the full length (7 ft) and add 3 #8 bars at approximately 16 in. c.c. on each side of strap bars (see details, Fig. 7.16).

Transverse Steel   In the transverse direction, we note a rather disproportionate percentage of the column load P_1 assumed by the left side of footing 1. Hence, we shall concentrate more of the transverse steel in that segment of the footer (as we did for similar cases of rectangular and trapezoidal footings). That is not quite the case for footing 2; it resembles more of an isolated (spread) footing condition, and therefore will be treated that way.

Thus, for footing 1, we shall assume the shaded width s, as 12 in. + 0.75d = 2 ft, as shown in Fig. 7.15. Correspondingly,

$$q = 214/(2)(9) = 11.26 \text{ ksf}$$
$$M = 11.26(4)^2/2 = 90.1 \text{ k-ft} = 1081 \text{ k-in.}$$
$$A_s(16 - 0.84A_s) = 1081/(0.9)(60) = 20$$

Solving,

$$A_s = 1.33 \text{ in.}^2/\text{ft} = 2.66 \text{ in.}^2/2 \text{ ft width}$$

Use four #8 at 6 in. c.c. (see Fig. 7.16). For the remaining 5 ft width, use q = 3.85 ksf.
Hence, $M = 3.85 (4)^2/2 = 30.8$ k-ft = 369.6 k-in.

$$A_s = 0.435 \text{ in.}^2/\text{ft} = 2.17 \text{ in.}^2/5 \text{ ft}$$

based on p, A_s = 3.15 in.^{2}/5 ft. Therefore, use four #8 bars at 16 in. c.c. (see Fig. 7.16).

Footing 2. Based on wide-beam shear,

$$(9.2)(14.44)d = 164.22 - 35.42d$$
$$130d = 164.22 - 35.42d$$
$$d = 1 \text{ ft } (12 \text{ in.})$$

When comparing column loads, the $d = 1$ ft appears low for footing 2 when $d = 16$ in. was needed for footing 1. Thus, check diagonal tension using $d = 16$ in. as for #1

$$A = (16 + 16)^2 = 1024 \text{ in.}^2 = 7.11 \text{ ft}^2$$
$$\text{Perimeter} = 4(16 + 16) = 128 \text{ in.} = 10.67 \text{ ft}$$
$$V = 355 - 7.11(3.85) = 327.63 \text{ kips}$$
$$v_c = 327.63/(10.67)(1.33) = 23.09 < 4\phi\sqrt{f'_c}; \qquad \text{O.K.}$$

Use $d = 16$ in. for both footings

$$p = (0.00333)(16)(108) = 5.75 \text{ in.}^2$$

Max $-M$ in footing 1 is 558.54 k-ft versus 546.83 k-ft at footing–strap juncture. Thus, extending the seven #8 bars into footing 1 and arbitrarily adding six more (three on each side) is more than adequate to account for the slightly larger moment.

In the transverse direction,

$$M = 3.85(3.93)^2/2 = 29.8 \text{ k-ft} = 357.68 \text{ k-in.}$$
$$A_s = 0.42 \text{ in.}^2/\text{ft} = 3.86 \text{ in.}^2/9.2 \text{ ft}$$
$$< p_{min} = 5.83 \text{ in.}^2$$

Thus, use eight #8 bars at 14 in. c.c.

7.5 COEFFICIENT OF SUBGRADE REACTION

The coefficient of subgrade reaction (also known as **subgrade modulus** or **modulus of subgrade reaction**) is a mathematical constant that denotes the foundation's stiffness. It is used in the analysis of foundation components (e.g., beams, semirigid footings, mats, piles) that interact with an elastic support system (e.g., soil). The common symbol for this coefficient is k_s; it is defined as:

$$k_s = q/\delta \tag{7-3}$$

where k_s = coefficient of subgrade reaction, kcf; (MN/m^3)
$\quad\quad\quad q$ = soil pressure at a given point, ksf; (kN/m^2)
$\quad\quad\quad \delta$ = settlement of the foundation at the same point, ft; (m)

Figure 7.17a is a schematic illustration of a deflected foundation of finite length (e.g., beam). For short, relatively stiff beams, the deflection (δ) is fairly linear over the length of the beam. Also, for short distances, k_s may be assumed as reasonably constant, thus, ensuring a linear (perhaps uniform) value for q. However, since we recognize that, generally, soil is neither homogeneous nor isotropic, and that $q - \delta$ is nonlinear, the magnitude (values) for k_s may vary appreciably over a building site. Furthermore, the values for k_s appear to be affected by footing size, shape, and depth. Terzaghi (22) proposed the following expressions:

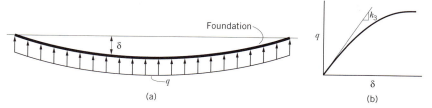

Figure 7.17 Interaction between a flexible foundation and an elastic support. (a) Flexible foundation on elastic soil, (b) Typical $q - \delta$ relationship.

For cohesive soils,

$$k_s = \frac{k_1}{B} \tag{7-4}$$

For granular soils,

$$k_s = k_1 \left[\frac{(B+1)}{(2B)} \right]^2 \tag{7-5}$$

Accounting for depth,

$$k_s = k_1 \left[\frac{(B+1)}{2B} \right]^2 \left[\frac{1+2D}{B} \right] \tag{7-6}$$

$$\leqq 2k_1 \left[\frac{(B+1)}{2B} \right]^2$$

For a rectangular footing, $L \times B$, on granular soil

$$k_s = k_1 \frac{(1+0.5B/L)}{1.5} \tag{7-7}$$

where L, B, D = footing length, width, depth, respectively
   k_s = value for a full-sized footing
   k_1 = value obtained from 1 ft $\times$ 1 ft load tests

Also, $k_1 = 12N$ (ksf) appears to be a reasonable approximation, particularly for slab analysis; N is the standard penetration number (see Sec. 3.5).

Figure 7.17b gives a graphical depiction of the general $q - \delta$ relationship for soils. One notes a distinct resemblance of this relationship to a typical stress–strain ($\sigma - \varepsilon$) curve. In turn, this infers a relationship between k_c and E_s. Vesic (26) proposed Eq. 7-8.

$$k_s = [0.65 E_s / (1 - \mu^2)] \sqrt[12]{E_s B^4 / EI} \tag{7-8}$$

where E_s = modulus (of elasticity) of the soil
   μ = Poisson ratio of the soil
   E = modulus of elasticity of the beam
   I = moment of inertia of beam's cross section

Ideally, k_1 is determined from field data, usually based on 1 ft $\times$ 1 ft plates, which are loaded, with the corresponding δ readings recorded. For very rigid plates, the δ is assumed uniform over the area of the plate. The data plotted, however, lack the linear (straight-line) relationship, even for very rigid plates. Thus, k_s is necessarily

Table 7.2 Approximate Range in Values for Coefficient of Subgrade Reaction, k_1, Based on 1 ft × 1 ft Plate Tests, k_3/ft^3 (values in parenthesis represent equivalent SI range in MN/m^3)

Relative Density (Granular Soil)	Loose	Medium Dense	Dense
Dry or moist	40–120	120–600	600–1200
	(6–18)	(18–90)	(90–180)
Submerged	50	160	600
	(8)	(24)	(90)

Consistency (Cohesive Soil)	Stiff	Very Stiff	Hard
	60–120	120–250	250+
	(9–18)	(18–38)	(38+)

estimated by selecting a slope that appears to represent the "early" segment of the $q - \delta$ curve. Averaging the results of several tests conducted within a small distance from one another may improve the accuracy. Table 7.2 provides a range of values for k_1 (for 1 ft × 1 ft).

7.6 BEAMS ON ELASTIC FOUNDATIONS

Let us consider the relationship of the flexural rigidity of a foundation and the elasticity of the soil supporting it; this is the case of a beam on an elastic foundation. The underlying assumption is that a beam is supported along its entire length by a continuous elastic support system, such that the intensity of reaction at any point from this system is proportional to the deflection at that point.

Many investigators have focused appreciable attention on the study of beams on elastic foundations (2, 3, 10, 11, 17, 18, 23, 24). E. Winkler (29) developed an approach back in 1867 that still receives a wide acceptance. Although the method assumes some support boundary conditions that are difficult to account for in the analysis, it is relatively easy to use and gives acceptable results. It views the foundation as resting on a series of springs. Hence, the vertical reaction, q, at any point is $q = k_s y$, where y is the deflection, and k_s is the subgrade modulus (see Eq. 7-3). The subgrade modulus, k_s, must be multiplied by the beam or strip width, b, when used in the analysis; $k = k_s b$. Units are FL^{-2}.

The basic differential equation for the deflection curve of the beam is

$$EI\frac{d^4y}{dx^4} = q = -ky \qquad (7\text{-}9)$$

Using the notation $\lambda = \sqrt[4]{k/4EI}$, the general solution of Eq. 7-9 can be expressed as

$$y = e^{\lambda x}(C_1 \cos \lambda x + C_2 \sin \lambda x) + e^{-\lambda x}(C_3 \cos \lambda x + C_4 \sin \lambda x) \qquad (7\text{-}10)$$

where C_1 through C_4 are determined from known loading and end conditions.

Table 7.3 Solutions to Some Select Cases of Beams on Elastic Foundations

Infinite Beams

Concentrated load P at point o:

$$y = \frac{P\lambda}{2k} \cdot \phi_{\lambda x}$$

$$M = \frac{P}{4\lambda} \cdot \psi_{\lambda x}$$

$$V = \frac{-P}{2} \cdot \theta_{\lambda x}$$

Max. at "0":

$$\delta_{\mathbb{C}} = \frac{P\lambda}{2k}$$

$$M_{\mathbb{C}} = \frac{P}{4\lambda}$$

$$V_{\mathbb{C}} = \frac{-P}{2}$$

Moment M_o:

$$y = \frac{M_o\lambda^2}{k} \cdot \beta_{\lambda x}$$

$$M = \frac{M_o}{2} \cdot \theta_{\lambda x}$$

$$V = -\frac{M_o\lambda}{2} \cdot \psi_{\lambda x}$$

Distributed load q over length l:

$$y_c = \frac{q}{2k}\left[2 - e^{-\lambda a}\cos\lambda a - e^{-\lambda b}\cos\lambda b\right]$$

$$M_c = \frac{q}{4\lambda^2}\left(e^{-\lambda a}\sin\lambda a + e^{-\lambda b}\sin\lambda b\right)$$

$$V_c = \frac{q}{4\lambda}\left[e^{-\lambda a}(\cos\lambda a - \sin\lambda a) - e^{-\lambda b}(\cos\lambda b - \sin\lambda b)\right]$$

when $a = b = l/2$

$$\delta_{max} = \frac{q}{4k}\left(1 - e^{-\lambda a}\cos\lambda a\right)$$

$$M_{max} = \frac{q}{2\lambda^2} \cdot e^{-\lambda a}\sin\lambda a$$

Semi-Infinite Beams

P_0 at end:

$$y = \frac{2P_o\lambda}{k} \cdot \theta_{\lambda x}$$

$$M = \frac{-P_o}{\lambda} \cdot \beta_{\lambda x}$$

$$V = -P_o \cdot \psi_{\lambda x}$$

M_0 at end:

$$y = \frac{-2\lambda^2 M_o}{k} \cdot \psi_{\lambda x}$$

$$M = M_o \cdot \phi_{\lambda x}$$

$$V = -2\lambda M_o \cdot \beta_{\lambda x}$$

Finite Beams

P at center:

$$y_c = \frac{P\lambda}{2k}\,\frac{\cosh\lambda l + \cos\lambda l + 2}{\sinh\lambda l + \lambda\sin\lambda l}$$

$$M_c = \frac{P}{4\lambda}\,\frac{\cosh\lambda l - \cos\lambda l}{\sinh\lambda l + \sin\lambda l}$$

$\phi_{\lambda x} = e^{-\lambda x}(\cos\lambda x + \sin\lambda x)$; $\psi_{\lambda x} = -e^{-\lambda x}(\sin\lambda x - \cos\lambda x)$; $\theta_{\lambda x} = e^{-\lambda x}\cos\lambda x$; $\beta_{\lambda x} = e^{-\lambda x}\sin\lambda x$

Table 7.4 Values for ϕ, β, ψ, and θ Factors

λx	ϕ	β	ψ	θ	λx	ϕ	β	ψ	θ
0.0	1.000	0.000	1.000	1.000	5.1	−.003	−.006	0.008	0.002
0.1	0.991	0.090	0.810	0.900	5.2	−.002	−.005	0.007	0.003
0.2	0.965	0.163	0.640	0.802	5.3	−.001	−.004	0.007	0.003
0.3	0.927	0.219	0.489	0.708	5.4	−.001	−.003	0.006	0.003
0.4	0.878	0.261	0.356	0.617	5.5	0.000	−.003	0.006	0.003
0.5	0.823	0.291	0.241	0.532	5.6	0.001	−.002	0.005	0.003
0.6	0.763	0.310	0.143	0.453	5.7	0.001	−.002	0.005	0.003
0.7	0.700	0.320	0.060	0.380	5.8	0.001	−.001	0.004	0.003
0.8	0.635	0.322	−.009	0.313	5.9	0.002	−.001	0.004	0.003
0.9	0.571	0.318	−.066	0.253	6.0	0.002	−.001	0.003	0.002
1.0	0.508	0.310	−.111	0.199	6.1	0.002	−.000	0.003	0.002
1.1	0.448	0.297	−.146	0.151	6.2	0.002	−.000	0.002	0.002
1.2	0.390	0.281	−.172	0.109	6.3	0.002	0.000	0.002	0.002
1.3	0.336	0.263	−.190	0.073	6.4	0.002	0.000	0.001	0.002
1.4	0.285	0.243	−.201	0.042	6.5	0.002	0.000	0.001	0.001
1.5	0.238	0.223	−.207	0.016	6.6	0.002	0.000	0.001	0.001
1.6	0.196	0.202	−.208	−.006	6.7	0.002	0.000	0.001	0.001
1.7	0.158	0.181	−.205	−.024	6.8	0.002	0.001	0.000	0.001
1.8	0.123	0.161	−.199	−.038	6.9	0.001	0.001	0.000	0.001
1.9	0.093	0.142	−.190	−.048	7.0	0.001	0.001	0.000	0.001
2.0	0.067	0.123	−.179	−.056	7.1	0.001	0.001	−.000	0.001
2.1	0.044	0.106	−.168	−.062	7.2	0.001	0.001	−.000	0.000
2.2	0.024	0.090	−.155	−.065	7.3	0.001	0.001	−.000	0.000
2.3	0.008	0.075	−.142	−.067	7.4	0.001	0.001	−.000	0.000
2.4	−.006	0.061	−.128	−.067	7.5	0.001	0.001	−.000	0.000
2.5	−.017	0.049	−.115	−.066	7.6	0.001	0.000	−.000	0.000
2.6	−.025	0.038	−.102	−.064	7.7	0.001	0.000	−.000	0.000
2.7	−.032	0.029	−.089	−.061	7.8	0.000	0.000	−.000	0.000
2.8	−.037	0.020	−.078	−.057	7.9	0.000	0.000	−.000	−.000
2.9	−.040	0.013	−.067	−.053	8.0	0.000	0.000	−.000	−.000
3.0	−.042	0.007	−.056	−.049	8.1	0.000	0.000	−.000	−.000
3.1	−.043	0.002	−.047	−.045	8.2	0.000	0.000	−.000	−.000
3.2	−.043	−.002	−.038	−.041	8.3	0.000	0.000	−.000	−.000
3.3	−.042	−.006	−.031	−.036	8.4	0.000	0.000	−.000	−.000
3.4	−.041	−.009	−.024	−.032	8.5	0.000	0.000	−.000	−.000
3.5	−.039	−.011	−.018	−.028	8.6	0.000	0.000	−.000	−.000
3.6	−.037	−.012	−.012	−.025	8.7	−.000	0.000	−.000	−.000
3.7	−.034	−.013	−.008	−.021	8.8	−.000	0.000	−.000	−.000
3.8	−.031	−.014	−.004	−.018	8.9	−.000	0.000	−.000	−.000
3.9	−.029	−.014	−.001	−.015	9.0	−.000	0.000	−.000	−.000
4.0	−.026	−.014	0.002	−.012	9.1	−.000	0.000	−.000	−.000
4.1	−.023	−.014	0.004	−.010	9.2	−.000	0.000	−.000	−.000
4.2	−.020	−.013	0.006	−.007	9.3	−.000	0.000	−.000	−.000
4.3	−.018	−.012	0.007	−.005	9.4	−.000	0.000	−.000	−.000
4.4	−.015	−.012	0.008	−.004	9.5	−.000	−.000	−.000	−.000
4.5	−.013	−.011	0.009	−.002	9.6	−.000	−.000	−.000	−.000
4.6	−.011	−.010	0.009	−.001	9.7	−.000	−.000	−.000	−.000
4.7	−.009	−.009	0.009	−.000	9.8	−.000	−.000	−.000	−.000
4.8	−.007	−.008	0.009	0.001	9.9	−.000	−.000	−.000	−.000
4.9	−.006	−.007	0.009	0.001	10.0	−.000	−.000	−.000	−.000
5.0	−.005	−.006	0.008	0.002					

$\phi_{\lambda x} = e^{-\lambda x}(\cos \lambda x + \sin \lambda x)$

$\beta_{\lambda x} = e^{-\lambda x}(\sin \lambda x)$

$\psi_{\lambda x} = e^{-\lambda x}(\sin \lambda x - \cos \lambda x)$

$\theta_{\lambda x} = e^{-\lambda x}(\cos \lambda x)$

For design, typical needs include moments, shears, and deflections. Table 7.3 provides the closed-form solutions of the basic differential equations, based on Winkler's approach, for some select cases of beams on elastic foundations. Table 7.4 provides numerical values for the factors ϕ, β, ψ, and θ, defined in Table 7.3.

Example 7.1

Given   A long concrete beam rests on a stiff to very stiff silt formation; assume a $k_s = 90$ pci (155 k/ft³). Per a 12-in. width, the moment of inertia of the beam is 1.8×10^4 in.⁴. Four equal concentrated loads, P, spaced 180 in. apart, act on the 12-in. beam width, as shown in Figure 7.18. Assume $E_c = 4 \times 10^6$ psi.

Find   The moment and deflection under the first and second loads.

Procedure

$$\lambda = \sqrt[4]{\frac{k}{4E_c I}} = \sqrt[4]{\frac{90 \times 12}{4(4 \times 10^6)(1.8 \times 10^4)}} = \frac{1}{128}$$

From Table 7.4, we obtain approximate values for ψ and ϕ at corresponding points:

Function \ Point	1	2	3	4
λx	0	$\frac{180}{128} = 1.406$	2.81	4.22
ψ	1	-0.201	-0.076	0.006
ϕ	1	0.284	-0.037	-0.020

$$M_1 = \frac{P}{4\lambda}\psi = \frac{P}{4\lambda}(1 - 0.201 - 0.076 + 0.006)$$

$$= 0.73\frac{P}{4\lambda} = 0.73\frac{P}{4(1/128)}$$

(i.e., 27% less than that produced by a single load).

$$M_2 = \frac{P}{4\lambda}[1 - 2(0.201) - 0.076] = \frac{0.522P}{4(1/128)}$$

$$M_3 = M_2; \qquad M_4 = M_1, \text{ etc. (from symmetry)}$$

$$y_1 = \delta_1 = \frac{P\lambda}{2k}\phi = \frac{P\lambda}{2k}(1 + 0.284 - 0.037 - 0.020)$$

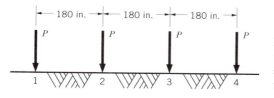

— 180 in. — — 180 in. — — 180 in. —

P P P P

1   2   3   4

Figure 7.18 The four concentrated loads acting on a beam supported by an elastic foundation.

$$y_1 = 1.227 \frac{P(1/128)}{2(90 \times 12)}$$

$$y_2 = \frac{P\lambda}{2k} [1 + 2(0.284) - 0.037]$$

$$y_2 = 1.531 \frac{P(1/128)}{2(90 \times 12)}$$

$$y_3 = y_2; \qquad y_4 = y_1, \text{ etc. (from symmetry)}$$

7.7 MAT OR RAFT FOUNDATIONS

Mat or **raft** foundations are thickened concrete slabs that support a number of columns or walls; hence, one may view mats as large combined footings. Mats may be preferred over spread footings on strata that are erratic or have low bearing capacities, or they have a heterogeneous character, or where large differential settlement is likely if one were to use spread footings. The use of mat foundations may be also advantageous where the foundation is below the water table and there is need to eliminate water infiltration into basement-type installations.

In cases where large settlements are anticipated, piles may be the viable option for the mat support. Piles may also provide a restraining force to overcome buoyancy in situations of high water table.

Figure 7.19 shows some common types of mat foundations. The following are some of the general features of each:

• **flat-plate** shown in Fig. 7.19a is perhaps the most common of the mat types. It generally requires less labor effort in placing the steel and less steel details, and is relatively easy to construct. Thus, although it may be somewhat thicker than other types, overall, it may be more economical.

• **thickened flat-plate** shown in Fig. 7.19b depicts a somewhat altered configuration of a flat-plate design, intended to accommodate large columns loads. The mat is thickened in the proximity of the columns so as to increase the punching shear capacity; the objective is to localize the thickness, without increasing the thickness for the entire mat.

• **waffle-plate** shown in Fig. 7.19c features an increase in mat thickness via the construction of a monolithic beam–slab composite in a grid fashion. Again, the intent is to provide rigidity with a minimum of concrete thickness, thus limiting or controlling the structural weight.

• **wall-plate** design shown in Fig. 7.19d incorporates the feature of an integral slab–wall interaction, thereby, resulting in a stiffer mat foundation.

All the mats shown in Fig. 7.19 are continuous, single-unit monolithic slabs, usually cast in one operation (single pour). The mat is reinforced with both positive and negative steel, adequately reinforced to resist moments in both the x and y directions. Generally, this makes for a rather large volume of steel reinforcements. Also, the effort related to the detailing and placing of such reinforcement is signifi-

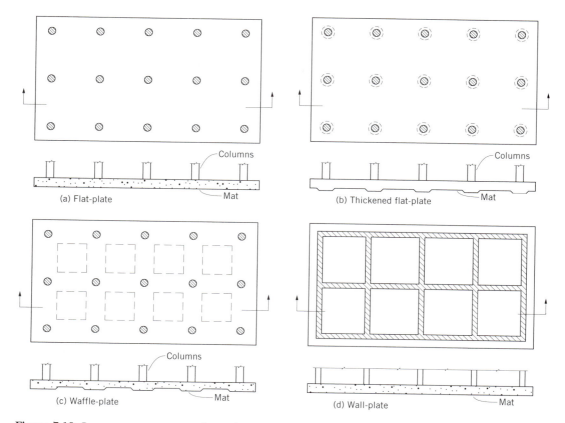

Figure 7.19 Some common types of mat foundations.

cant. Frequently, one is confronted with other concerns related to a large single pour: special techniques and machinery (e.g., cranes and special buckets for pouring of the concrete, concrete pumps), intense labor preparations, and high-concrete-volume delivery arrangements. Thus, a single large pour may make a mat design rather expensive. Hence, for soils with good bearing capacity, isolated (spread) footings are usually more economical.

Figure 7.20 shows an arrangement where a series of isolated footings are substituted for a single mat. Each footing is poured monolithically, not structurally connected to any other, and is designed as a single spread footing (discussed in Chapter 6). In order to limit vertical movement of one footing relative to the other, shear keys could be incorporated into the design. Furthermore, one may elect to provide a water-stop installation at the joints or interfaces of the footings so as to control upward infiltration of water.

Allowable Bearing Pressure for Mat Foundations

The ultimate bearing capacity of the soil under a mat may be computed using Eq. 4-9, repeated here.

$$q_u = cN_c d_c s_c i_c + \gamma D N_q d_q s_q i_q + \tfrac{1}{2}\gamma B N_\gamma s_\gamma i_\gamma \qquad (4\text{-}9)$$

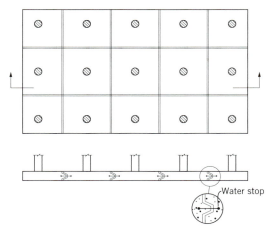

Figure 7.20 Mat foundations consisting of a series of spread footings.

The allowable bearing capacity, q_a, is obtained by dividing q_u by a reasonable safety factor. Since mat foundations typically provide a bridging effect over intermittent soft or erratic spots in the soil, mats are usually not prone to the magnitude of settlement, particularly differential settlement, associated with spread footings. Hence, the allowable bearing value to be used for design of mat foundations may be somewhat higher than that for spread (isolated) footings. For example, a factor of safety of 3 is usually sufficient; that is, $q_a = q_u/3$ is deemed suitable for most mat installations.

Based on the Standard Penetration Test (SPT) numbers, the allowable bearing capacity may be estimated from:

Fps units: $$q_a = \frac{N}{4}\left(\frac{B+1}{B}\right)^2 k_d' \qquad \text{kips/ft}^2 \qquad (7\text{-}11a)$$

SI units: $$q_a = 12N\left(\frac{B+0.305}{B}\right)^2 \qquad \text{kN/m}^2 \qquad (7\text{-}11b)$$

where $k_d = 1 + 0.33(D/B)$ (after Meyerhof).

The above expressions are predicated on an estimated or anticipated settlement of 2.5 cm. Correspondingly, a higher allowable q_a may be feasible for larger tolerable settlement (e.g., perhaps for steel buildings, or for buildings of only two or three stories). As always, judgment is considered an indispensable ingredient in the overall design effort.

Mat Settlements

As for virtually all types of foundations, mat settlements must also be limited to tolerable amounts. Both the amount of settlement and degree of tolerance are complicated phenomena to determine. The interaction of the superstructure and soil are frequently complicated by variations in soil parameters within a given site and by the structural rigidity of the structure. Furthermore, while general settlement may be a problem, differential settlement appears to be almost always a primary concern to the designer. The amount of total or differential settlement that is tolerable depends mostly on the effects to the superstructure; auxiliary facilities such as sewer, gas, or water lines and elevators can generally be designed to accommodate settlements without too much difficulty. (In fact, utilities and piping are more easily installed and maintained if they are placed above the mat). As a general rule for mats, one may aim for a maximum total settlement of about 2 in. and a differential settlement of about 1.5 in. However, the building designer is perhaps the best suited professional to make the judgment regarding tolerable settlement for that particular design.

Committee 336 of American Concrete Institute (1) proposed a procedure for existing differential settlement, using a relative stiffness factor, K_r, defined as

$$K_r = \frac{E'I_B}{E_s B^3} \tag{7-12}$$

$E'I_B$ is expressed as

$$E'I_B = E'I_f + \sum E'I_B + E'\frac{t\,h^3}{12}$$

where $E'I_B$ = flexural rigidity of superstructure and mat, perpendicular direction to B
E_s = modulus of elasticity of soil
B = width of mat
$E'I_f$ = flexural rigidity of footing or mat
$E'\frac{t\,h^3}{12}$ = flexural rigidity of shear walls perpendicular to B
t = wall thickness
h = wall height

The ratio of differential settlement, δ, to total settlement, ΔH, is

K_r	δ
0	0.35 ΔH for square mats
	0.5 ΔH for long mats
0.5	0.1 ΔH
>0.5	0; treated as rigid mats

Some factors that may be considered toward the objective of limiting settlement are

1. Use lower design values of soil pressures q_a. Typically, this calls for a larger size footing (longer and wider, etc.), an option that is frequently dictated by space limitations.

2. Improve the soil stratum supporting the mat (e.g., engineered fill). This option may be frequently dictated by the economics associated with the excavation of the existing fill, availability and the placing of the new fill, and sometimes by the problems of placing the new fill in places of relatively high water table.

3. Minimize the added load to a soil by removing (excavating) soil and placing the mat at a lower elevation. The beneficial effect is a weight reduction on the supporting soil equal to the weight of the removed soil. Of course, as always with an excavation, one must be cognizant of stability associated with such construction. For cases where excavation is appreciably below the water table and the strata consists of a fine sand, a major concern may be **boiling** (discussed in Sec. 2.6). The installation of sheet piles designed to overcome the boiling may or may not be feasible, particularly if the installation is near adjoining structures that may be affected by installation vibrations or by the lowering of the water table via dewatering. In the case of clay formations, one may experience a heave of the bottom (a rise in the bottom of the excavation), somewhat similar to the rebounding effect from consolidation tests. Such heave is frequently difficult to account for, since the imposed load from the mat frequently does not negate the heave effects at the same rate when the heave occurred. Furthermore, expansion of some clay (lateral flow) is also a phenomenon that sometimes occurs in clay formations, particularly in highly saturated formations.

Design of Mat Foundations

Mats may be designed as either rigid bodies or as flexible slabs supported by an elastic stratum, or as a combination of both. The combination approach is common in current practice. The coverage in this text is limited to merely a partial description of the Strength Design Method (SDM). A detailed evaluation is beyond the scope of this text.* However, a large number of computer programs are commercially available for mat analysis.

Using the Strength Design Method (SDM), the following procedure is summarized for the design of mat foundations:

1. Proportion the mat using unfactored loads and overtuning moments. Thus, the pressure distribution will be based on:

$$q = \frac{R}{A} \pm \frac{M_x y}{I_x} \pm \frac{M_y x}{I_y}; \qquad \begin{array}{l} M_x = Re_x \\ M_y = Re_y \end{array}$$

where R = resultant = ΣP = sum of all loads on the mat
   A = area of mat
   x, y = coordinates of any point on the mat where q is to be determined
   e_x, e_y = eccentricity of the resultant, R, with respect to the x- and y-axes
   I_x, I_y = moment of inertia of mat area with respect to the x- and y-axes.

The "ultimate" soil pressure, q_{ult}, can be determined as $q_{ult} = q_a$ (factored loads/ unfactored loads).

2. Determine the mat thickness (minimum) based on punching shear at critical columns based on column load and shear perimeter. Avoid the use of shear reinforcement; thus, the thickness of the mat will be determined based on the punching shear strength of the unreinforced concrete.

3. Design the reinforcing steel for bending by treating the mat as a rigid body and isolating strips in perpendicular directions, provided the following criteria are met: (a) column spacing is less than $1.75/\lambda$, or the mat is very thick, and (b) variation in column loads and spacing is not over 20%. These strips are analyzed as independent, continuous, or combined footings with multiple column loads, supported by soil pressure on the strip, and column reactions equal the factored (or unfactored)

---

* For a more thorough, detailed description of the procedure, see Refs. (1) and (4).

loads obtained from the superstructure analysis. Consideration of the shear strength transfer between strips may be necessary in order to satisfy the summation of forces in the vertical direction.

The designer may also proceed with a computer analysis, typically via computer programs that divide the mat into a number of discrete, finite elements using grid lines. The procedure may be based on either a **finite difference, finite grid,** or **finite element** method. Perhaps of the three, the finite element method is the more popular.

Regardless of the method used, one must be cognizant of inherent difficulties in obtaining an exact solution to the problem of designing the mat. Among these obstacles are

1. The difficulty in establishing elastic parameters.
2. Predicting the subgrade response, particularly when confronted with variations in soil properties under the mat.
3. The mat shape.
4. Variations and assumptions in the superstructure loads, behavior, and the stiffening effect of superstructure on mat.

Example 7.5 _____

Given   The mat dimensions and columns shown in Fig. 7.21 and design data.

$$q_a = 1 \text{ ksf}; \quad f_y = 50 \text{ ksi}; \quad f_c' = 3 \text{ ksi } (E = 3,100 \text{ ksi} = 450,000 \text{ ksf});$$
$$\text{L.F.} = 1.6; \quad k_s = 18 \text{ kcf}$$

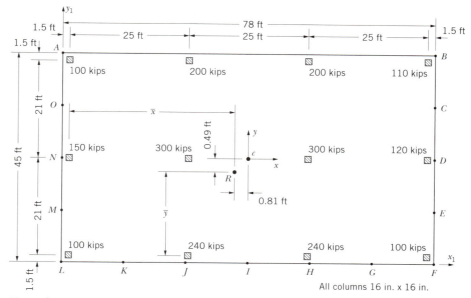

Figure 7.21 Column locations and mat size.

Find Mat design via USD (conventional method).

Procedure Resultant $R = \Sigma P = 2160$ kips; $A = 3510$ ft^2

$$\bar{x} = \frac{25(200 + 300 + 240) + 50(200 + 300 + 240) + 75(110 + 120 + 100)}{2160}$$

$$= 36.69 \text{ ft}$$

Hence,

$e_x = 0.81$ ft (left of mat centroid, negative as shown)

$$\bar{y} = \frac{21(150 + 300 + 300 + 120) + 42(100 + 200 + 200 + 110)}{2160} = 20.51 \text{ ft}$$

Hence,

$e_y = 0.49$ ft (below centroid of mat, negative as shown)

$I_x = 78(45)^3/12 = 5.92 \times 10^5$ ft^4

$I_y = 45(78)^3/12 = 17.80 \times 10^5$ ft^4

$M_x = Re_x = (2160)(0.81) = 1750$ k-ft

$M_y = Re_y + (2160)(0.49) = 1058$ k-ft

$$q = \frac{2160}{3510} \pm \frac{(1750)y}{5.92 \times 10^5} \pm \frac{(1058)x}{17.8 \times 10^5}$$

$$q = -0.615 \pm (2.95 \times 10^{-3})y \pm (0.59 \times 10^{-3})x$$

Thus, q values may be determined at any point defined by the x and y coordinates. Table 7.5 gives q values for some selected points.

Next, we shall find the depth (mat thickness) based on diagonal tension at the 240-kip column and the 110-kip column (see Fig. 7.21), based on a three-

Table 7.5 Soil Pressures, q, at Some Selected Points Under the Mat (negative sign denotes compression)

Point	R/A	Y coord.	$2.956 \times 10^{-3}y$	X coord.	$0.59 \times 10^{-3}x$	q (ksf)
A	0.615	+22.5	+0.066	−39	−0.023	−0.572
B	0.615	+22.5	+0.066	+39	+0.023	−0.526
C	0.615	+10.5	+0.031	+39	+0.023	−0.561
D	0.615	—	—	+39	+0.023	−0.592
E	0.615	−10.5	−0.031	+39	+0.023	−0.623
F	0.615	−22.5	−0.066	+39	+0.023	−0.658
G	0.615	−22.5	−0.066	+25	+0.015	−0.666
H	0.615	−22.5	−0.066	+12.5	+0.007	−0.674
I	0.615	−22.5	−0.066	0	0	−0.681
J	0.615	−22.5	−0.066	−12.5	−0.007	−0.688
K	0.615	−22.5	−0.066	−25.5	−0.015	−0.696
L	0.615	−22.5	−0.066	−39	−0.023	−0.704
M	0.615	−10.5	−0.031	−39	−0.023	−0.673
N	0.615	0	0	−39	−0.023	−0.638
O	0.615	+10.5	+0.031	−39	−0.023	−0.607

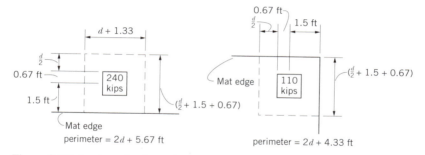

Figure 7.22 Sections for investigating diagonal shear stress.

sided and two-sided shear perimeter, respectively, as shown in Fig. 7.22. For $f'_c = 3$ ksi, $v_c = 26.81$ ksf, and neglecting soil pressure (it's relatively small), for the 240-kip column, we have

$$d(\text{perimeter}) v_c = (\text{L.F.})(\text{column load})$$
$$d(2d + 5.67)26.81 = 1.6(240)$$
$$2d^2 + 5.67d - 14.32 = 0$$
$$d = 1.99 \text{ ft} = 22.72 \text{ in.; say } d = 23 \text{ in. and } D = 27 \text{ in.}$$

For 110-kip column

$$d(d + 4.33) - 6.56 = 0$$

Solving,

$$d = 1.19 \text{ ft} = 14.27 \text{ in. (say } D = 18 \text{ in.)}$$

Thus, $d = 23$ in. controls.

At this point perhaps we should check to see if we comply with the ACI recommended limit on column spacing and loads.

$$\lambda = \sqrt[4]{\frac{k_s B}{4EI}}$$

$$k_s = 18 \text{ ksf;} \qquad B = 12 \text{ ft;} \qquad E = 450{,}000 \text{ ksf;} \qquad D = 2.25 \text{ ft}$$

$$I = \frac{12(2.25)^3}{12} = 15.2 \text{ ft}^4$$

$$\lambda = \sqrt[4]{\frac{18 \times 12}{4(450{,}000)(15.2)}}$$

$$\lambda = 0.00532$$

Thus, max. spacing $= 1.75/\lambda = 32.9$ ft > 25 ft by 7.9 ft $> 20\%$ of 25 ft. Thus, we're exceeding both the spacing and column load variations recommended for a "rigid" condition analysis. The load variation between exterior and interior columns most frequently exceeds the 20% recommendation, thus posing a frequent dilemma in the rigid analysis. For the sake of illustrating the procedure, however, we shall proceed on the assumption of a rigid condition.

Next, we shall determine the reinforcement requirements. One approach consists of dividing the mat into strips, say 3 strips along the directions and 4 in the y direction. For the 3 strips along the x direction, we have:

Strips	Width (ft)	$q_{avg.}$ (ksf)	$q_{avg}A$ (kips)	ΣP (kips)
A B C O	12	0.579	542	610
C F M O	21	0.604	989.6	870
E F L M	12	0.667	624.4	680
Σ			2156	2160

Sample calculations, strip ABCO:

$$q_{avg} = \frac{0.572 + 0.526 + 0.561 + 0.654}{4} = 0.579 \text{ ksf}$$

$$(q_{avg})A = (0.579)(12 \times 78) = 542 \text{ kips}$$
$$\Sigma P = 100 + 200 + 200 + 110 = 610 \text{ kips}$$

One notes that the $(q_{avg})A \neq \Sigma P$ for the respective strips. Hence, before we can draw the shear and moment diagrams, we shall adjust q_{avg} and ΣP. Again, the illustration is for only one strip; the others, including those in the y direction, are left as an exercise for the student. Strip ABCO.

$$\text{Average load} = \frac{542 + 610}{2} = 576 \text{ kips}$$

$$\text{Adjusted load reduction factor} = \frac{576}{610} = 0.944$$

$$\text{Adjusted } q_{avg} = (0.579)\left(\frac{576}{542}\right) = 0.615 \text{ ksf}$$

The resultant load, R_1, is 0.61 ft to the right of the center line (℄), as shown in Fig. 7.23. Thus, q is not uniform [i.e., $q = (R/A) + (Re/I)$]. This adjustment in q will permit the moment diagram to close.

Next, find the reinforcement requirements via

$$M_u = \phi A_s f_y (d - a/2)$$

where $M_u = (M)(\text{L.F}) = 1.6M \text{ (ft-lb)}/\text{ft}$
$\phi = 0.9; \quad d = 23 \text{ in.}$
$a = A_s f_y / 0.85 f_c' b = 50 A_s / 0.85 \cdot (3) \cdot (12)$

M = the respective max. positive and max. negative moments per ft of width. For example, in strip ABCO, the maximum moments are

$$+M = 148.5/12 \text{ ft} = 12.375 \text{ (ft-lb.)}/\text{ft}$$
$$-M = 540.2/12 = 45.02 \text{ (ft-lb.)}/\text{ft}$$

Hence, one may determine A_s for strip ABCO. A like approach should follow for all strips in both longitudinal and transverse directions. Subsequently one proceeds with the detailing of the foundation. These aspects are left as an exercise for the student.

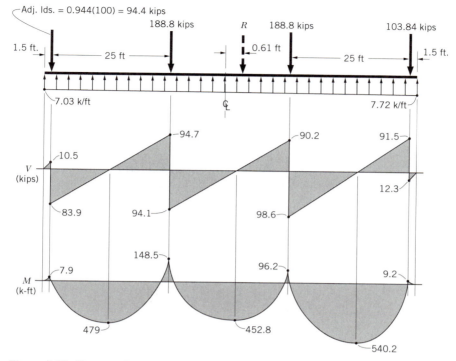

Adj. lds. = 0.944(100) = 94.4 kips

188.8 kips

R

188.8 kips

103.84 kips

1.5 ft.

25 ft

0.61 ft

25 ft

1.5 ft.

7.03 k/ft

℄

7.72 k/ft

94.7

90.2

91.5

V
(kips)

10.5

83.9

94.1

98.6

12.3

148.5

96.2

M
(k-ft)

7.9

9.2

479

452.8

540.2

Figure 7.23 Shear and moment diagrams for the strip ABCO.

Problems

7.1 Design a rectangular-shaped, combined footing, assuming a rigid footing and uniform soil pressure, for the conditions shown. See Tables P7.1 and P7.2 and Fig. P7.1.

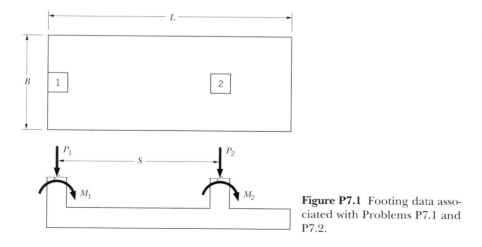

L

B

1

2

P_1

P_2

S

M_1

M_2

Figure P7.1 Footing data associated with Problems P7.1 and P7.2.

Table P7.1 Data for Probs. P7.1 and P7.3

Prob.	Col. no.	Size (in.)	Rebars	D.L. (kips)	L.L. (kips)	M_1 (kip-ft)	M_2* (kip-ft)	S (ft)	q_a (ksf)	f'_c (ksi)	f_y (ksi)
(a)	1	12 × 12	6 #8	130	80	0					
	2	14 × 14	8 #8	160	100		0	15	3.0	3	50
(b)	1	16 × 16	8 #8	180	110	0					
	2	20 × 20	10 #8	220	150		0	16	3.0	3	60
(c)	1	12 × 12	6 #8	130	100	0					
	2	16 × 16	8 #8	150	130		0	16	3.5	3	60
(d)	1	14 × 14	8 #8	120	140	0					
	2	18 × 18	10 #8	200	170		0	18	3.5	3	50
(e)	1	15 × 15	8 #8	175	150	0					
	2	20 × 20	12 #8	230	160		0	18	4.0	3	50
(f)	1	16 × 16	8 #8	160	130	230					
	2	21 × 21	12 #8	230	160		230	20	4.0	3	60
(g)	1	18 × 18	10 #8	180	150	245					
	2	21 × 21	12 #8	200	180		−300	20	3.0	3	60
(h)	1	18 × 18	10 #8	190	190	150					
	2	22 × 22	10 #10	200	200		250	20	3.0	3	60

* A minus sign represents a direction opposite that shown. Assume M_1 and M_2 are already factored moments.

Table P7.2 Data for Probs. P7.1, 7.2, 7.4

Prob.	Col. no.	Size (mm)	Rebars (SI)	D.L. (kN)	L.L. (kN)	M_1 (kN-m)	M_2* (kN-m)	S (m)	q_a (kPa)	f'_c (MPa)	f_y (MPa)
(a)	1	300 × 300	6 25	550	350	0					
	2	360 × 360	8 25	700	450		0	4.5	150	21	350
(b)	1	400 × 400	8 25	800	500	0					
	2	500 × 500	10 25	950	600		0	5	150	21	400
(c)	1	300 × 300	6 25	550	450	0					
	2	400 × 400	8 25	600	550		0	5	175	21	400
(d)	1	360 × 360	8 25	540	625	0					
	2	460 × 460	10 25	900	750		0	5.5	175	21	350
(e)	1	380 × 380	8 25	780	600	0					
	2	500 × 500	12 25	1020	700		0	5.5	200	21	350
(f)	1	400 × 400	8 25	700	550	140					
	2	530 × 530	12 25	1020	700		240	6	200	21	400
(g)	1	460 × 460	10 25	800	600	260					
	2	530 × 530	12 25	900	800		−400	6	150	21	400
(h)	1	460 × 460	10 25	850	850	170					
	2	560 × 560	10 25	900	900		270	6	150	21	400

* A minus sign represents a direction opposite that shown. Assume M_1 and M_2 are already factored moments.

7.2 Design a trapezoid-shaped combined footing for the data given in Table 7.1, assuming the footing length extends only to the outside column dimensions, as shown in Fig. P7.2.

Figure P7.2 Footing data associated with Problems P7.2 and P7.3.

7.3 Design a trapezoid-shaped combined footing for the data given in Table P7.2, assuming the footing length extends only to the outside column dimensions, as shown in Fig. P7.2.

7.4 Design strap footings for the data given in Table P7.2 and clearance restriction for column 1, as shown in Fig. P7.4.

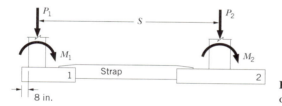

Figure P7.4 Footing data associated with Problem P7.4.

7.5 Complete the mat design in Example 7.5.

BIBLIOGRAPHY

1. American Concrete Institute Committee 336 (1987), "Suggested Design Procedures for Combined Footings and Mats," ACI Committee 336 Report.
2. Beaufait, F. W., "Numerical Analysis of Beams on Elastic Foundations," *Proc. ASCE*, 103, EM1, 205–209, Feb. 1977.
3. Biot, M. A., "Bending of an Infinite Beam on an Elastic Foundation," *Trans. ASME, J. Appl. Mech.*, vol. 59, A1–A7, 1937.
4. Bowles, J. E., "Mat Design," *J. ACI*, vol. 83, no. 6, Nov.–Dec. 1986, pp. 1010–1017.
5. Bowles, J. E., *Foundation Analysis and Design*, 4th ed., New York, McGraw Hill, 1988.
6. Brown, P. T., "Strip Footings with Concentrated Loads on Deep Elastic Foundations," University of Sydney, School of Civil Engineering Research Report No. R225, Sept. 1973.
7. Filonenko-Borodich, M. M., "Some Approximate Theories of Elastic Foundations," *Uch. Zap. Mosk. Gos.*, Univ. Mekh., vol. 46, pp. 3–18, 1940 (in Russian).
8. Fischer, J. A., et al., "Settlement of a Large Mat on Sand," *5th PSC, ASCE*, vol. 1, part 2, pp. 997–1018, 1972.

9. Fletcher, D. Q., and L. R. Herrmann, "Elastic Foundation Representation of Continuum," *Proc. ASCE,* vol. 97, EM1, pp. 95–107, Feb. 1971.

10. Hayashi, K., *Theory of Beams on Elastic Foundation,* Berlin, Springer-Verlag, 1921 (in German).

11. Hetenyi, Miklos, *Beams on Elastic Foundation,* Ann Arbor, Mich., University of Michigan Press, 1946.

12. Hetenyi, M., "Series Solutions for Beams on Elastic Foundations," *J. Appl. Mech.,* pp. 507–514, June 1971.

13. Hogg, A. H., "Equilibrium of Thin Plate Symmetrically Loaded Resting on an Elastic Foundation of Infinite Depth," *Philosophical Magazine,* vol. 25, pp. 576–582, 1938.

14. Kameswara, Rao, N. S. P., Das, Y. C., and M. Anandakrishnan, "Variational Approach to Beams on Elastic Foundations," *Proc. ASCE,* vol. 97, EM, pp. 271–294, Apr. 1971.

15. Kay, J. N., and R. L. Cavagnaro, "Settlement of Raft Foundations," *ASCE J. Geotech. Eng. Div.,* vol. 109, no. 11, pp. 1367–1381, Nov. 1983.

16. Kobayashi, H., and K. Sonoda, "Timoshenko Beams on Linear Viscoelastic Foundations," *ASCE J. Geotech. Eng. Div.,* vol. 109, June 1983.

17. Malter, Henry, "Numerical Solutions for Beams on Elastic Fondations," *J. Soil Mech. Found. Div.,* vol. LXXXIV, no. ST2, Part I, Mar. 1958.

18. Popov, E. P., "Successive Approximations for Beams on Elastic Foundations," *Trans. ASCE,* vol. CXVI, pp. 1083, 1951.

19. Scott, R. F., *Foundation Analysis,* Englewood Cliffs, N.J., Prentice-Hall, pp. 119–200, 1981.

20. Shukla, S. N., "A Simplified Method for Design of Mats on Elastic Foundation," *JACI Proc.,* vol. 81, no. 5, pp. 469–475, Sept.–Oct. 1984.

21. Swiger, W. F., "Evaluation of Soil Moduli," *6th PSC, ASCE,* vol. 2, pp. 79–92, 1974.

22. Terzaghi, K., "Evaluation of Coefficient of Subgrade Reaction," *Geotechnique,* vol. 5, no. 4, pp. 297–326, Dec. 1955.

23. Timoshenko, S., *Strength of Materials,* 3rd ed., Part II, Advanced Theory and Problems, Robert E. Krieger Publishing Co., Huntington, N.Y. 1976, pp. 1–25.

24. Vesic, A. S., "Bending of Beams Resting on Isotropic Elastic Solid," *ASCE J. Eng. Mech. Div.,* vol. 87, EM2, pp. 35–53, Apr. 1961.

25. Vesic, A. S., "Beams on Elastic Subgrade and the Winkler's Hypothesis," *5th ICSMFE,* vol. 1, pp. 845–850, 1961.

26. Vesic, A. S., "Bending of Beams Resting on Isotropic Elastic Solid," *Proc. ASCE,* vol. 87, EM2, pp. 35–51, Apr. 1961.

27. Vesic, A. S., and W. H. Johnson, "Model Studies of Beams Resting on a Silt Subgrade," *JSMDF, ASCE,* vol. 89, SM 1, pp. 1–31, Feb. 1963.

28. Vesic, A. S., "Slabs on Elastic Subgrade and Winkler's Hypothesis," 8th International Conference on Soil Mechanics and Foundation Engineering, Moscow 1973.

29. Winkler, E., *Theory of Elasticity and Strength,* Prague, H. Dominicus, 1867 (in German).

30. Woinowsky-Krieger, S., "The State of Stress in Thick Elastic Plates," *Ingenieur-Archiv,* vol. 4, pp. 305–331, 1931 (in German).

CHAPTER 8

Lateral Earth Pressure

8.1 INTRODUCTION

The problem associated with *lateral earth pressure* and retaining-wall stability is one of the most common in the civil engineering field, and a segment of soil mechanics that has been receiving widespread attention from engineers for a long time. Historical records indicate that efforts to devise procedures and formulate methodologies for the analysis of, and designing for, the effects of earth pressure date back to over three centuries—and perhaps much longer. It may be interesting to note that many of the theories developed by some of these early investigators still serve as the basis for present-day analysis of retaining walls.

The typical structures whose primary or secondary purpose is to resist earth pressures may include various types of retaining walls, sheet piling, braced sheeting of pits and trenches, bulkheads or abutments, and basement or pit walls. These may be self-supporting (e.g., gravity or cantilever-type concrete walls) or they may be laterally supported by means of bracing or anchored ties. The latter are discussed in some detail in Chapter 10.

The lateral earth pressure depends on several factors (1, 4): (1) the physical properties of the soil; (2) the time-dependent nature of soil strength, (3) the interaction between the soil and the retaining structure at the interface; (4) the general characteristics of the deformation in the soil–structure composite; and (5) the imposed loading (e.g., height of backfill, surcharge loads).

Two basic types of soil pressures are evaluated in this chapter, active and passive. If the soil mass pushes against a retaining wall, such as to push it away, the soil becomes the actuating element and the pressure resulting thereby is known as *active pressure*. On the other hand, if the wall pushes against the soil (e.g., see Fig.

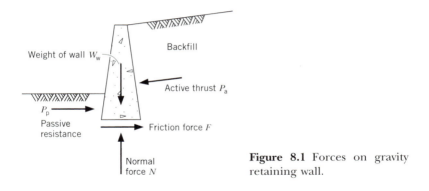

Weight of wall W_w

Backfill

Active thrust P_a

P_p
Passive resistance

Friction force F

Normal force N

Figure 8.1 Forces on gravity retaining wall.

8.1) the resulting pressure is known as *passive pressure*. In this case the actuating element is the retaining wall itself.

Although much research has been performed and appreciable advancement made during the past two centuries regarding the distribution of earth pressures and on the analysis of a wide range of earth-retaining structures, some of the theories formulated by such persons as Coulomb in 1776 and Rankine in 1857 still remain as the fundamental approaches to the analysis of most earth-supporting structures, particularly for sandy soils. Furthermore, although some research data and experience indicate that assumptions related to pressure distributions on retaining walls, or on the failure surface of backfills, are not quite those depicted by these early investigators, substantial evidence exists that the analysis and design efforts based on their theories give acceptable results for most cases of cohesionless-type backfills. The results are significantly less dependable for the more cohesive soils.

8.2 ACTIVE AND PASSIVE EARTH PRESSURES

Figure 8.1 shows some of the forces acting on a typical gravity retaining wall. Frictional forces that may be developed on the front and back faces of the retaining wall are not shown. The lateral force induced by the backfill pushes against the wall with a resultant pressure P_a. In turn, the retaining wall resists the lateral force of the backfill, thereby retaining its movement. In this case, it is readily apparent that the soil becomes the actuating force. The thrust P_a is the resultant of the active pressure, or simply the *active thrust*. The resistance to the active thrust is provided by the frictional force at the bottom of the wall and by the soil in front of the wall. For the sake of illustration, assume that the wall was pushed to the left by the active thrust P_a. In this case, relative to the soil in front of the wall, the wall becomes the actuating force, with the soil in front of the wall providing the passive resistance to movement. This resistance is known as the *passive earth pressure*, with the resultant of this pressure denoted by P_p.

The magnitude of the lateral force varies considerably as the wall undergoes lateral movement resulting in either tilting or lateral translation, or both. This phenomenon was focused on by Terzaghi (21) through his classic experiments in 1929–1934 and by others (1, 3, 7, 8, 23).

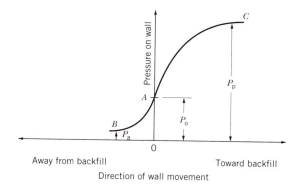

Figure 8.2 Relationship between earth pressure and wall movement.

Figure 8.2 depicts the relationship between the earth pressure and the wall movement. P_0 represents the magnitude of pressure when no movement of the retaining wall takes place; it is commonly referred to as *earth pressure at rest*. As the wall moves toward the backfill, the pressure increases, reaching a maximum value of P_p at point *C*. On the other hand, if the wall moves away from the backfill, the force decreases, reaching a minimum value of P_a at point *B*.

The relative magnitude of the active and passive earth pressures may perhaps be better illustrated with the aid of Fig. 8.3. For the sake of simplicity, several assump-

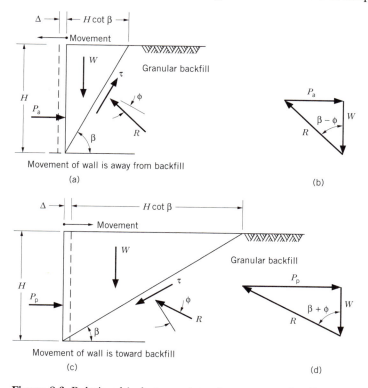

Figure 8.3 Relationship between lateral pressure and wall movement for cohesionless soil mass and plain failure surfaces. (a), (b) Development of active pressure P_a. (c), (d) Development of passive pressure P_p.

tions are made. Specific deviations from these will be discussed in the next three sections:

1. Frictional forces between backfill and retaining wall are assumed negligible.
2. The wall is vertical, and the surface of the backfill is horizontal.
3. The backfill is a homogeneous, granular material.
4. The failure surface is assumed to be a plane.

Figures 8.3a and 8.3b show the active case, where the wall moves away from the backfill, together with the corresponding force polygon. Similarly, the case of passive resistance, together with its force polygon, is shown in Figs. 8.3c and 8.3d.

The magnitudes of the active and passive forces P_a and P_p could be derived from the basic condition of static equilibrium as follows.

Case of Active Pressure   From Fig. 8.3b,

$$P_a = W \tan(\beta - \phi) \tag{a}$$

From Fig. 8-3a,

$$W = (\tfrac{1}{2}\gamma H)(H \cot \beta) = \tfrac{1}{2}\gamma H^2 \cot \beta \tag{b}$$

Substituting Eq. (b) into Eq. (a), we obtain

$$P_a = \tfrac{1}{2}\gamma H^2 \cot \beta \tan(\beta - \phi) \tag{c}$$

The maximum P_a may be obtained as follows. Let $\partial P_a / \partial \beta = 0$ and solve for β_{cr}. Then substitute β_{cr} in Eq. (c); hence,

$$\frac{\partial P_a}{\partial \beta} = \tfrac{1}{2}\gamma H^2 [\cot \beta \sec^2(\beta - \phi) + \tan(\beta - \phi)(-\csc^2 \beta)] = 0$$

or

$$\frac{\cos \beta \sin \beta - \sin(\beta - \phi) \cos(\beta - \phi)}{[\cos(\beta - \phi) \sin \beta]^2} = 0 \tag{d}$$

From the trigonometric identities we have

$$\cos \beta \sin \beta = \tfrac{1}{2} \sin 2\beta$$
$$\sin(\beta - \phi) = \sin \beta \cos \phi - \cos \beta \sin \phi$$
$$\cos(\beta - \phi) = \cos \beta \cos \phi + \sin \beta \sin \phi$$

Substituting in Eq. (d) and rearranging terms, we have

$$\frac{\cos(2\beta - \phi) \sin \phi}{[\cos(\beta - \phi) \sin \beta]^2} = 0 \tag{e}$$

Equation (e) is satisfied when $\cos(2\beta - \phi) = 0$. Thus,

$$2\beta - \phi = 90$$

or

$$\beta_{cr} = 45 + \frac{\phi}{2} \tag{f}$$

Substituting in Eq. (c), we obtain

$$P_a = \tfrac{1}{2}\gamma H^2 \cos\left(45 + \frac{\phi}{2}\right)\tan\left(45 - \frac{\phi}{2}\right)$$

But $\cot(45 + \phi/2) = \tan(45 - \phi/2)$. Hence, we obtain

$$P_a = \tfrac{1}{2}\gamma H^2 \tan^2\left(45 - \frac{\phi}{2}\right) \tag{8-1}$$

Case of Passive Pressure   From Fig. 8.3d,

$$P_p = W\tan(\beta + \phi) \tag{g}$$

From Fig. 8.3c,

$$W = \tfrac{1}{2}\gamma H(H\cot\beta) = \tfrac{1}{2}\gamma H^2 \cot\beta \tag{h}$$

Thus, Eq. (g) becomes

$$P_p = \tfrac{1}{2}\gamma H^2 \cot\beta\,\tan(\beta + \phi) \tag{i}$$

As previously, letting $\partial P_p/\partial\beta = 0$, we get

$$\frac{\partial P_p}{\partial\beta} = \frac{\gamma}{2}H^2[\cot\beta\,\sec^2(\phi + \beta) + (-\csc^2\beta)\tan(\phi + \beta)] = 0$$

But,

$$\frac{\gamma}{2}H^2 \neq 0$$

Thus,

$$[\cot\beta\,\sec^2(\phi + \beta) + (-\csc^2\beta)\tan(\phi + \beta)] = 0$$

Rearranging terms,

$$\left[\left(\frac{\cos\beta}{\sin\beta}\right)\left(\frac{1}{\cos(\phi + \beta)}\right)^2\right]\frac{\sin\beta}{\sin\beta} + \left[\left(\frac{-1}{\sin^2\beta}\right)\left(\frac{\sin(\phi + \beta)}{\cos(\phi + \beta)}\right)\right]\frac{\cos(\phi + \beta)}{\cos(\phi + \beta)} = 0$$

or

$$\frac{\cos\beta\sin\beta - \sin(\phi + \beta)\cos(\phi + \beta)}{[\sin^2\beta][\cos^2(\phi + \beta)]} = 0$$

Since $\sin^2\beta\cos^2(\phi + \beta) \neq 0$

$$\cos\beta\sin\beta - \sin(\phi + \beta)\cos(\phi + \beta) = 0$$

From trigonometric identities,

$$\cos\beta\sin\beta = \tfrac{1}{2}\sin 2\beta$$
$$\sin(\phi + \beta) = \sin\phi\cos\beta + \cos\phi\sin\beta$$
$$\cos(\phi + \beta) = \cos\phi\cos\beta - \sin\phi\sin\beta$$
$$\sin\phi\cos\phi = \tfrac{1}{2}\sin 2\phi$$
$$\sin^2\phi = \tfrac{1}{2}(1 - \cos 2\phi)$$
$$\cos^2\phi = \tfrac{1}{2}(1 + \cos 2\phi)$$
$$\cos 2\phi = \cos^2\phi - \sin^2\phi = 1 - 2\sin^2\phi = -1 + 2\cos^2\phi$$

Hence, we have

$$\tfrac{1}{2}\sin 2\beta - (\sin\phi\cos\beta + \cos\phi\sin\beta)(\cos\phi\cos\beta - \sin\phi\sin\beta) = 0$$

Expanding,

$$\tfrac{1}{2}\sin 2\beta - [\sin\phi\cos\phi\cos^2\beta - \sin^2\phi\cos\beta\sin\beta \\ + \cos^2\phi\sin\beta\cos\beta - \cos\phi\sin\phi\sin^2\beta] = 0$$

or

$$\tfrac{1}{2}\sin 2\beta - [\sin\phi\cos\phi(1 - \sin^2\beta) - \sin^2\phi\cos\beta\sin\beta \\ + (1 - \sin^2\phi)\sin\beta\cos\beta - \cos\phi\sin\phi\sin^2\beta] = 0$$

Combining terms,

$$\tfrac{1}{2}\sin 2\beta - [\sin\phi\cos\phi(1 - 2\sin^2\beta) + \sin\beta\cos\beta(1 - 2\sin^2\phi)] = 0$$

or

$$\tfrac{1}{2}\sin 2\beta - \sin\phi\cos\phi\cos 2\beta + \tfrac{1}{2}\sin 2\beta(2\sin^2\phi - 1) = 0$$

Rearranging,

$$\tfrac{1}{2}\sin 2\beta[1 + 2\sin^2\phi - 1) - \sin\phi\cos\phi\cos 2\beta = 0$$

or

$$\sin 2\beta\sin^2\phi - \sin\phi\cos\phi\cos 2\beta = 0$$

Factoring $\sin\phi$, we get

$$\sin\phi[\sin 2\beta\sin\phi - \cos\phi\cos 2\beta] = 0$$

From trigonometric identity

$$\cos\phi\cos 2\beta - \sin 2\beta\sin\phi = \cos(\phi + 2\beta)$$

Thus, we have

$$-\sin\phi\cos(\phi + 2\beta) = 0$$

This is satisfied when

$$\cos(\phi + 2\beta) = 0$$

or

$$\phi + 2\beta = 90°$$

Thus,

$$\beta_{cr} = 45 - \frac{\phi}{2} \tag{j}$$

Therefore,

$$P_p = \frac{\gamma}{2}H^2\cot\beta_{cr}\tan(\phi + \beta_{cr})$$

or

$$P_p = \frac{\gamma}{2} H^2 \cot\left(45 - \frac{\phi}{2}\right) \tan\left(\phi + 45 - \frac{\phi}{2}\right)$$

But

$$\cot\left(45 - \frac{\phi}{2}\right) = \tan\left(45 + \frac{\phi}{2}\right)$$

Thus,

$$P_p = \tfrac{1}{2}\gamma H^2 \tan^2\left(45 + \frac{\phi}{2}\right) \qquad (8\text{-}2)$$

Equations 8-1 and 8-2 are frequently written as

$$P_a = \tfrac{1}{2}\gamma H^2 K_a \qquad (8\text{-}1a)$$

and

$$P_p = \tfrac{1}{2}\gamma H^2 K_p \qquad (8\text{-}2a)$$

where

$$K_a = \tan^2\left(45 - \frac{\phi}{2}\right)$$

and

$$K_p = \tan^2\left(45 + \frac{\phi}{2}\right)$$

K_a and K_p are generally referred to as *coefficients* for *active* and *passive pressure*, respectively. They are constants for any given soil where ϕ = constant.

It is readily apparent that the coefficient K_p is significantly larger than K_a. However, it is sometimes the practice to assume a value of K_p approximately 10 times K_a. Though this is true for only a rather small range of ϕ values, in many instances the value for ϕ for sand does fall within this range. From the above expression, however, we note a more fundamental relationship:

$$K_a = \frac{1}{K_p}$$

The coefficient of earth pressure at rest K_0 has been shown experimentally to approximate, after J. Jaky, (5)*

$$K_0 = 1 - \sin \overline{\phi}$$

where $\overline{\phi}$ is the effective angle of internal friction.

This expression is acceptable for normally consolidated soils, both cohesionless and cohesive. For overconsolidated clays, the value of K_0 is slightly larger than that given by the above expression.

---

* K_0 is somewhat higher for (a) finer grain soils, (b) loose cohesionless soils, (c) soils of small ϕ values, and (d) overconsolidated soils. K_0 is somewhat smaller with increase in overburden pressures.

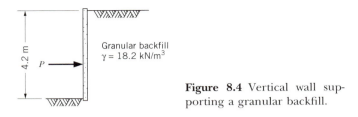

Figure 8.4 Vertical wall supporting a granular backfill.

Example 8.1

Given   A retaining wall as shown in Fig. 8.4.

Find   (a) K_a, K_p, and K_p/K_a for $\phi = 15, 20, 25, 30, 35,$ and $40°$.

   (b) P_a and P_p for $\phi = 30°$.

Procedure   (a) $K_a = \tan^2(45 - \phi/2)$ and $K_p = \tan^2(45 + \phi/2)$. The coefficients are given in the following table.

	15°	20°	25°	30°	35°	40°
K_a	0.589	0.490	0.406	0.333	0.271	0.217
K_p	1.698	2.040	2.464	3.000	3.690	4.599
K_p/K_a	2.883	4.163	6.069	9.009	13.616	21.193

   (b) $P_a = \frac{1}{2}\gamma H^2 K_a = \frac{1}{2}(18.2)(4.2)^2(0.333)$
   $P_b = \frac{1}{2}\gamma H^2 K_p = \frac{1}{2}(18.2)(4.2)^2(3.000)$

Answer $P_a = 53.45 \text{ kN}$ and $P_b = 481.57 \text{ kN}$

8.3 RANKINE'S THEORY

Although historical evidence exists that shows that as early as in 1687 Marquis Sebastian de Prestre de Vauban, a French military engineer, formulated some guidelines for designing some earth-retaining structures, it was not until 1776 that Charles Augustin Coulomb published his now famous and fundamental earth pressure theory. Since then, a number of investigators, including William John Macquorn Rankine, Jean Victor Poncelot, Karl Culmann, and numerous more current investigators, have refined and contributed much toward the solution of problems related to earth pressure. Coulomb's and Rankine's are perhaps the two best known theories and are frequently referred to as the *classical earth pressure theories*.

The theory proposed by Rankine in 1857 is based on the assumption that a conjugate relationship exists between vertical and lateral pressures on vertical planes within a mass of homogeneous, isotropic, and cohesionless material behind a smooth retaining wall. Rankine's theory reflects a simplification of Coulomb's method.

Cohesionless Backfill and Level Surface

The basic concept behind Rankine's theory can be depicted via Mohr's circle. Consider the element shown in Fig. 8.5a subjected to the geostatic stresses shown. The value for σ_1 could be approximated as the product of the average unit weight times depth, namely, $\sigma_1 \cong \gamma h$. If the wall were to move to the left, thereby creating a case of active stress, the value for σ_1 would become the major principal stress. The corresponding Mohr circle for this case is depicted by circle 1 in Fig. 8.5b. On the other hand, if the wall were to push against the backfill, a case of passive pressure would be developed. The vertical stress would then become the minor principal stress, and the lateral stress would thus become the major principal stress. The Mohr circle for this condition is depicted by circle 2 in Fig. 8.5b.

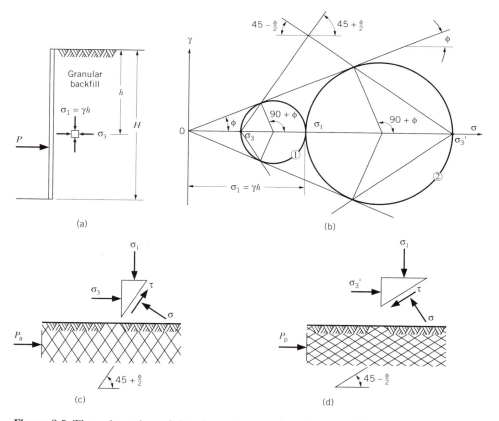

Figure 8.5 The orientation of slip planes in granular soil mass with a level surface under active and passive states of stress. (a) Element in granular soil. (b) Mohr's circles for active (1) and passive (2) states of stress in element shown in (a). (c) Slip planes for active case. (d) Slip planes for passive case.

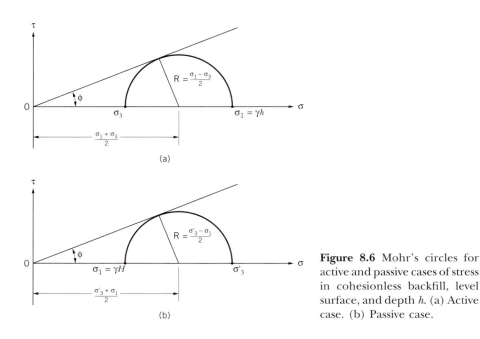

(a)

(b)

Figure 8.6 Mohr's circles for active and passive cases of stress in cohesionless backfill, level surface, and depth h. (a) Active case. (b) Passive case.

For clarity, Fig. 8.6 isolates the corresponding Mohr circles for the active and passive cases depicted in Fig. 8.5b. From Fig. 8.6a we note that

$$\sin \phi = \frac{(\sigma_1 - \sigma_3)/2}{(\sigma_1 + \sigma_3)/2} = \frac{\sigma_1 - \sigma_3}{\sigma_1 + \sigma_3}$$

or, rearranging terms and solving for σ_3,

$$\sigma_3 = \sigma_1 \frac{1 - \sin \phi}{1 + \sin \phi} = \gamma h \frac{1 - \sin \phi}{1 + \sin \phi} \tag{a}$$

or, since $(1 - \sin \phi)/(1 + \sin \phi) = \tan^2(45 - \phi/2)$, Eq. (a) becomes

$$\sigma_3 = \gamma h \tan^2 \left(45 - \frac{\phi}{2} \right) \tag{8-3}$$

For the passive case (Fig. 8.6b) we have

$$\sin \phi = \frac{(\sigma_3' - \sigma_1)/2}{(\sigma_3' + \sigma_1)/2} = \frac{\sigma_3' - \sigma_1}{\sigma_3' + \sigma_1}$$

Thus, after rearranging terms and solving for σ_3',

$$\sigma_3' = \sigma_1 \frac{(1 + \sin \phi)}{(1 - \sin \phi)} = \gamma h \frac{1 + \sin \phi}{1 - \sin \phi} \tag{b}$$

or

$$\sigma_3' = \gamma h \tan^2 \left(45 + \frac{\phi}{2} \right) \tag{8-4}$$

We note that $\tan^2(45 - \phi/2)$ and $\tan^2(45 + \phi/2)$ in Eqs. 8-3 and 8-4 are constants for constant values of ϕ. Hence, the corresponding pressures against the retaining wall vary linearly with depth, as indicated by Fig. 8.7. The corresponding resultant pressures, active and passive, can be calculated for a unit length of retaining wall as

$$P_a = \tfrac{1}{2}\gamma H^2 \tan^2\left(45 - \frac{\phi}{2}\right) = \tfrac{1}{2}\gamma H^2 K_a \tag{8-5}$$

$$P_p = \tfrac{1}{2}\gamma H^2 \tan^2\left(45 + \frac{\phi}{2}\right) = \tfrac{1}{2}\gamma H^2 K_p \tag{8-6}$$

which correspond to Eqs. 8-1 and 8-2, respectively.

The corresponding slip planes for the active and passive cases are shown in Figs. 8.5c and 8.5d.

Cohesionless Backfill and Inclined Surface

Let us now consider a cohesionless mass with a sloping surface behind a smooth vertical retaining wall. Assume this condition to be depicted by Fig. 8.8a. The lateral stresses acting on the vertical faces of the element (i.e., the faces parallel to the wall) are parallel to the inclined surface. Thus, any such planes experience not only normal but also shear stresses. Needless to say, they are no longer principal planes, as was the case for horizontal surfaces.

The corresponding resultant pressure on the wall could be determined with the aid of Mohr's circle. Figure 8.8c symbolizes an active state of stress. The magnitude of the vertical stress is depicted by the distance $0C$; the lateral stress, acting parallel to the sloped surface, is represented by the distance $0A$. Hence, from Fig. 8.8c we have

$$\sigma_h = (0A) \tag{a}$$

$$0A = \left(\frac{0B - AB}{0B + AB}\right)(0C) = \left(\frac{0B - AB}{0B + AB}\right)\gamma h \cos i \tag{b}$$

and

$$\left.\begin{array}{l} 0B = (0D)\cos i \\ r = (0D)\sin\phi \\ BD = (0D)\sin i \\ AB = \sqrt{r^2 - (BD)^2} = \sqrt{(0D\sin\phi)^2 - (0D\sin i)^2} \end{array}\right\} \tag{c}$$

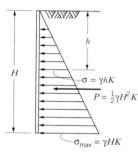

$\sigma = \gamma h K$

$P = \tfrac{1}{2}\gamma H^2 K$

$\sigma_{max} = \gamma H K$

Figure 8.7 Pressure distribution.

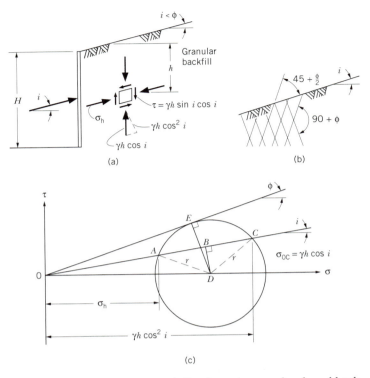

Figure 8.8 Lateral pressure and slip planes in granular sloped back-fill under active state of stress. (a) Inclined granular backfill. (b) Orientation of slip planes. (c) Mohr's circle for active state of stress.

Substituting into Eq. (b), we get

$$0A = \left(\frac{(0D)\cos i - (0D)\sqrt{\sin^2\phi - \sin^2 i}}{(0D)\cos i + (0D)\sqrt{\sin^2\phi - \sin^2 i}} \right) \gamma h \cos i \quad (d)$$

Factoring and canceling $(0D)$ and substituting $\sin^2\phi = 1 - \cos^2\phi$ and $\sin^2 i = 1 - \cos^2 i$, Eq. (d) becomes

$$0A = \left(\frac{\cos i - \sqrt{1 - \cos^2\phi - 1 + \cos^2 i}}{\cos i + \sqrt{1 - \cos^2\phi - 1 + \cos^2 i}} \right) \gamma h \cos i$$

or

$$0A = \left(\frac{\cos i - \sqrt{\cos^2 i - \cos^2\phi}}{\cos i + \sqrt{\cos^2 i - \cos^2\phi}} \right) \gamma h \cos i \quad (8\text{-}7)$$

or

$$\sigma_h = 0A = \gamma h K_a \quad (8\text{-}8)$$

and

$$P_a = \tfrac{1}{2}\gamma H^2 K_a = \tfrac{1}{2}\gamma H^2 \left(\frac{\cos i - \sqrt{\cos^2 i - \cos^2\phi}}{\cos i + \sqrt{\cos^2 i - \cos^2\phi}} \right) \cos i \quad (8\text{-}9)$$

For the passive case,

$$\sigma_h = OC = (OB + AB)\left(\frac{OA}{OB - AB}\right) = \left(\frac{OB + AB}{OB - AB}\right)(OA)$$

or

$$\sigma_h = \left(\frac{OB + AB}{OB - AB}\right)(\gamma h \cos i)$$

But,

$$OB = (OD)\cos i; \; BD = (OD)\sin i; \; r = ED = (OD)\sin \phi$$

Thus,

$$AB = [r^2 - (BD)^2]^{1/2} = [(OD \sin \phi)^2 - (OD \sin i)^2]^{1/2}$$

Hence,

$$OC = \left[\frac{(OC)\cos i + OD(\sin^2\phi - \sin^2 i)^{1/2}}{(OD)\cos i - OD(\sin^2\phi - \sin^2 i)^{1/2}}\right]\gamma h \cos i$$

From trigonometric identities,

$$\sin^2\phi = 1 - \cos^2\phi \text{ and } \sin^2 i = 1 - \cos^2 i$$

we have

$$OC = \left[\frac{\cos i + (1 - \cos^2\phi - 1 + \cos^2 i)^{1/2}}{\cos i - (1 - \cos^2\phi - 1 + \cos^2 i)^{1/2}}\right]\gamma h \cos i$$

or,

$$\sigma_h = OC = \gamma h \cos i \left[\frac{\cos i + \sqrt{\cos^2 i - \cos^2\phi}}{\cos i - \sqrt{\cos^2 i - \cos^2\phi}}\right]$$

But,

$$P_p = \tfrac{1}{2}H(OC)$$

Hence,

$$P_p = \tfrac{1}{2}\gamma H^2 K_p = \tfrac{1}{2}\gamma H^2 \left(\frac{\cos i + \sqrt{\cos^2 i - \cos^2\phi}}{\cos i - \sqrt{\cos^2 i - \cos^2\phi}}\right)\cos i \qquad (8\text{-}10)$$

Equation 8-9 is Rankine's expression for the active lateral pressure at depth H for a backfill's unit weight γ. Equation 8-10 is Rankine's expression for the passive case.

For a given sloped surface and uniform soil properties K_a becomes a constant. Thus, the intensity of load, or stress, varies linearly with depth. Hence, as before, the total resultant active force may be given by Eq. 8-9. Again, one notes that the direction of the resultant is parallel to the sloped surface.

For the case of level surface, Eqs. 8-9 and 8-10 reduce to Eqs. 8-5 and 8-6, respectively.

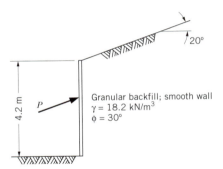

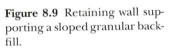

Granular backfill; smooth wall
$\gamma = 18.2$ kN/m^3
$\phi = 30°$

Figure 8.9 Retaining wall supporting a sloped granular backfill.

Example 8.2

Given   A retaining wall as shown in Fig. 8.9.

Find P_a and P_p.

Procedure   From Eqs. 8-9 and 8-10, respectively,

$$P_a = \tfrac{1}{2}\gamma H^2 K_a = \tfrac{1}{2}(18.2)(4.2)^2 \left[\frac{\cos 20 - \sqrt{\cos^2 20 - \cos^2 30}}{\cos 20 + \sqrt{\cos^2 20 - \cos^2 30}} \right] \cos 20$$

or

$$P_a = 160.52 K_a = 160.52 \left[\frac{0.94 - \sqrt{0.883 - 0.750}}{0.94 + \sqrt{0.883 - 0.750}} \right] \cos 20$$

$$P_a = 160.52 \frac{0.575}{1.305} (0.94)$$

and

$$P_p = 160.52(0.940) \left(\frac{1.305}{0.575} \right)$$

Answer $P_a = 66.484$ kN/m   and $P_p = 342.452$ kN/m

8.4 COULOMB'S EQUATION

In 1776 Coulomb introduced an expression for determining the active thrust on retaining walls. Although Coulomb was cognizant of the effects of both strength parameters c and ϕ, and of the likely probability that the sliding surface is not a plane, he elected to base his analysis on the assumption that the sliding surface is a plane and the backfill is granular ($c = 0$). He did this in order to simplify somewhat the mathematically complex problem introduced when cohesion and nonplane sliding surfaces are considered. He did, however, account for the effects of friction interaction between the soil backfill and the face of the retaining wall, and he considered the more general case of a sloped backface of the retaining wall. In this respect it is a more general approach than the Rankine cases considered in the previous section.

According to Coulomb's theory, the thrust is induced by the sliding wedge, as shown in Fig. 8.10a. For this reason, it is sometimes referred to as the *sliding wedge analysis*. The corresponding force polygon is shown in Fig. 8.10b. The development of Coulomb's equation follows from this basic relationship. From Fig. 8.10b using the Law of Sines, we have

$$\frac{P_a}{\sin(\alpha - \phi)} = \frac{W}{\sin(-\alpha + \phi + \beta + \delta)}$$

or

$$P_a = W\frac{\sin(\alpha - \phi)}{\sin(-\alpha + \phi + \beta + \delta)} \tag{8-11}$$

The weight W of the wedge can be obtained from Fig. 8.11.

$$W = \frac{Lh}{2}\gamma = \frac{(\overline{AD} + \overline{CD})h}{2}\gamma \tag{a}$$

But

$$AB = \frac{H}{\cos(\beta - 90)} = \frac{h}{\sin(\beta - \alpha)}$$

or

$$h = H\left[\frac{\sin(\beta - \alpha)}{\cos(\beta - 90)}\right]$$

Also,

$$\frac{CD}{\sin(90 - \alpha + i)} = \frac{h}{\sin(\alpha - 1)}$$

$$CD = h\frac{\sin(90 - \alpha + i)}{\sin(\alpha - i)}$$

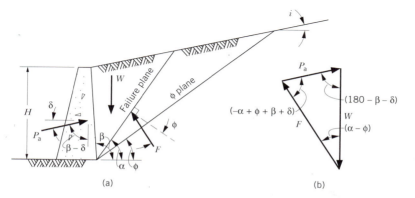

Figure 8.10 Coulomb's theory, general case with cohesionless backfill, active state. (a) Coulomb's sliding wedge. (b) Force polygon.

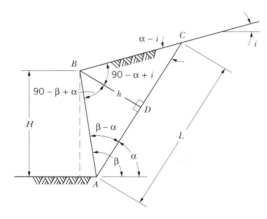

Figure 8.11 General cross section of sliding wedge.

and

$$\frac{AD}{\sin(90 - \beta + \delta)} = \frac{h}{\sin(\beta - \alpha)}$$

$$AD = h\frac{\sin(90 - \beta + \delta)}{\sin(\beta - \alpha)}$$

Substituting into Eq. (a),

$$W = \frac{h^2}{2}\gamma\left[\frac{\sin(90 - \beta + \alpha)}{\sin(\beta - \alpha)} + \frac{\sin(90 - \beta + \delta)}{\sin(\alpha - i)}\right] \qquad \text{(b)}$$

or, in terms of H,

$$W = \frac{H^2}{2}\gamma\left[\frac{\sin(\beta - \alpha)}{\cos(90 - \beta + \delta)}\right]^2\left[\frac{\sin(90 - \alpha + i)}{\sin(\alpha - i)} + \frac{\sin(90 - \beta + \delta)}{\sin(\beta - \alpha)}\right] \qquad \text{(c)}$$

Hence, substituting into Eq. 8-11,

$$P_a = \left(\frac{H^2\gamma}{2}\right)\frac{\sin(\alpha - \phi)}{\sin(-\alpha + \phi + \beta + \delta)}\left[\frac{\sin(\beta - \alpha)}{\cos(90 - \beta + \delta)}\right]^2$$
$$\times\left[\frac{\sin(90 - \alpha + i)}{\sin(\alpha - i)} + \frac{\sin(90 - \beta + \delta)}{\sin(\beta - \alpha)}\right] \qquad \text{(8-12)}$$

To determine the orientation of the failure plane that produces a maximum P_a, the following condition must be satisfied:

$$\frac{\partial P_a}{\partial \alpha} = 0 \qquad \text{(d)}$$

The solution to this equation is tedious and beyond the scope of this text. Müller-Breslau solved this general problem in 1906 (12). Hence,

$$P_a = \frac{H^2\gamma}{2}\left[\frac{\csc\beta\sin(\beta - \phi)}{\sqrt{\sin(\beta + \delta)} + \sqrt{\dfrac{\sin(\phi + \delta)\sin(\phi - i)}{\sin(\beta - i)}}}\right]^2 = \frac{H^2\gamma}{2}K_a \qquad \text{(8-13)}$$

where

$$K_a = \left[\frac{\csc \beta \sin(\beta - \phi)}{\sqrt{\sin(\beta + \delta)} + \sqrt{\frac{\sin(\delta + \phi)\sin(\phi - i)}{\sin(\beta - i)}}} \right]^2$$

The corresponding passive thrust is expressed by Eq. 8-14:

$$P_p = \frac{H^2\gamma}{2} \left[\frac{\csc \beta \sin(\beta + \phi)}{\sqrt{\sin(\beta - \delta)} - \sqrt{\frac{\sin(\phi + \delta)\sin(\phi + i)}{\sin(\beta - i)}}} \right]^2 = \frac{H^2\gamma}{2} K_p \quad (8\text{-}14)$$

where

$$K_p = \left[\frac{\csc \beta \sin(\beta + \phi)}{\sqrt{\sin(\beta - \delta)} - \sqrt{\frac{\sin(\phi + \delta)\sin(\phi + i)}{\sin(\beta - i)}}} \right]^2$$

We note that for a vertical smooth backface (that is, $\beta = 90°$ and $\delta = 0$), Eq. 8-13 reduces to Eq. 8-9 derived on the basis of Rankine's theory.

Coulomb arbitrarily placed the resultant thrust P_a at the third point from the bottom. Correspondingly, he assumed the pressure distribution to vary linearly with depth. Although this assumption appears to give results acceptable for very rigid walls and granular backfill, it is not valid for relatively flexible bulkheads, for cohesive backfills, or where the retaining wall rotates about points not close to the bottom.

Example 8.3

Given A retaining wall as shown in Fig. 8.12

Find P_a and P_p.

Procedure From Eqs. 8-13 and 8-14,

$$P_a = \frac{H^2\gamma}{2} \left[\frac{\csc \beta \sin(\beta - \phi)}{\sqrt{\sin(\beta + \delta)} + \sqrt{\frac{\sin(\phi + \delta)\sin(\phi - i)}{\sin(\beta - i)}}} \right]^2$$

$$P_a = \frac{(4.2)^2(18.2)}{2} \left[\frac{\csc 98 \sin(98 - 30)}{\sqrt{\sin(98 + 20)} + \sqrt{\frac{\sin(30 + 20)\sin(30 - 20)}{\sin(98 - 20)}}} \right]^2$$

Answer
$$P_a = 72.94 \text{ kN/m}$$

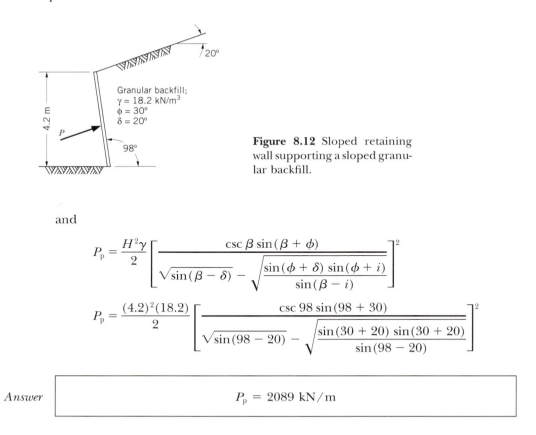

Figure 8.12 Sloped retaining wall supporting a sloped granular backfill.

Granular backfill;
$\gamma = 18.2$ kN/m³
$\phi = 30°$
$\delta = 20°$

4.2 m

P

98°

20°

and

$$P_p = \frac{H^2\gamma}{2}\left[\frac{\csc\beta\,\sin(\beta+\phi)}{\sqrt{\sin(\beta-\delta)}-\sqrt{\dfrac{\sin(\phi+\delta)\,\sin(\phi+i)}{\sin(\beta-i)}}}\right]^2$$

$$P_p = \frac{(4.2)^2(18.2)}{2}\left[\frac{\csc 98\,\sin(98+30)}{\sqrt{\sin(98-20)}-\sqrt{\dfrac{\sin(30+20)\,\sin(30+20)}{\sin(98-20)}}}\right]^2$$

Answer

$$P_p = 2089 \text{ kN/m}$$

8.5 LATERAL EARTH PRESSURES IN PARTIALLY COHESIVE SOILS

The Mohr circle may be used to determine the lateral thrust on retaining walls from backfill soil for which the shear strength may be expressed by $s = c + \sigma \tan \phi$. In this case, the values of c and ϕ imply effective cohesion and effective angle of internal friction, respectively. As a special case, let us assume a vertical and smooth retaining wall and a $(c\text{-}\phi)$ soil backfill with a horizontal surface. At a given depth h on the element shown in Fig. 8.13a, the vertical stress σ_1 is equal to γh. The lateral stress is designated by σ_3. Both of these are principal stresses. The corresponding Mohr circle for this case is shown in Fig. 8.13b.

From Fig. 8.13b we note that

$$\sin\phi = \frac{(\sigma_1-\sigma_3)/2}{(\sigma_1+\sigma_3)/2+c\cot\phi} = \frac{\sigma_1-\sigma_3}{\sigma_1+\sigma_3+2c\cot\phi}$$

Rearranging terms,

$$\sigma_3(1+\sin\phi) = \sigma_1(1-\sin\phi) - 2c\sin\phi\cot\phi$$

or

$$\sigma_3 = \sigma_1\left(\frac{1-\sin\phi}{1+\sin\phi}\right) - 2c\left(\frac{\cos\phi}{1+\sin\phi}\right) \tag{a}$$

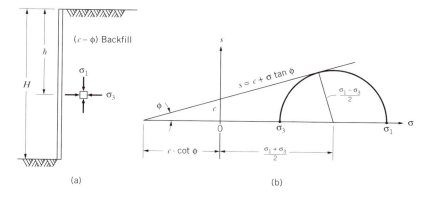

Figure 8.13 Cohesive (c-ϕ) backfill, active state.

But

$$\frac{\cos \phi}{1 + \sin \phi} = \frac{\sqrt{1 - \sin^2 \phi}}{1 + \sin \phi} = \frac{\sqrt{(1 - \sin \phi)(1 + \sin \phi)}}{\sqrt{(1 + \sin \phi)(1 + \sin \phi)}}$$

or

$$\frac{\cos \phi}{1 + \sin \phi} = \sqrt{\frac{1 - \sin \phi}{1 + \sin \phi}}$$

Thus, substituting into Eq. (a),

$$\sigma_3 = \sigma_1 \frac{1 - \sin \phi}{1 + \sin \phi} - 2c \sqrt{\frac{1 - \sin \phi}{1 + \sin \phi}} \qquad \text{(b)}$$

But $\sigma_1 = \gamma h$ and $(1 - \sin \phi)/(1 + \sin \phi) = \tan^2(45 - \phi/2)$. Hence,

$$\sigma_3 = \gamma h \tan^2 \left(45 - \frac{\phi}{2} \right) - 2c \tan \left(45 - \frac{\phi}{2} \right) \qquad \text{(c)}$$

or

$$\sigma_3 = \gamma h K_a - 2c \sqrt{K_a} \qquad \text{(d)}$$

The maximum horizontal stress (or pressure) for the active case occurs when $h = H$. The pressure distribution is shown in Fig. 8.14b. The corresponding resultant P_a is

$$P_a = \tfrac{1}{2} \gamma K_a H^2 - 2c \sqrt{K_a} H \qquad \text{(8-15)}$$

For the passive state, the major principal stress would be the horizontal stress, shown as σ_3 in Fig. 8.13a; the minor stress is then σ_1. Thus, Fig. 8.13b would show σ_1 and σ_3 in reverse positions. Then

$$\sin \phi = \frac{\sigma_3 - \sigma_1}{\sigma_3 + \sigma_1 + 2c \cot \phi}$$

Simplifying and solving for σ_3, we get

$$\sigma_3 = \sigma_1 \frac{(1 + \sin \phi)}{(1 - \sin \phi)} + 2c \frac{\cos \phi}{1 - \sin \phi} \qquad \text{(e)}$$

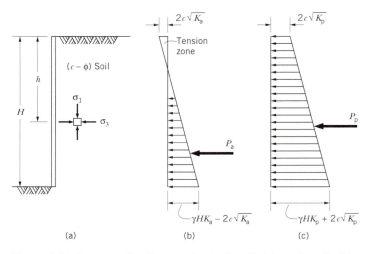

Figure 8.14 Pressure distributions in (c-ϕ) soil. (a) (c-ϕ) soil. (b) Active state. (c) Passive state.

But

$$\frac{\cos \phi}{1 - \sin \phi} = \frac{\sqrt{1 - \sin^2 \phi}}{1 - \sin \phi} = \sqrt{\frac{1 + \sin \phi}{1 - \sin \phi}}$$

Hence,

$$\sigma_3 = \sigma_1 \left(\frac{1 + \sin \phi}{1 - \sin \phi} \right) + 2c \sqrt{\frac{1 + \sin \phi}{1 - \sin \phi}} \tag{f}$$

or

$$\sigma_3 = \sigma_1 \tan^2 \left(45 + \frac{\phi}{2} \right) + 2c \tan \left(45 + \frac{\phi}{2} \right) \tag{g}$$

or

$$\sigma_3 = \gamma h K_p + 2c \sqrt{K_p} \tag{h}$$

The stress (pressure) distribution is shown in Fig. 8.14c. From this the pressure resultant P_p is

$$P_p = \tfrac{1}{2} \gamma K_p H^2 + 2c \sqrt{K_p} H \tag{8-16}$$

8.6 UNSUPPORTED CUTS IN (c-ϕ) SOIL

Unsupported excavations would theoretically be possible in (c-ϕ) soils if the lateral pressure (σ_3 for the active case) would not exceed the strength of the soil. The general expression for the horizontal stress σ_3 in a (c-ϕ) soil for the active case was given by Eq. (d) in Section 8.5. For convenience it is reproduced here:

$$\sigma_3 = \gamma h K_a - 2c \sqrt{K_a} \tag{d}$$

At ground surface, $h = 0$. Thus,

$$\sigma_3 = -2c\sqrt{K_a} \qquad (\text{tension})$$

This implies the formation of a crack as depicted in Fig. 8.15a. The corresponding pressure distribution based on Eq. (d) is shown in Fig. 8.15b.

The theoretical depth of the crack h_t can be determined by recognizing that, at the bottom of the crack, $\sigma_3 = 0$. Thus, from Eq. (d),

$$0 = \gamma h_t K_a - 2c\sqrt{K_a}$$

or

$$h_t = \frac{2c}{\gamma\sqrt{K_a}} \qquad (8\text{-}17)$$

The theoretical maximum depth H_c of unsupported excavation may be calculated as the point where the tension forces equal the cohesive strength. Hence, from Fig. 8.15b, $H_c = 2h_c$. This could also be obtained from Eq. (d) if $\sigma_3 = 2c\sqrt{K_a}$, when $h = H_c$:

$$2c\sqrt{K_a} = \gamma H_c K_a - 2c\sqrt{K_a}$$

or

$$H_c = \frac{4c}{\gamma\sqrt{K_a}} = 2h_t \qquad (8\text{-}18)$$

Though Eq. 8-18 provides a theoretical depth to which an excavation may be made without lateral support, it should be used cautiously. Surface moisture that may enter the crack may induce hydrostatic stresses or may decrease the shear strength of the soil. Hence, the unsupported excavation to such depths should be for short duration at best. Even then, judgment reflecting on the potential consequences from unsupported excavation is indeed warranted (13).

In general, it is advisable to minimize the use of cohesive backfill whenever possible. With changes in moisture content, the pressure induced by highly cohesive soil may change significantly. For example, as the clay dries up and shrinks, the

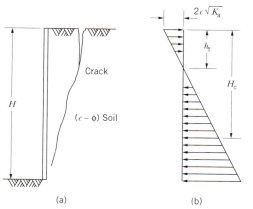

Figure 8.15 Tension crack in $(c\text{-}\phi)$ soil and corresponding pressure distribution, active state. (a) Tension crack. (b) Pressure distribution.

pressure on the wall is likely to be significantly reduced. On the other hand, a dry soil subjected to moisture may swell significantly, thereby increasing the pressure by a significant amount. Furthermore, repeated shrinking and swelling cycles may result in cumulative movement in the retaining wall itself. Yet to predict forces and subsequent effects with any degree of accuracy may be rather difficult.

Example 8.4 _____

Given   An unsupported cut as shown in Fig. 8.16.

Find   (a) Stress at top and bottom of cut.
   (b) Maximum depth of potential tension crack.
   (c) Maximum unsupported excavation.

Procedure   (a) From Eq. (c), Section 8.5,

$$\sigma_3 = \gamma h \tan^2 \left(45 - \frac{\phi}{2} \right) - 2c \tan \left(45 - \frac{\phi}{2} \right)$$

at the top, $h = 0$,

$$\sigma_3 = 0 - 2(25) \tan(45 - \tfrac{10}{2})$$

Answer

$$\sigma_3 = -41.95 \text{ kN/m}^2$$

and at the bottom, $h = 4.2$ m,

$$\sigma_3 = 18.2(4.2) \tan^2(45 - \tfrac{10}{2}) - 2(25) \tan(45 - \tfrac{10}{2})$$
$$\sigma_3 = 53.82 - 41.95$$

Answer

$$\sigma_3 = 11.87 \text{ kN/m}^2$$

(b) From Eq. 8-17,

$$h_t = \frac{2c}{\gamma \sqrt{K_a}} = \frac{2(25)}{(18.2) \sqrt{\tan^2(45 - \tfrac{10}{2})}}$$

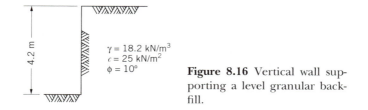

$\gamma = 18.2$ kN/m^3
$c = 25$ kN/m^2
$\phi = 10°$

Figure 8.16 Vertical wall supporting a level granular backfill.

Answer

$$h_t = 3.27 \text{ m}$$

(c) From Eq. 8-18,

$$H_c = 2h_t$$

Answer

$$H_c = 6.54 \text{ m}$$

8.7 EFFECTS OF SURCHARGE LOADS

Concentrated or uniformly distributed loads, commonly referred to as *surcharge loads*, acting on the surface of the backfill, will (1) increase the lateral pressure against the retaining wall and (2) move the point of application of the resultant pressure upward.

Figure 8.17 shows a uniformly distributed surcharge q (kN/m^2) acting on the surface of the backfill. The vertical stress σ_1 on an element within the soil backfill at a depth h is equal to $\gamma h + q$. Correspondingly, the lateral stress σ_3, from Eq. (c), Section 8.5, for a (c-ϕ) soil, is

$$\sigma_3 = (\gamma h + q) \tan^2\left(45 - \frac{\phi}{2}\right) - 2c \tan\left(45 - \frac{\phi}{2}\right)$$

The intensity of pressure at any given depth can be determined by superimposing the effects of the backfill and of the surcharge, as indicated in Fig. 8.18. The effect could be envisioned as that of an imaginary equivalent soil layer of thickness q/γ on top of the backfill. Thus, the effect is additive, resulting in the trapezoidal shape shown in Fig. 8.18. Furthermore, the resultant thrust of the two superimposed effects acts at a point between the resultants of the two pressure blocks shown in Fig. 8.18. Hence, the surcharge increases both the lateral thrust and the overturning moment.

For concentrated surcharge loads Q, such as may be induced by a continuous footing, railroad tracks, and the like, running parallel to the wall, it is possible,

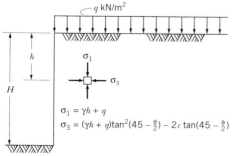

$\sigma_1 = \gamma h + q$

$\sigma_3 = (\gamma h + q)\tan^2(45 - \frac{\phi}{2}) - 2c \tan(45 - \frac{\phi}{2})$

Figure 8.17 Uniform surcharge on level backfill surface.

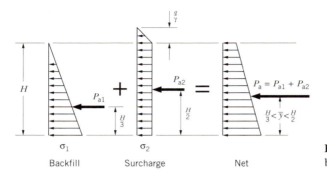

Figure 8.18 Stresses induced by backfill surcharge.

although rather laborious, to estimate the increased stresses on the wall based on Boussinesq's equation, consistent with the theory of elasticity for a semi-infinite homogeneous soil mass. However, graphical methods are more expedient for this purpose (see Examples 8.5 and 8.6).

Experimental data indicate that Boussinesq's formula for lateral stress gives acceptable results where the wall movement is compatible to soil deformations within the backfill. On the other hand, if the retaining wall is totally rigid such that the soil deformation is greatly restricted by the rigid boundary, the horizontal stress approaches a value twice that given by Boussinesq's equation. This effect becomes less noticeable as the distance of Q relative to the wall increases.

The line of action of the active thrust induced by the concentrated surcharge loads Q is commonly based on empirical procedures. Such a procedure gives acceptable results for cohesionless backfill; it may be illustrated with the aid of Fig. 8.19. When the concentrated surcharge load is located to the left of point C, the

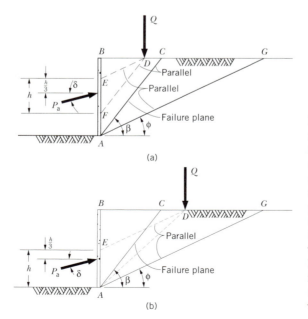

Figure 8.19 Procedure for estimating the line of action of the resultant active thrust P_a, caused by the concentrated surcharge Q on the left and right of the potential slip planes. (a) Location of active thrust induced by concentrated load Q acting between wall and failure plane. (b) Location of active thrust P_a, induced by concentrated load Q acting outside the sliding wedge.

active thrust P_a may be determined by drawing lines ED and FD parallel to lines AG and AC, respectively. The point of application of P_a is one-third the distance EF from point E, as indicated in Fig. 8.19a. When load Q is located to the right of the failure plane, as indicated in Fig. 8.19b, line ED is parallel to line AG. The point of application of P_a is one-third the distance EA from point E, as shown in Fig. 8.19b. We note that the line of action of the resultant thrust moves up the wall as the load Q approaches the wall. Furthermore, the lateral thrust as well as the overturning effect decreases as the load Q moves away from the wall.

8.8 CULMANN'S METHOD

The following graphical procedure was devised by Karl Culmann over a century ago (1875). It is used to determine the magnitude and the location of the resultant earth pressures, both active and passive, on retaining walls. This method is applicable with acceptable accuracy to cases where the backfill surface is level or sloped, regular or irregular, and where the backfill material is uniform or stratified. Also, it considers such variables as wall friction, cohesionless soils, and, with some procedural modifications, cohesive soils and surcharge loads, both concentrated and uniformly distributed. It does, however, require that the angle of internal friction of the soil be a constant for the total backfill. The procedure presented here is limited to cohesionless soils.

Reference is made to Fig. 8.20 in describing the procedure for determining the active pressure for a case of cohesionless soil by Culmann's method:

1.   Select a convenient scale to show a representative configuration of retaining wall and backfill. This should include height and slope of the retaining

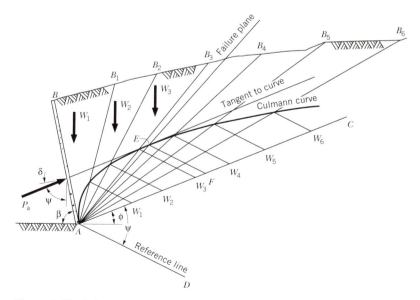

Figure 8.20 Culmann's active earth pressure for cohesionless soils.

wall, surface configuration of the backfill, location and magnitude of concentrated (line) surcharge loads, uniformly distributed surcharge, and so on.

2.   From point A draw line AC, which makes an angle of ϕ with the horizontal.

3.   Draw line AD at an angle of ψ from line AC. Figure 8.20 shows the angle ψ to be the angle between the vertical and the resultant active pressure.

4.   Draw rays AB_1, AB_2, AB_3, and so on, that is, assumed failure surfaces.

5.   Determine the weight of each wedge, accounting for variations in densities if the backfill is a layered system, for variable moisture content, and so on.

6.   Select a convenient scale and plot these weights along line AC. For example, the distance from A to W_1 along line AC equals W_1; similarly, the distance from W_1 to W_2 along line AC equals W_2, and so on.

7.   From each of the points located on line AC, draw lines parallel to line AD to intersect the corresponding assumed failure surfaces; that is, the line from W_1 will intersect line AB_1, the one from W_2 will intersect line AB_2, and so on.

8.   Connect these points of intersection with a smooth line, *Culmann's curve.*

9.   Parallel to line AC, draw a tangent to Culmann's curve. In Fig. 8.20 point E represents such a tangent point. More than one tangent is possible if the Culmann line is irregular.

10.   From the point of tangency, draw line EF parallel to line AD. The magnitude of EF, based on the selected scale, represents the active pressure P_a. If several tangents to the curve are possible, the largest of such values becomes the value of P_a. The failure surface passes through E and A, as shown in Fig. 8.20.

Surcharge loads and their respective effects on the location of the resultant could be accounted for as described in the previous section. Examples 8.5 and 8.6 may further enhance this explanation.

Figure 8.21 illustrates the procedure for determining the *passive* resistance via Culmann's method. The approach is similar to that for the active pressure, with some notable differences: (1) line AC makes an angle of ϕ degrees below rather than above the horizontal; (2) the reference line makes an angle of ψ with line AC, with ψ measured as indicated in Fig. 8.21. For the assumed sliding wedges, the weights W_1, W_2, and so on are plotted along line AC. From these points, lines are drawn parallel to the reference line to intersect the corresponding rays, as shown in Fig. 8.21. The Culmann line represents a smooth curve connecting such points of intersection. A tangent to the Culmann curve parallel to line AC is drawn, with the resultant earth pressure being the scaled value of line EF, as shown in Fig. 8.21.

Example 8.5 _____

Given   A retaining wall with backfill as shown in Fig. 8.22.

Find   The active thrust via Culmann's method.

Procedure   Figure 8.22 shows a 7-m vertical wall supporting a granular backfill whose ϕ value equals 30°. The wall is assumed smooth. A line load of 100 kN/m runs

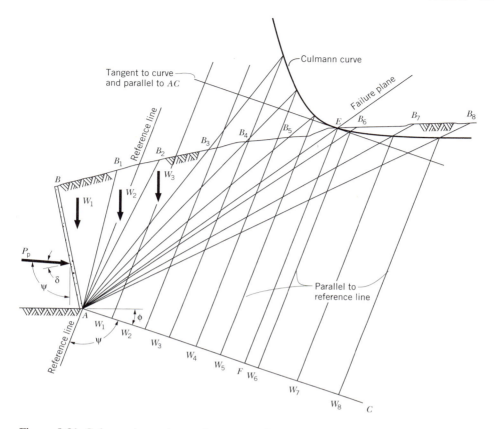

Figure 8.21 Culmann's passive earth pressure for cohesionless soils.

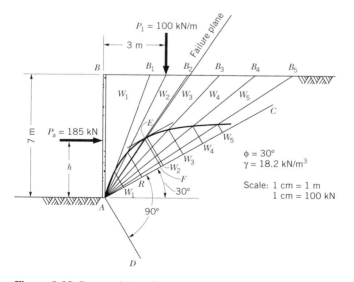

Figure 8.22 Determining the active thrust on a wall via Culmann's method.

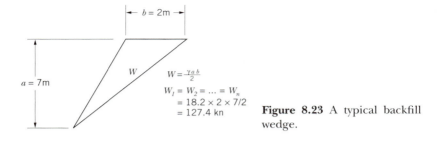

Figure 8.23 A typical backfill wedge.

parallel to the wall. For an arbitrary scale of 1 cm = 1 m the given data are plotted to scale.

For convenience the bases for all the wedges are the same. Hence, the weight of all the wedges equals 127.4 kN, as shown in Fig. 8.23. The corresponding points along AC are shown in Fig. 8.22 for an arbitrary scale of 1 cm = 100 kN. From these points lines are drawn parallel to line AD so as to intersect rays AB_1, AB_2, AB_3, and so on. Note that a similar line is drawn for the line load P_1. By connecting these points of intersection with a smooth curve (Culmann's curve) and drawing a tangent to this curve parallel to AC, we obtain the value for P_a, which is equal to the corresponding scaled value EF. The scaled value for $EF =$ 1.85 cm for 185 kN.

The point where P_a acts is determined as described in the preceding section and as shown in Fig. 8.24. Line EG is parallel to line AC, and line GF is parallel to the failure plane. P_a therefore acts at one-third distance EF from point E, or a total of 4.25 m above point A.

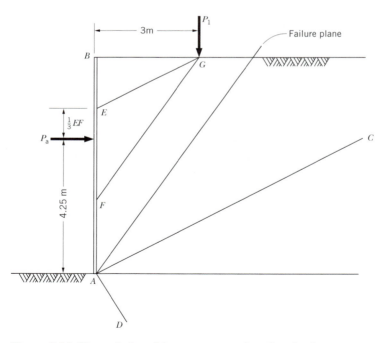

Figure 8.24 Thrust induced by concentrated surface load.

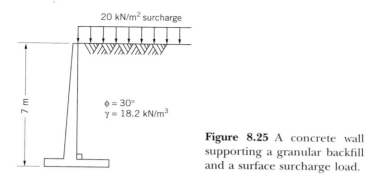

Figure 8.25 A concrete wall supporting a granular backfill and a surface surcharge load.

Example 8.6

Given   A surcharge on a backfill as shown in Fig. 8.25.

Find   The active thrust via Culmann's method.

Procedure   The uniform surcharge shown in Fig. 8.25 is transformed into an equivalent effective weight as shown in Fig. 8.26. From there on, the procedure is very similar to that given in Example 8.5. Note, however, that the soil on top of the heel of the retaining wall cannot form a wedge during failure, provided the retaining structure remains intact.

Hence, line *AB* forms an imaginary rigid surface of the backfill. Furthermore, the point where P_a acts may be determined by assuming that the pressure distribu-

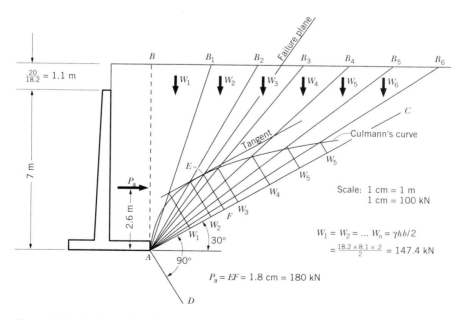

Figure 8.26 Culmann's solution.

tion and the location of the resultant pressure are as shown in Fig. 8.18. In our case the resultant thrust turns out to be located 2.6 m from the bottom of the wall.

Problems

8.1   For the condition outlined in Section 8.2, the expressions for the coefficients for active and passive pressures were $K_a = \tan^2 (45 - \phi/2)$ and $K_p = \tan^2 (45 + \phi/2)$.
   (a) For 5° increments, determine the value of K_a and K_p.
   (b) Plot the relationships determined in part (a).
   (c) Is the ratio of K_p/K_a a constant?
   (d) At what value of ϕ would the passive coefficient approach infinity?

8.2   A vertical retaining wall retains a granular backfill whose angle of internal friction and unit weight are 30° and 18.5 kN/m³, respectively. For $H = 5$ m:
   (a) Determine the active thrust P_a per meter length of wall by Eq. 8-1a.
   (b) Determine P_a by Rankine's equation 8-9 for $i = 0, 10, 15, 20,$ and 25°.
   (c) What is the percentage error when calculating P_a by assuming a level backfill instead of the actual slopes given in part (b)?
   (d) Plot percentage error versus i for the results of part (c). Is a trend implied? Explain.

8.3   A 2-m high, vertical retaining wall pushes against a granular backfill whose angle of internal friction is 30°. Assume the unit weight of the fill to be 17.6 kN/m³.
   (a) Determine the total passive resistance P_p via. Eq. 8-2a.
   (b) Determine P_p via Rankine's equation 8-10 for $i = 0, 10, 15, 20,$ and 25°.
   (c) What is the percentage error when calculating P_p by assuming a level backfill instead of the actual slopes given?
   (d) Plot percentage error versus i for the results of part (c). Is there a distinct trend? Explain.

8.4   A smooth, vertical retaining wall, 5 m high, retains a granular backfill whose angle of internal friction is 30°. Assume the unit weight of the backfill to be 18.5 kN/m³.
   (a) Determine the active thrust P_a by Eq. 8-1a.
   (b) Determine P_a by Coulomb's equation 8-13 for $i = 0, 10, 15, 20,$ and 25°.
   (c) What is the percentage error when calculating P_a by assuming a level backfill instead of the actual slopes given in part (b)?
   (d) Plot percentage error versus i for the results of part (c). What might one conclude? Explain.

8.5   A smooth vertical retaining wall, 2 m high, pushes against a granular backfill whose angle of internal friction is 30°. Assume the unit of weight of the fill to be 17.6 kN/m³.

    (a) Determine the total passive resistance P_p via Eq. 8-2a.

    (b) Determine P_p via Coulomb's equation 8-14 for $i = 0, 10, 15, 20,$ and 25°.

    (c) What is the percentage error when calculating P_p by assuming a level backfill instead of the actual slopes given in part (b)?

    (d) Plot percentage error versus i for the results of part (c). What might one conclude? Explain.

8.6   A smooth vertical retaining wall, 5 m high, retains a granular backfill whose angle of internal friction is equal to 30°. Assume the unit weight of the backfill to be 18.5 kN/m³. For a backfill slope $i = 0, 10, 15, 20,$ and 25°:

    (a) Determine K_a via the Rankine equation and Coulomb equation.

    (b) Determine the corresponding differences in the values of K_a.

    (c) Plot the differences from part (a) versus i.

8.7   For the retaining wall shown in Fig. P8.7 determine the active thrust via Coulomb's theory for the following given data: $i = 22°$, $\gamma = 17.3$ kN/m³, $\phi = 32°$, $\delta = 20°$, $\beta = 100°$, and $H = 5$ m.

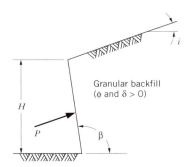

Granular backfill
(ϕ and $\delta > 0$)

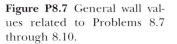

Figure P8.7 General wall values related to Problems 8.7 through 8.10.

8.8   Rework Problem 8.7 assuming $\delta = 0$ (a smooth wall). What is the percentage error introduced by assuming a smooth wall when the angle δ actually is 20°?

8.9   Rework Problem 8.7 by assuming that the angle $i = 0°$. What percentage error is introduced by assuming a value of 0° when i actually equals 22°?

8.10   Rework Problem 8.7 assuming the angle $\beta = 90°$. What percentage error is introduced by assuming a value of 90° when it actually equals 100°?

8.11   The depth of the vertical unsupported cut shown in Fig. P8.11 is 3.7 m. Assume $c = 22$ kN/m², $\phi = 12°$, $\gamma = 17.3$ kN/m³. Determine:

    (a) The stress at the top and bottom of the cut.

    (b) The maximum depth of the potential tension crack.

    (c) The maximum unsupported excavation depth.

8.12   Figure P8.11 shows a vertical unsupported cut of 3.7-m depth. Assume $c = 22$ kN/m², $\gamma = 17.3$ kN/m³, and $\phi = 6°$. Determine:

    (a) The stress at the top and bottom of the cut.

    (b) The maximum depth of the potential tension crack.

    (c) The maximum unsupported excavation depth.

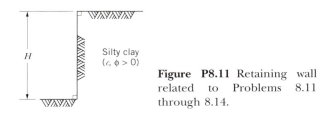

Figure P8.11 Retaining wall related to Problems 8.11 through 8.14.

8.13 Figure P8.11 shows a vertical unsupported cut of 3.7 m. Assume $\gamma = 17.3$ kN/m^3 and $\phi = 12°$. Cohesion is suspected to vary from 20 to 30 kN/m^2. For the range of 20–30 kN/m^2, in increments of 2 kN/m^2, plot c versus the following quantities:

    (a) The stress at the top and bottom of the cut.

    (b) The maximum depth of the potential tension crack.

    (c) The maximum unsupported excavation depth.

8.14 Rework Problem 8.13 for $\phi = 0°$.

For Problems 8.15 through 8.20 refer to Fig. P8.15 and determine the active thrust for the respective data given. Assume a frictionless wall.

8.15 $H = 5$ m; $\phi = 30°$; $\beta = 90°$; $h_w = 0$; $\gamma = 18$ kN/m^3; $q = 0$.

8.16 $H = 5$ m; $\phi = 30°$; $\beta = 90°$; $h_w = 0$; $\gamma = 18$ kN/m^3; $q = 250$ kN/m^2.

8.17 $H = 5$ m; $\phi = 30°$; $\beta = 90°$; $h_w = 2$ m; $\gamma = 18$ kN/m^3; $q = 0$.

8.18 $H = 5$ m; $\phi = 30°$; $\beta = 90°$; $h_w = 2$ m; $\gamma = 18$ kN/m^3; $q = 250$ kN/m^2.

8.19 $H = 5$ m; $\phi = 30°$; $\beta = 75°$; $h_w = 0$; $\gamma = 18$ kN/m^3; $q = 0$.

8.20 $H = 5$ m; $\phi = 30°$; $\beta = 95°$; $h_w = 0$; $\gamma = 18$ kN/m^3; $q = 0$.

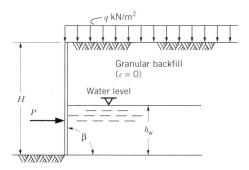

Figure P8.15 Retaining wall related to Problems 8.15 through 8.20.

Referring to Fig. P8.15, solve for P_a via Coulomb's method for the data given in Problems 8.21 through 8.26. Assume γ is a constant for the whole depth and $c = 0$.

8.21 $H = 5$ m; $\phi = 30°$; $\beta = 100°$; $h_w = 0$; $\gamma = 18$ kN/m^3; $q = 0$.

8.22 $H = 5$ m; $\phi = 30°$; $\beta = 100°$; $h_w = 0$; $\gamma = 18$ kN/m^3; $q = 250$ kN/m^2.

8.23 $H = 5$ m; $\phi = 30°$; $\beta = 90°$; $h_w = 2$ m; $\gamma = 18$ kN/m^3; $q = 0$.

8.24 $H = 5$ m; $\phi = 30°$; $\beta = 90°$; $h_w = 2$ m; $\gamma = 18$ kN/m^3; $q = 250$ kN/m^2.

8.25 $H = 5$ m; $\phi = 30°$; $\beta = 95°$; $h_w = 0$; $\gamma = 18$ kN/m³; $q = 0$.

8.26 $H = 5$ m; $\phi = 30°$; $\beta = 95°$; $h_w = 0$; $\gamma = 18$ kN/m³; $q = 250$ kN/m².

8.27   Solve for the active thrust via Coulomb's method for the data given in Fig. P8.27, assuming $P_1 = P_2 = 300$ kN/m, and $q = 0$.

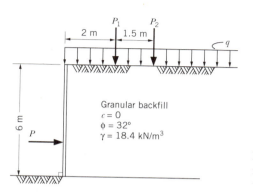

Granular backfill
$c = 0$
$\phi = 32°$
$\gamma = 18.4$ kN/m³

Figure P8.27 Retaining wall related to Problems 8.27 through 8.30.

8.28   Referring to Fig. P8.27, for $q = 250$ kN/m² and $P_1 = P_2 = 0$, determine the active thrust on the cantilever retaining wall via Coulomb's method.

8.29   Referring to Fig. P8.27, if $P_1 = P_2 = 300$ kN/m and $q = 250$ kN/m², determine the active thrust on the retaining wall via Coulomb's method.

8.30   Referring to Fig. P8.27, for $P_1 = P_2 = 500$ kN/m and $q = 250$ kN/m², determine the active thrust of the retaining wall via Coulomb's method.

BIBLIOGRAPHY

1.   Andrawes, K. Z., and M. El-Sohby, "Factors Affecting Coefficient of Earth Pressures K_0," *ASCE J. Geotech. Eng. Div.*, vol. 99, July 1973.

2.   Brooker, E. W., and H. O. Ireland, "Earth Pressures at Rest Related to Stress History," *Can. Geotech. J.*, vol. 2, no. 1, 1965.

3.   Edil, T. B., and A. W. Dhowian, "At-Rest Lateral Pressure of Peat Soils," *ASCE J. Geotech. Eng. Div.*, vol. 107, Feb. 1981.

4.   Huntington, W. C., *Earth Pressures and Retaining Walls*, Wiley, New York, 1957.

5.   Jáky, J., "Earth Pressure in Silos," *Proceedings of the Second International Conference on Soil Mechanics and Foundation Engineering*, Rotterdam, vol. I, 1948.

6.   Lee, I. K., and J. R. Herington, "Effect of Wall Movement on Active and Passive Pressures," *ASCE J. Geotech. Eng. Div.*, vol. 98, June 1972.

7.   Mackey, R. D., and D. P. Kirk, "At Rest, Active and Passive Earth Pressures," *Proc. Southeast Asian Reg. Conf. Soil Eng.*, Bangkok, 1967.

8.   Massarsch, K. R., "New Method for Measurement of Lateral Earth Pressure in Cohesive Soils," *Can. Geotech. J.*, vol. 12, Feb. 1975.

9.   Massarsch, K. R., and B. B. Broms, "Lateral Earth Pressures at Rest in Soft Clay," *ASCE J. Geotech. Eng. Div.*, vol. 102, Oct. 1976.

10.   Mayne, P. W., and F. H. Kulnawy, "K_0-OCR Relationships in Soil," *ASCE J. Geotech. Eng. Div.*, vol. 108, June 1982.

11.   Moore, P. J., and G. K. Spencer, "Lateral Pressures from Soft Clay," *ASCE J. Geotech. Eng. Div.*, vol. 98, Nov. 1972.

12. Müller-Breslau, H., *Erddruck auf Stutzmauern,* Alfred Kroner Verlag, Stuttgart, Germany, 1906, 1947.

13. Murphy, D. J., G. W. Clough, and R. S. Woolworth, "Temporary Excavation in Varved Clay," *ASCE J. Geotech. Eng. Div.,* vol. 101, Mar. 1975.

14. Peck, R. B., "Earth Pressure Measurements in Open Cuts, Chicago, Illinois, Subway," *Trans. ASCE,* vol. 108, 1943.

15. Peck, R. B., et al., "A Study of Retaining Wall Failures," *2nd Int. Conf. Soil Mech. Found. Eng.,* vol. 3, 1948.

16. Reese, L. C., and R. C. Welch, "Lateral Loading of Deep Foundations in Stiff Clay," *ASCE J. Geotech. Eng. Div.,* vol. 101, July 1975.

17. Rehman, S. E., and B. B. Broms, "Lateral Pressures on Basement Wall: Results from Full-Scale Tests," *5th Eur. Conf. Soil Mech. Found. Eng.,* vol. 1, 1972.

18. Rowe, P. W., and K. Peaker, "Passive Earth Pressure Measurements," *Geotechnique,* vol. 15, Mar. 1965.

19. Shields, D. H., and A. Z. Tolunay, "Passive Pressure Coefficients by Method of Slices," *ASCE J. Soil Mech. Found. Div.,* vol. 99, no. SM12, Dec. 1973.

20. Spangler, M. G., and J. Mickle, "Lateral Pressure on Retaining Walls Due to Backfill Surface Loads," *Highw. Res. Board Bull.,* vol. 141, 1956.

21. Terzaghi, K., "Large Retaining Wall Test," *Eng. News Rec.,* Feb. 1, 1922; Mar. 8, 1920; Apr. 19, 1934.

22. Terzaghi, K., and R. B. Peck, *Soil Mechanics in Engineering Practice,* Wiley, New York, 1967.

23. Tschebotarioff, G. P., *Foundations, Retaining and Earth Structures,* 2nd ed., McGraw-Hill, New York, 1973.

CHAPTER 9

Retaining Walls

9.1 INTRODUCTION

A retaining wall may be defined as a structure whose primary purpose is to prevent lateral movement of earth or some other material. For some special cases (e.g., basement walls of a home or certain bridge abutments), a retaining wall may also have the function of supporting vertical loads.

Figure 9.1 shows several common types of retaining walls, and some of the terms used in connection with retaining walls. The type of wall chosen for any one purpose depends on the soil conditions, construction conditions, economy, and function. Gravity walls are most frequently used for heights less than 5 m, cantilever walls for up to about 8 m, and counterforts for greater heights.

- **Gravity** retaining wall, shown in Figure 9.1a, is usually built of plain concrete, although stone is often used as a substitute. This type of wall depends only on its own weight for stability, and hence, its height is subject to some definite practical limits.

- **Semi-gravity** wall is in essence a gravity wall that has been given a wider base (a toe or heel or both) to increase its stability. Some reinforcement is usually necessary for this type of wall.

- **T-shaped wall** (Fig. 9.1b) is perhaps the most common cantilever wall. For this type of wall, the weight of the earth in the back of the stem (the backfill) contributes to its stability.

- **L-shaped wall** (Fig. 9.1c) is frequently used when property-line restrictions forbid the use of a T-shaped wall. On the other hand, when it is not feasible (e.g., due to construction restrictions) to excavate for a heel, a **reverse L-shape** may serve the need.

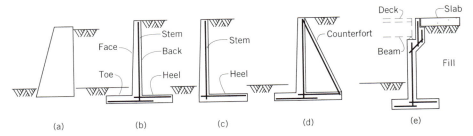

Figure 9.1 Types of retaining walls.

- **Counterfort wall** (Fig. 9.1d) consists of three main components: base, stem, and intermittent vertical ribs, called counterforts, which tie the base and the stem together. These ribs, which act as tension ties, transform the stem and heel into continuous slabs supported on three sides—at two adjacent counterforts and at the base of the stem.

- **Buttressed wall** is constructed by placing the ribs on the front face of the stem where they act in compression.

- **Bridge abutment** is a retaining wall, generally short (in plan view) and typically accompanied by wing walls. For example, the length of the wall is generally only slightly more than the width of the road, with the wing walls assuming the stability for the earth embankment on each side of the roadway. Figure 9.1e shows a schematic arrangement of the typical abutment retaining wall. Figure 9.2 is an actual bridge abutment-retaining wall composite.

9.2 SOME ASPECTS OF SOIL-RETAINING WALL INTERACTION

Figure 9.3 depicts a general force system that acts on a retaining wall. The active thrust, P_a, is the resultant force induced by the backfill and any surcharge that may be present on the surface of the backfill. It tends to move the wall to the left and tends to overturn the wall structure about point O. The force F, which acts on the bottom of the base, resists the horizontal translation; it is helped by the passive resistance, P_p. Resistance to overturning is provided by the weight, W (wall and backfill over the heel), and by any vertical component of P_a. The overturning resistance contributed by P_p is typically insignificant, and thus is usually neglected in the overturning evaluation.

Generally, the solution to a retaining wall problem entails:

1. *Evaluation of conditions that may affect the retaining structure and overall force system*   These may include the character and degree of compaction of the backfill, surcharge loads, seepage pressures, possible ice formations, changes as a result of wall movement, bearing capacity of the soil, and fluctuation of the water table.

2. *Development of reasonable external force system*   This encompasses an estimate of the strength parameters c and ϕ and magnitudes or orientations of various force resultants acting on the wall, or both.

Figure 9.2 Retaining walls (also note upper wall) and bridge abutment, Allegheny County, Pennsylvania, Rt. 764-10. (Courtesy of Portland Cement Assoc.)

3. *Design of the concrete components*   Included in this step is establishment of a viable safety factor, selection of the type of wall, proportioning of components, reinforcement, detailing, possible use of piles, specifications, and possible special requirements for construction.

Usually one constructs a retaining wall by excavating to the base elevation for a width necessary for concrete forming and construction. The back of the excavation

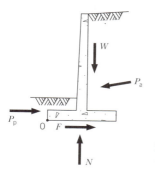

Figure 9.3 General force system on a retaining wall.

is usually sloped. Subsequent to the construction of the wall, one proceeds with the backfilling operations.

Both the type and the manner and degree of compaction of the backfill are major influences on the magnitude of the active thrust P_a. Generally, it is desirable to have a clean, granular backfill; this permits good drainage and a more predictable behavior of the backfill, and usually results in a lower P_a. When clay is used as backfill, say because a "select" granular fill is not a viable option, then it is advisable to investigate its potential effect on the structure, as a result of varying conditions, including wall movement and variable water contents. Effects of moisture in the form of swelling of the clay, poor drainage, and ensuing hydrostatic forces, or ice-related thrust forces, are factors that may result in subsequent increase in the active thrust.

As the wall yields (e.g., via some translation and/or some rotation), the active thrust is reduced, usually only temporarily until the backfill reassumes an active state; simultaneously, the passive resistance P_p also varies with time and with the change in P_a. Thus, the inherent difficulty is to determine the pressure at various stages of wall movement.

Studies conducted for the purpose of establishing pressures over the height of the wall, using small-scale models for mostly granular backfill, indicate that pressure distributions do not quite assume a hydrostatic shape (linear) and, therefore, the resultant force is not located quite at one-third of the height from the bottom. For example, studies conducted by the author on a small laboratory model, using sand backfill on a movable steel-plate wall (4 ft wide and 7 ft high) displayed a pressure increase with depth somewhat close to linear for approximately 80% of the height; a significant decrease was detected near the bottom. An explanation for this phenomenon may be the effects of base friction, as the wall permits some translation. On the other hand, pressures near the bottom exceeding twice the theoretical values were reported by some investigators for instrumented full-scale rigid walls; pressures close to theoretical values (linear) were reported for the upper half.

Generally, the pressure distributions appear to vary with the conditions of backfill material and with wall rigidity. Indeed, the results appear to vary so significantly that no reliable assessment could be formalized for the distribution to account for all such conditions. Thus, in view of this dilemma, it is expedient and, based on performance history of many designs, acceptably accurate to assume a linear pressure distribution (hydrostatic shape) over the height of the wall. If surcharges are involved, it is common practice to assume the pressure to be uniform over the height of the wall. The net affect would be the superposition of the two forces, as depicted in Fig. 9.4.

When determining the values of K_a it is common practice to assume the value of ϕ of approximately 30° for granular backfill; from laboratory testing, it is common practice to adjust the angle of internal friction obtained in a triaxial test to a plane–strain value by increasing it by approximately 10% ($\phi_{ps} = 1.1\phi_{tr}$). A common value for density of cohesionless soil is about 17 kN/m³. Depending on the water conditions and wall rigidity, for silt or clay backfill, the lateral pressure may approach that for hydrostatic conditions (i.e., $K_a \cong 1$).

Compacting the backfill excessively may indeed be detrimental; it may induce excessive pressures and create stability problems of the wall—points to keep in mind when developing specifications for this purpose. Conversely, reducing water

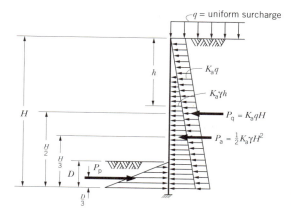

Figure 9.4 Commonly assumed pressure distributions on retaining walls.

built up via drainage is desirable as an effective means in reducing long-term pressures (including hydrostatic and ice formation) for both cohesive and cohesionless backfill. Drainage behind walls is generally accomplished by providing weepholes, accompanied by granular backfill. Fabric filters (geofabrics) are also available to provide such drainage and to filter fines in order to prevent plugging up the weepholes.

Most retaining wall failures are gradual, with the wall usually providing early indications of tilting or movement. Occasionally, a failure may occur suddenly and abruptly, where the wall and the soil rotate in a slope-failure manner. In most instances, however, wall failures are caused by inadequate bearing capacity, usually in clay formations. When the bearing capacities are questionable or inadequate, then it is typical for walls to be supported via piles (piles are discussed in detail in Chapter 11). If excessive sliding occurs, it is common practice to increase the passive resistance by projecting the base deeper into the soil, via a shear key. In the case of pile foundations, this could be accomplished by providing a batter for the piles.

It is common practice to assume a safety factor of 1.6 to 1.8 for retaining walls; ACI Article 9.924 recommends a load factor of 1.7 for lateral earth pressure design.

It is somewhat debatable as to how much friction is developed between the backfill soil and the backface of the wall (stem). Generally, the friction force has a stabilizing effect on the wall. Thus, if one were to design ignoring the friction developed between the stem and the soil, the result would be an inherent added factor of safety against tilting.

9.3 SLIDING AND OVERTURNING

The structural elements of the wall should be so proportioned that the following minimum safety factors are realized:

$$\text{(a)} \quad F_{\text{sliding}} = \frac{\text{Resisting forces}}{\text{Active forces}} \qquad \begin{aligned} &= 1.5 \text{ (for granular backfill)} \\ &= 2.0 \text{ (for cohesive backfill)} \end{aligned}$$

and against overturning about the toe (point O in Fig. 9.3):

(b) $F_{overturning} = \dfrac{\text{Stabilizing moments}}{\text{Overturning moments}}$ $\begin{aligned} &= 1.5 \text{ (for granular backfill)} \\ &= 2.0 \text{ (for cohesive backfill)} \end{aligned}$

The horizontal component of the lateral pressure (Fig. 9.3) tends to force the wall to slide along its base. The resisting force is provided by the horizontal force, F, composed of friction and adhesion, and by the passive resistance P_p of the soil in front of the wall. The passive resistance should not be counted on for this purpose if there is a chance that the soil in front of the wall may be eroded or excavated during the life of the wall.

Commonly assumed values for coefficient of friction μ and base cohesion c_b are given by the following expressions:

(c) $\tan \phi > \mu > (2/3)\tan \phi$

(d) $0.5c \leq c_b \leq 0.75c$

Additional passive resistance, and therefore additional sliding resistance, may be derived from the use of the shear key constructed under the base of the wall, as shown in Figure 9.5b. The rear location of the key results in a somewhat larger sliding plane than if the key is located toward the front of the base. The need for reinforcement of this key should be viewed in the context of the stresses induced in it; the key is normally poured monolithically (concurrently) with the base of the wall.

If the bearing capacity dictates the need for piles (i.e., low-bearing capacity), the horizontal component for resisting sliding is typically derived from the horizontal component of the battered (sloped) piles. No frictional resistance and no cohesion under the base should be assigned to the resistance force when piles are used.

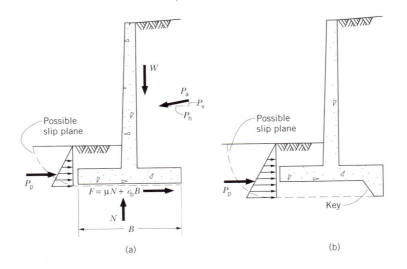

Figure 9.5 Forces resisting sliding of retaining walls.

9.4 TILTING

Some tilting is almost inevitable and somewhat desirable. In the process of tilt movement, the active thrust is reduced, although usually for only a short time, particularly for granular backfill. That is, the thrust is reduced only until the backfill creeps or slides forward and reassumes the original contact with the backface of the wall.

Usually, tilting is the result of rotation about the toe. This may be attributed to an inadequate backfill weight (the use of lightweight material, a short heel dimension, etc.) or by a foundation failure (bearing failure) in the zone of the toe. This phenomenon is shown in Fig. 9.6a.

Tilting may also occur in the opposite direction. A shear failure, as depicted by Fig. 9.6b, may occur if the layer of soil for a depth of $1.5H$ to $2H$ under the base of the wall is weak. Generally, this is considered to be a shallow failure. On the other hand, if a weak soil underlies a better soil, a deep failure may occur; the stability analysis should include this possibility. A safety factor of 2 against any shear failure is common.

The procedure for investigating shear failures consists of determining the location of the critical sliding surface, similar to the analysis for slope stability. The sliding surface is generally assumed to be circular. A number of trial circles are made before the critical one is determined; the critical surface for the sliding is that which gives the smallest factor of safety.

9.5 ALLOWABLE BEARING CAPACITY

A retaining wall must be so proportioned that it has an adequate safety factor against bearing failure. Usually, the safety factor used for the purpose is 2 for granular soils and 3 for cohesive soil. When the soil possesses a low bearing capacity or when a larger base is not a viable option, the use of piles is the typical alternative.

The allowable soil pressure can be computed by using the expression given by Eq. 9-1 for the ultimate bearing capacity, then dividing that result by a suitable factor of safety.

$$q_u = cN_c d_c i_c + \gamma D N_q d_q i_q + \tfrac{1}{2}\gamma B N_\gamma i_\gamma \tag{9-1}$$

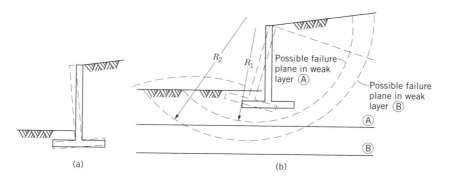

Figure 9.6 Two modes of wall tilting.

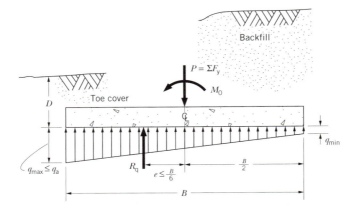

Figure 9.7 Soil pressure on base.

It is also common practice to utilize the results from the standard penetration test (SPT) given by the following:

$$q_a = \frac{N}{6}\left(\frac{B+1}{B}\right)^2, \quad \text{ksf}$$

The common range for the allowable bearing capacity as a function of the ultimate bearing capacity is as follows:

$$\frac{q_u}{4} \le q_a \le \frac{q_u}{3}$$

The actual bearing pressure on the base is a combination (superposition) of normal forces and the effects of overturning, as depicted by Fig. 9.7. The pressure q at any given point is given by the following expression:

$$q_{\substack{max \\ min}} = \frac{P}{A} \pm \frac{M_0\,(B/2)}{I} = P\left[\frac{1}{A} \pm \frac{e(B/2)}{I}\right] \le q_a$$

Note that the q_{max} must not exceed the allowable q_a; typically, q_a has an inherent factor of safety against failure of 3 to 4. Quite obviously, the overturning moment creates a rather uneven distribution from the toe to the heel of the base; a larger base and the proportions of toe and heel may be options to evaluate to provide a more uniform distribution for the design. Ideally, the pressure is uniform over the whole base; actually, this is virtually never the case. Hence, the challenge to the designer is to provide a balance between an economical and a safe design.

9.6 SETTLEMENT

As was pointed out in Sec. 2.8, settlement in granular soil takes place in relatively short time spans, while that in cohesive soil is a time-related phenomenon. Thus, retaining walls resting on granular soil generally undergo most, if not all, of the anticipated settlement within the time of the wall construction. On the other hand, those that rest on cohesive soils may settle for a long period of time after the completion of the wall. In that case, the amount of consolidation may be estimated using the consolidation theory developed in Sec. 2.8.

Since most of settlement from consolidation is related not only to the strata characteristics (thickness, compressibility, etc.), but also to the induced forces, it is perhaps apparent that for minimum differential settlement over the width of the base, a uniform pressure under the base is desirable. Also, a q_{max} much larger than q_{min} may result not only in excessive tilting, but also bearing problems. That is, the smaller the eccentricity e in Fig. 9.7, the less the pressure difference between q_{max} and q_{min} and, correspondingly, the less tendency for tilting. Of course, "backward" tilting is also a possibility, if, say, the substrata consists of a thick compressible layer, and the weight of the backfill is excessive, and thus creates a large soil pressure on the heel portion of the base.

For long walls, differential settlements (and some tilting) are not only possible, but frequently inevitable. This occurs in spite of the fact that the rigidity of the wall, coupled with that of the footing, provides reasonably stiff structural members (large sections modulus) to bridge over weak formations. This may occur, for example, when there is a sudden change in soil conditions, from say a rock to a weak soil formation. The typical consequence in that case is a vertically oriented crack in the wall. In turn, the usual corrective approach is to provide a vertical joint for the full height of the wall. Indeed, this is practical for reasons of minimizing cracking as well as to account for effects of temperature changes and shrinkage, and for construction conveniences. Such a joint may be merely a grooved configuration, similar to a shear key, between the base and the stem. The groove–key configuration provides a reasonable alignment between adjoining structural components. Such a joint could also serve as an expansion joint by minimizing the bond on one side of the joint via greasing the bars and the like; thus, expansion is permitted and cracking minimized, if not totally eliminated; see Fig. 9.8.

9.7 DRAINAGE

As indicated in Sec. 9.2, an ideal backfill is a granular soil such as sand and gravel with preferably no clay or silt. A cohesive backfill has poor drainage characteristics, may create excessive hydrostatic pressures or expand, or may create an ice formation and, thus, additional active thrust against the wall. Indeed, when the use of cohesive backfill is inevitable, then one should design the retaining wall for larger lateral pressures—a more expensive design.

Drainage could be accomplished by a relatively free-draining soil, consisting of generally less than 10% fines (silt and clay) forming a column over the heel of the retaining wall (Fig. 9.9). In addition, weepholes (generally about 2.5 to 3 in. in diameter, and 10 to 15 ft apart) are commonly used to permit discharge of the water in the backfill. Perforated drain pipes, about 4 to 6 in. in diameter, running parallel to the length of the wall, are also used for draining the backfill water. Vertical strips of filter material are further means of improving backfill drainage.

To keep the weepholes and the perforated drain pipe from clogging, a clean gravel, usually of the size of $\frac{3}{8}$ in., plus or minus, is used over the drain pipe and around the weepholes, as shown in Fig. 9.9. Such filter cover (clean gravel) and

Figure 9.8 The vertical joint in the retaining wall provides alignment between adjoining wall lengths and may be adapted to serve as an expansion and construction joint as well. (Courtesy of Portland Cement Assoc.)

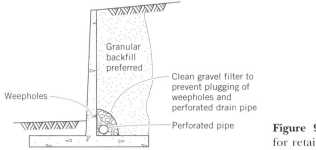

Figure 9.9 Drainage features for retaining walls.

periodic cleaning are commonly incorporated specifications in the drain-pipe design.

9.8 PROPORTIONS OF RETAINING WALLS

One proceeds with a design of a retaining wall by first selecting tentative dimensions. These trial dimensions are analyzed for stability and structural integrity. These sections are then revised as needed to obtain the most suitable proportion in the context of economy and safety. Figure 9.10 provides some common ranges of dimensions that may be helpful during the early trial phase of the design.

Gravity/Semi-gravity Walls

Figure 9.10a shows a typical proportion designation for a semi-gravity retaining wall. Typically, the walls are trapezoidal-shaped, with the base having back and front protrusions in order to distribute the base pressure toward a more uniform shape. Although the protrusions could be equal, shown in Fig. 9.10a, the toe projection may be somewhat larger than the heel in order to reduce the toe pressure. Generally, the wall is a massive structure; it experiences relatively low concrete stresses, is typically lightly reinforced, and is constructed of low-strength concrete.

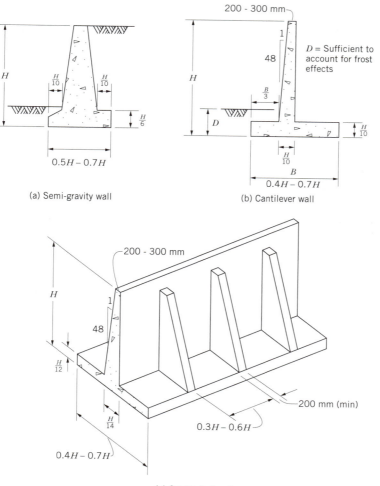

Figure 9.10 Range of proportions for some common retaining walls.

Cantilever Walls

Compared to semi-gravity walls, cantilever walls have slender components (stem and base), are reinforced to resist both flexure and thermal stresses, and are generally carefully formed and constructed to reasonably close tolerances. The components are stress-analyzed for structural adequacy and are selected for adequate stability. A small batter on the front face (typically 1 to 48) provides a slight compensation for possible forward tilting and also provides a certain degree of esthetic appeal; a vertical face or perhaps a slight forward tilting may give the illusion of instability.

Figure 9.10b provides a common range of relative proportions for cantilever walls. The base is selected so that the resultant forces fall within the middle third of the base; otherwise, a large eccentricity may result in an undesirable toe pressure. The top of the stem is generally a minimum of 0.2 m, but preferably 0.3 m in width (thickness) in order to facilitate the placement of the concrete; a narrow stem top may restrict the pouring of the concrete. The stem, the toe, and the heel are analyzed as beams, satisfying both flexure and shear requirements. Typically, a load factor of 1.7 to 1.8 is used in the analysis of these components. Concretes are usually of 3000 psi or higher; steel reinforcement is of a grade of 50 ksi or higher.

Counterfort Walls

Typical proportions for counterfort retaining walls are shown in Fig. 9.10c. These proportions, however, may vary to a greater extent than for cantilever walls since the base dimensions and stem are interrelated to the spacing of the counterforts. The space of the counterforts is commonly about $\frac{2}{3}$ of the height of the wall for moderate heights, and about $\frac{1}{2}$ of the height for walls in excess of about 8 m. The ribs (counterforts) may extend to the full height of the wall or may be cut-off at some point near the top, thus permitting the upper part of the wall to act as a cantilever. The top of the wall is usually 0.2 to 0.3 m thick, again in order to facilitate the pouring of the concrete.

Example 9.1

Given   The data shown via Fig. 9.11; also, assume

$$f_c' = 3 \text{ ksi}; \qquad f_y = 60 \text{ ksi}; \qquad \text{L.F.} = 1.8; \qquad \text{ACI Code}$$

Assume stem thickness at top (d_{top} + 3.5 in. cover) = recommended min. thickness = 300 mm = 11.8 in., say 12 in.

Find   A suitable design.

Procedure   Based on $\phi = 36°$, $K_a = \tan^2(45 - \phi/2) = 0.26$. Thus, the pressure distribution on the stem is shown in Fig. 9.12.

Referring to Fig. 9-13,

$$P = \tfrac{1}{2}K_a\gamma H^2 = 0.5(0.26)(110)(18)^2 = 4633 \text{ lb/ft of wall}$$
$$P' = K_a(q_s)(H) = 0.26(400)(18) = 1872 \text{ lb/ft of wall}$$
$$v_c = 2\phi\sqrt{f_c'} = 2(.85)\sqrt{3000} = 93.11 \text{ psi}$$

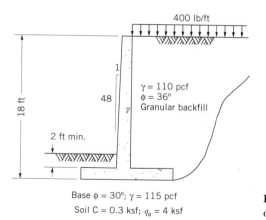

Base φ = 30°; γ = 115 pcf
Soil C = 0.3 ksf; q_a = 4 ksf

Figure 9.11 Retaining wall data.

Stem thickness at bottom, d_{bot}, based on shear, is

$$d_{bot} = \frac{(P + P')(\text{L.F.})}{(v_c)(12\text{ in.})} = \frac{(4633 + 1872)(1.8)}{(93.11)(12)} = 10.5\text{ in.}$$

Based on a stem thickness at top = 300 m = 11.8 in.,

$$d_{top} = 11.8\text{ in.} - 3.5\text{ in. cover} = \text{say 8 in.}$$

For 1:48 slope on stem face (front), the d at bottom is

$$\left(8 + \frac{18}{4}\right) = 12.5\text{ in.,} \qquad \text{say 13 in.}$$

Thus, stem thickness at bottom is 13 in. + 3.5 in. cover = 16.5 in., or rounded to 17 in.; stem thickness at top would likewise be rounded to 12 in. The variation of d at various points along the height is shown in Fig. 9.14a.

Maximum moment at bottom of steel is

$$M_{stem} = (4633)(6) + (1872)(9) = 44{,}644\text{ ft-lb} = 535{,}752\text{ in.-lb}$$

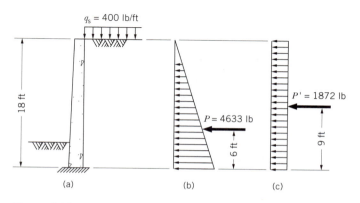

Figure 9.12 Soil pressures on stem.

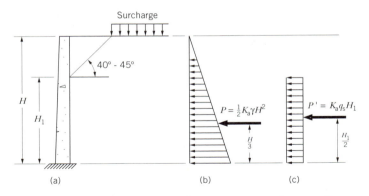

Figure 9.13 Recommended surcharge pressure when surcharge begins at some distance from backface of wall.

The variation of moment along the stem is shown in Fig. 9.13b. Now to determine A_s:

$$a = A_s f_y / 0.85 f_c' b = A_s (60) / 0.85 (3) (12) = 1.96 A_s$$

$$M_u = \phi A_s f_y \left(d - \frac{a}{2} \right)$$

or

$$A_s \left(d - \frac{a}{2} \right) = M_u / \phi f_y$$

At bottom of stem:

$$A_s \left(13 - \frac{1.96}{2} A_s \right) = (535.8) 1.8 / 0.9 (60); \qquad 1.8 = \text{load factor}$$

$$0.98 A_s^2 - 13 A_s = 17.86$$

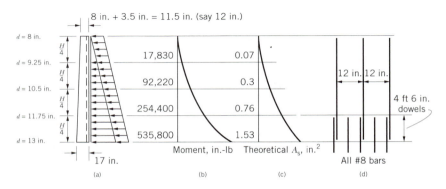

Figure 9.14 Force and moment distribution on stem.

or

$$A_s^2 - 13.27A_s + 18.33 = 0$$

$$A_s = \frac{13.27 \pm \sqrt{176.09 - 73.32}}{2} = \frac{13.27 \pm 10.14}{2} = 1.56 \text{ in.}^2$$

Now check A_s based on

$$p_{max} = 0.016; \qquad p_{min} = 200/f_y = 0.0033$$

Actual $p = A_s/bd = 1.56/(12)(13) = 0.0100 > 0.003 \qquad$ O.K.
$$< 0.016$$

Use #8 bars at a spacing of $(0.79/1.56) \times 12 = 6.08$ in., say 6 in. c.c.

Development length $= 0.04 d_b f_y / \sqrt{f_c'} = 0.04(1)60{,}000/\sqrt{3{,}000} = 43.82$ in.
$$= 3.65 \text{ ft}$$

Cut-off Points

Not all bars need to be extended to the full height of the stem. Figure 9.14 shows the load distribution (backfill) on the stem, the bending moments over the height, and values of A_s required (based on a load factor = 1.8) over the height of the stem. One-third (minimum) of A_s required at the bottom will need to be extended to the top. One notes that A_s for a #8 bar ($= 0.79$ in.^{2}) is equal to the required A_s at approximately 4.5 ft from the bottom of the stem (see Fig. 9.14c). Thus, one #8 bar, at 12 in. c.c. spacing, would suffice. Hence, if we extended the dowels to 4.5 ft above the base, and extended every second bar to the full stem height, we would meet the requirement of $\frac{1}{3}$ of A_s (actually it's $\frac{1}{2} A_s$), and would have ample A_s at bottom (1.57 in.^{2} vs. 1.53 in.^{2} required). The 4.5-ft dowel length satisfies the development length (slightly over that required); it is reasonably conducive to handling during steel placement (very long dowels are cumbersome to handle and to fix in place since the other ends of these dowels are also embedded in the footing). This expedites the labor efforts and is likely to result in some cost savings even though the A_s extended to the top of the stem exceeds the minimum of $\frac{1}{3} A_s$. The cut-off arrangement is shown in Fig. 9.14d.

Base of the Retaining Wall

It is desirable to construct a base to provide compression under the entire base. From the point of view of tilting, differential settlement, and bearing capacity, the most desirable pressure distribution would be a rectangular one, as shown in Figure 9.15a. From the standpoint of a viable design, the rectangular distribution is unattainable in view of the practical limit to base width. That is, the rectangular shape may be approached by simply adjusting the length of the base (including toe and heel ratio) so as to make the resultant pressure coincide with the resultant load at the center of the base. However, since this implies an impractically long toe, the most common shapes of pressure distributions are trapezoidal or triangular.

To calculate the size of the base of a retaining wall, we can proceed by either making an assumption regarding the base width l, and then checking the stability, pressure, or by calculating the base width as follows: We begin by assuming that

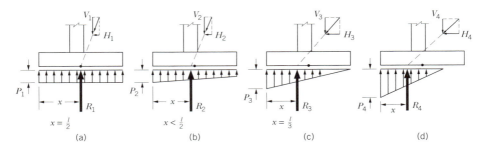

Figure 9.15 Pressure distributions corresponding to changes in the resultant load.

the soil pressure is distributed as shown in Fig. 9.15c, that the length of the toe is $(l/3)$, and that the material represented by the shaded area in Fig. 9.16 has a unit weight of 110 lb/ft³, including the concrete. For calculating the base width, the surcharge load was transformed into an equivalent soil height (i.e., $400/110 = 3.64$ ft) and "lumped" into W_2, as shown in Fig. 9.16. The actual dimensions were used for the stability analysis and heel design. This is a reasonable assumption for approximating the size of the base. Later, the actual weight will be used. Thus, $W_1 = (3.5)(l/3)(110)$ acting at a distance of $l/6$ from e, $W_2 = (19.5 + 3.64)(2l/3)110$ acting at a distance of $l/3$ from e, and $\Sigma M_e = 0$;

$$\left(3.5 \times \frac{110}{3}\right)(l)(1/6) + 2030\left(\frac{19.5}{2}\right) + 5437\left(\frac{19.5}{3}\right) = \left(2 \times 110 \times \frac{23.14}{3}\right) l \times l/3$$

$$21.4l^2 + (19{,}773) + 35{,}340 = 565.6l^2$$
$$l^2 = 55{,}113/544.20 = 101.3$$
$$l = 10.06 \text{ ft}, \quad \text{say 10 ft}$$

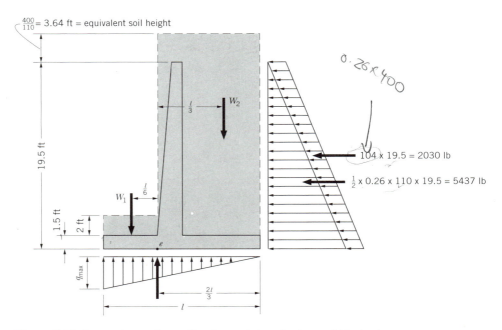

Figure 9.16 Approximate forces for determining the base width l.

We make a preliminary check of the soil pressure

$$R = W_1 + W_2 = q'_{max}\frac{10}{2} = \frac{1}{2}lq'_{max}$$

Therefore,

$$(3.5)\left(\frac{10}{3}\right)(110) + 23.14(2/3)(10)110 = 5\,q'_{max}$$

$$q'_{max} = \frac{1283 + 16,969}{5} = 3650\text{ lb}/\text{ft}^2$$

Hence, we see that with respect to pressure, the base width is satisfactory.

Stability; Overturning

For a more exact approach, the stability analysis must take into account the actual volumes and weights for concrete and soil, and it must cover overturning, sliding, and maximum bearing pressures. For our calculations, we use a base thickness of 1.5 ft (see Fig. 9.17) and the base length computed above.

$$\Sigma M_a = 0$$

Weights, lb		Moment Arm, ft	Righting Moment (RM), ft-lb
$W_1 = (2)(3.33)(110)$	$= 733$	1.67	1224
$W_2 = (1/2)(18)\left(\frac{5}{12}\right)(150)$	$= 563$	3.61	2032
$W_3 = (18)\left(\frac{12}{12}\right)(150)$	$= 2,700$		11,448
$W_4 = (21.68)(5.25)(110)$	$= 12,497$	7.37	92,103
$W_5 = (10)(1.5)(150)$	$= 2,250$	5.00	11,250
	$18,743$		118,057

Next we analyze the overturning effect.

Overturning force, lb	Moment arm, ft	Overturning moment (OM), ft-lb
$P' = 2028$	9.75	19,773
$P = 5437$	6.5	35,340
$\Sigma\quad 7465$		55,113

Let us denote by x the distance of the resultant force from point a;

$$x = \frac{RM - OM}{\Sigma W} = \frac{118,057 - 55,113}{18,743} = 3.36\text{ ft.}$$

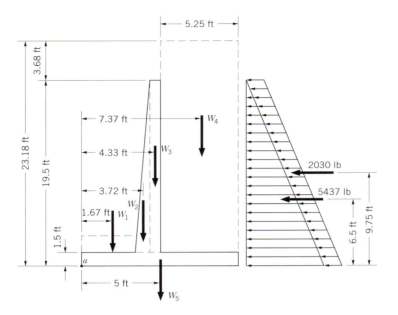

Figure 9.17 Forces acting on the wall.

The factor of safety (SF) against overturning is $118{,}057/55{,}113 = 2.14$; O.K.

$$e = \frac{l}{2} - x = \frac{10}{2} - 3.37 = 1.64 \text{ ft}$$

Sliding

Let us assume that the coefficient of friction of soil on concrete is 0.50. This is deemed reasonable since the bottom of the excavation is typically rather rough.

$$\text{Horizontal forces} = P + P' = 5437 + 2030 = 7465 \text{ lb}$$
$$\text{Resisting horizontal forces} = P_r = \mu\,(\Sigma N) = 0.5 \times 18{,}743 = 9{,}372 \text{ lb}$$
$$\begin{array}{c}\text{Safety factor against sliding}\\ \text{(neglecting } P_p \text{ at toe)}\end{array} = \text{SF} = 9372/7465 = 1.26$$

A shear key will have to be designed to increase the safety factor against sliding.

Base Pressure

Denoting the base pressure by q and assuming a straight-line pressure distribution, as shown in Fig. 9.18, we find that the formula

$$q = \frac{P}{A} \pm \frac{Mc}{I}$$

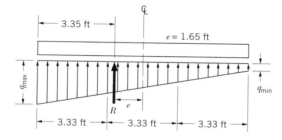

Figure 9.18 Soil pressure on base.

applies. In our case,

$$q = \Sigma W/A \pm \frac{We(L/2)}{I}; \qquad I = \frac{1 \times (10)^3}{12} = 83.33 \text{ ft}^4$$

$$q_{max} = \frac{18,743}{l \times 10} + \frac{(18,743 \times 1.64 \times 5)}{83.33} = 3719 \text{ lb/ft}^2$$

$$q_{min} = \frac{18,743}{l \times 10} - \frac{(18,743 \times 1.64 \times 5)}{83.33} \doteq 29.9 \text{ lb/ft}^2$$

Design of Heel

We begin by calculating the moment M at the back face.

$$M = -\frac{(12,497 + 1181)5.25}{2} + \frac{(29.9)(5.25)(5.25)}{2} + \frac{(1967 - 29.9)(5.25)^2}{6}$$

(upward reaction included)

$$M = -35,905 + 412 + 8899 = -26,594 \text{ ft-lb}$$
$$V = W_4 + W_c - \text{upward pressure}$$

$$V = 13,678 - \left(\frac{1967 + 29.9}{2}\right)(5.25) = 8436 \text{ lb}$$

Note that the plane of shear is not taken at a distance d from the face, since the plane of shear will look somewhat like that shown in Fig. 9.19.

$$d = \frac{8436 \times 1.8}{(93.11)(12)} = 13.59 \text{ in.}, \qquad \text{say } 13.75 \text{ in.}$$

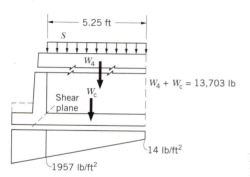

Figure 9.19 Pressure forces on heel.

Now to determine A_s. From

$$M_u = \phi A_s f_y \left(d - \frac{a}{2}\right); \qquad a = \frac{A_s f_y}{0.85 f'_c b} = 1.96 A_s$$

$$A_s \left(13.75 - \frac{1.96}{2} A_s\right) = (26.594 \times 12)(1.8)/0.9(60)$$

$$\text{or} \qquad A_s^2 - 14.03 A_s + 10.73 = 0$$

$$A_s = \frac{14.03 \pm \sqrt{196.8 - 42.92}}{2} = 0.81 \text{ in.}^2/\text{ft}$$

Assuming that #8 bars (same size as for stem and possibly for toe) are used, we find that the spacing is 12 in. c.c. (i.e., spacing = $(0.79/0.81)(12) = 11.7$ in., say 12 in.).

Checking

$p = A_s/bd = 0.81/(12)(13.75) = 0.0049 > 0.003$ O.K.
$l_d = 43.82$ in. $= 3.65$ ft (see stem calculations) < 5.25 ft available;   O.K.

Since the base will be poured on quite possibly uneven excavation, it is prudent to increase slightly the overall thickness of the base, by perhaps 1 in., beyond that required for cover. In our case, we have an overall thickness of, say 18 in.

Design of Toe

We base our calculation on Fig. 9.20. For the moment at the face of the stem, we obtain

$$M_{\text{toe at face of stem}} = \frac{(2489 - 445)(3.33)^2}{2} + (3719 - 2489)\frac{3.33}{2}\left(\frac{2}{3} \times 3.33\right)$$

$$M_{\text{toe}} = 11,333 + 4546 = 15,879 \text{ ft-lb} = 190,548 \text{ in.-lb}$$

A d for the toe near the stem, approximately the same as for the heel of 13.75 in., might be desirable when viewed in the context of construction expediency. Thus, via a brief comparison with the calculations for the heel stresses, the shear stresses for the toe are satisfactory. Similarly, A_s needed for the toe will be quite satisfied if we use the same A_s provided by the dowels; thus, we shall extend the dowels into the base, as shown in Fig. 9.25.

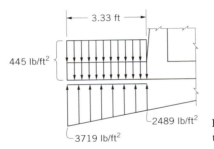

Figure 9.20 Pressure forces on toe.

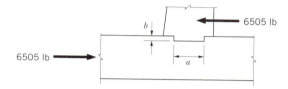

Figure 9.21 Design of stem shear key.

Shear Keys

We have to consider the stem key and the base key.

Stem Key   Since in pouring a retaining wall one first pours the base and then the stem, it is evident that there will be lack of homogeneity, and thus we lack shear strength at the junction of the base and the stem. To remedy this situation a shear key (stem key) is provided. In our case, the horizontal forces are 6505 lb. (Fig. 9.21).

Assuming a case of pure shear and a permissible shear stress of $2\phi\sqrt{f'_c} = 93$ psi, referring to Fig. 9.21, (also see Fig. 9.24) we find that the groove width $a = 6505$ lb/(12 in.)(93 psi) = 5.83 in., say 6 in. The bearing stress is conservatively assumed to be $0.10f'_c = 300$ psi. Thus, the groove depth

$$b = \frac{6505}{12 \times 300} = 1.81 \text{ in., say, 2 in.}$$

Base Key   The stability analysis showed that the wall is not sufficiently safe with respect to sliding. Although the toe fill (cover) and the earth in front of the toe would increase the stability against sliding, their contribution cannot always be relied on to develop the expected passive resistance. Creep, frost action, and erosion are phenomena that lead to this unreliability. In Fig. 9.22, the region $abcde$ is assumed as the failure plane; hence, the base resistance is approximately equal to $F_2 + F_3$.

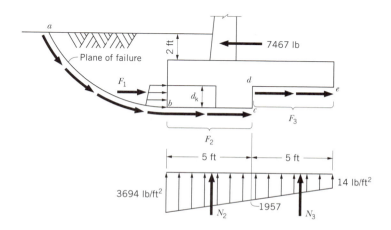

Figure 9.22 Forces acting on base to resist sliding.

$$F_2 = (\mu_{\text{soil on soil}}) N_2; \qquad F_3 = (\mu_{\text{conc on soil}}) N_3$$

or

$$F_2 = (\mu_{\text{ss}}) N_2; \qquad F_3 = (\mu_{\text{cs}}) N_3$$

By assuming that the coefficient μ of soil equals the coefficient μ of concrete on soil, and that both are 0.5 (conservative), we find that $F_2 + F_3 = 9382$ lb (see previous calculations for sliding). We may include a safety factor of 1.8 to safeguard against sliding.

The additional force to be taken by the shear key is $F_1 = (7465)(1.8) - 9372 = 4065$ lb. For $\phi = 30°$, $K_p \cong 3.0$. Thus, $P_p = K_p \gamma H = (3.0)(110) H = 330 H$.

From Fig. 9.23, we obtain $F_1 = (330/2)(H_2^2 - H_1^2)$. (For H, see Fig. 9.23.)

$$4065 = 165(H_2^2 - 3.5^2) \text{ approximately,}$$
$$4065 = 165 H_2^2 - 2021$$

$$H_2^2 = \frac{(4065 + 2021)}{165}, \qquad H_2 = 6.07 \text{ ft}$$

$$\text{key depth } d_k = H_2 - H_1 = 6.05 - 3.5 = 2.5 \text{ ft}$$

We next determine the thickness of the shear key. From the shear, we have

$$d = \frac{V}{vb} = \frac{4065}{(93)(12)} = 3.64 \text{ in.,} \qquad \text{say 4 in.}$$

Since some of the dowel embedment can satisfy A_s, we find, using a 3-in. cover, that the thickness of the key is 7 in. min., perhaps 10 in. or 12 in., as practical construction reasons (such as minimum backhoe bucket width).

Temperature and Shrinkage Reinforcement

The quantity of steel needed to compensate for expansion or contraction caused by temperature changes, or by shrinkage and creep, is governed partially by the ACI Code and partially by practical considerations related to the steel erection and forming.

From the various phases of construction of bridge abutments shown in Fig. 9.24, one notes a number of details (e.g. dowel steel, heel, toe, shear key) common in retaining wall design. Figure 9.25 depicts the design details of the retaining wall in our example problem.

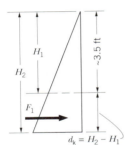

Figure 9.23 Forces on base key.

Figure 9.24 Lower right hand corner: Partial bridge abutment, Ohio Department of Transportation, Stark County project over Tuscarawas River, Massillon, Ohio. Note base (heel, toe, and shear key) details, and dowels sticking up from base. (Courtesy of Monotube Pile Corporation.)

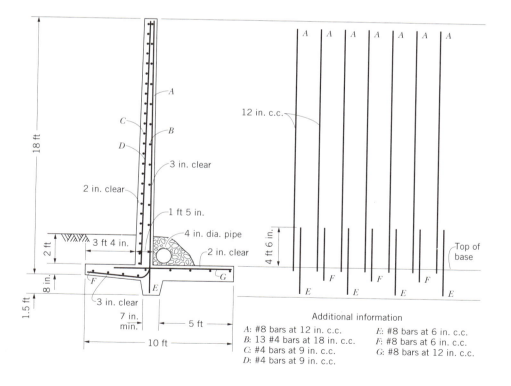

Additional information

A: #8 bars at 12 in. c.c.   E: #8 bars at 6 in. c.c.
B: 13 #4 bars at 18 in. c.c.   F: #8 bars at 6 in. c.c.
C: #4 bars at 9 in. c.c.   G: #8 bars at 12 in. c.c.
D: #4 bars at 9 in. c.c.

Figure 9.25 Final design.

9.9 REINFORCED EARTH

Figure 9.26 depicts the general features of a rather unique earth-retaining structure. It is a relatively new concept which has attracted appreciable interest in research and application since its introduction (2, 3, 4, 5, 8, 9, 10). The earth mass is reinforced with a series of strips, which are attached to a face "skin." The skin resists the lateral thrust, induced by the active earth pressure, through the anchorage provided by the strips. The strips derive their tensile capacity from the "bond" developed between the strips and the earth.

The strips may be of any material, such as metal, fiber, wood, or plastics, which satisfies the following requirements: (1) has adequate tensile strength, (2) is corrosive resistant, and (3) is suitable for friction or bond development. Also, the strips should provide adequate surface area and display sufficient flexibility to develop the frictional resistance required, and they should be compatible with the soil movement and deformation.

Granular backfill is generally desirable. It permits easy drainage and thus creates a state of reduced lateral pressure. Also, it develops better strip bond than do cohesive soils.

The lateral pressure on the wall is assumed to vary linearly with depth, as shown in Fig. 9.27a. The corresponding tensile forces in the strips are therefore computed as indicated in Fig. 9.27b. The length of the strips must be sufficient to extend beyond the potential failure plane shown in Fig. 9.27a. The cross section of the strips must be sufficient to satisfy tensile stress limits for the material used.

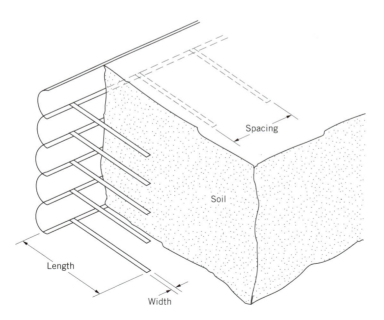

Figure 9.26 Reinforced-earth retaining wall.

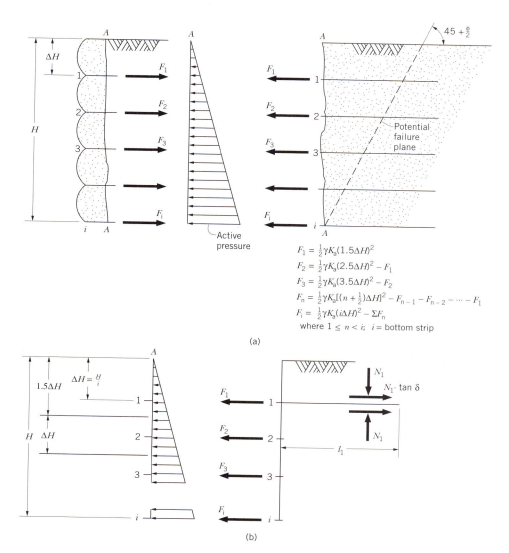

$$F_1 = \tfrac{1}{2}\gamma K_a(1.5\Delta H)^2$$
$$F_2 = \tfrac{1}{2}\gamma K_a(2.5\Delta H)^2 - F_1$$
$$F_3 = \tfrac{1}{2}\gamma K_a(3.5\Delta H)^2 - F_2$$
$$F_n = \tfrac{1}{2}\gamma K_a[(n+\tfrac{1}{2})\Delta H]^2 - F_{n-1} - F_{n-2} - \cdots - F_1$$
$$F_i = \tfrac{1}{2}\gamma K_a(i\Delta H)^2 - \Sigma F_n$$
where $1 \le n < i$; i = bottom strip

(a)

(b)

Figure 9.27 Forces on reinforced-earth wall. (a) Earth pressure and induced strip tensions. (b) Procedure for estimating strip tensions.

Example 9.2

Given Granular backfill; $\phi = 30°$; $\gamma = 18$ kN/m³; $H = 1.5$ m and $\Delta H = 1.5/5 = 0.3$ m; strip spacing = 1 m; $\sigma_s = 14 \times 10^4$ kN/m² (approximately 20,000 psi); $\delta = 20°$ (friction angle for soil strip).

Find Suitable strip sizes.

Procedure $K_a = 0.333$. Thus,

$$F_1 = \tfrac{1}{2}(18)(0.33)(0.45)^2 = 0.60 \text{ kN/m}$$
$$F_2 = \tfrac{1}{2}(18)(0.33)(0.75)^2 - F_1 = 1.07 \text{ kN/m}$$

$$F_3 = \tfrac{1}{2}(18)(0.33)(1.05)^2 - F_2 - F_1 = 1.06\,\text{kN/m}$$
$$F_4 = \tfrac{1}{2}(18)(0.33)(1.35)^2 - F_3 - F_2 - F_1 = 2.04\,\text{kN/m}$$
$$F_5 = \tfrac{1}{2}(18)(0.33)(1.5)^2 - F_4 - F_3 - F_2 - F_1 = 1.37\,\text{kN/m}$$

The cross-sectional area of the strips required is

$$A = \frac{F_{\max}}{\sigma_s} = \frac{2.04}{14 \times 10^4} = 0.249 \times 10^{-4}\,\text{m}^2 \approx 0.15\,\text{cm}^2$$

In order to develop the frictional resistance, a reasonable width must be available if we are to keep the length from becoming excessive. Furthermore, some reasonable thickness must be provided to compensate for some corrosion with time. Hence, somewhat arbitrarily, we select an 8-cm × 0.3-cm section; thus, the stress is

$$\sigma_s = \frac{2.04\,\text{kN}}{2.4 \times 10^{-4}\,\text{m}^2} = 0.85 \times 10^4\,\text{kN/m}^2 < 14 \times 10^4\,\text{kN/m}^2 \text{ allowable}$$

The length of each strip must extend past the potential failure surface for a sufficient distance to develop the necessary frictional resistance. Referring to Fig. 9.28, the distances from the face to the failure surface are:

$$L_1' = 4\Delta H \tan\left(45 - \frac{\phi}{2}\right) = 0.69\,\text{m}$$

$$L_2' = 3\Delta H \tan\left(45 - \frac{\phi}{2}\right) = 0.52\,\text{m}$$

$$L_3' = 2\Delta H \tan\left(45 - \frac{\phi}{2}\right) = 0.35\,\text{m}$$

$$L_4' = \Delta H \tan\left(45 - \frac{\phi}{2}\right) = 0.17\,\text{m}$$

$$L_5' = 0$$

The required lengths $L_1, L_2, \ldots, L_i$ are:

$$F_i \times (\text{safety factor}) = \gamma(n\Delta H)\tan\delta(L_n - L_i') \times \text{width} \times 2 \text{ faces}$$

Thus, assuming a safety factor of 1.2, we have

$$F_1(1.2) = 18(0.3)(0.36)(L_1 - 0.69)\left(\tfrac{8}{100}\right)(2)$$
$$0.6(1.2) = 0.35(L_1 - 0.69)$$

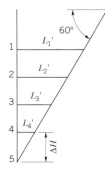

Figure 9.28 Strip lengths from face of wall to potential failure plane.

and

$$L_1 = 2.75 \text{ m}$$

Also,

$$1.07(1.2) = 0.7(L_2 - 0.52)$$
$$L_2 = 2.35 \text{ m}$$

Similarly,

$$L_3 = 2.85 \text{ m}$$
$$L_4 = 2.91 \text{ m}$$
$$L_5 = 2.40 \text{ m}$$

Answer | Thus, assume $L = 3$ m for each strip; the width for each strip is 8 cm and the thickness 0.3 cm.

Problems

Notation: H = stem height, ft; H_w = water depth, ft; P_i = ice thrust, lb; B = base width, ft; t = base thickness, in.; S = surcharge, psf; i = slope of backfill; ϕ = angle of internal friction; γ_s = soil weight, pcf; μ = coefficient of friction of concrete on soil; q_a = allowable soil pressure under base, psf; Use $\gamma_{conc.}$ = 150 pcf.

Note: The following problems are set up in such a way that the stability analysis of the first group is tied in with the force analysis of the second group, and both aid in the designs called for by the third group. For example, the calculations from Problem 9.1 may be used to determine the forces in Problem 9.6, which in turn are useful for the design required in Problem 9.11. A similar relationship exists among Problems 9.2, 9.7, and 9.12, and so on.

Stability Analysis: Sliding, Overturning, and Soil Pressure

Problems 9.1 through 9.5. Investigate the stability against sliding, overturning, and foundation soil pressure of the retaining walls described below. Assume a stem thickness of 1 in. for every foot of stem height, h. The base thickness, t, is uniform for the whole base.

9.1　For the wall in Fig. P9.1, $H = 16$ ft, $B = 9$ ft, $t = 15$ in., $S = 300$ psf, $\gamma_s = 100$ pcf, $\phi = 33°$, $\mu = 0.4$, $q_a = 4000$ psf.

9.2　For the wall on Fig. P9.2, $H = 15$ ft, $B = 9$ ft, $t = 14$ in., $i = 30°$, $\gamma_s = 100$ pcf, $\phi = 30°$, $\mu = 0.45$, $q_a = 5500$ psf.

9.3　For the wall in Fig. P9.3, $H = 18$ ft, $B = 11$ ft, $t = 16$ in., $\gamma_s = 110$ pcf, $\mu = 0.4$, $\phi = 30°$, $q_a = 5000$ psf, $H_w = 8$ ft.

9.4　For the wall in Fig. P9.4, $H = 16$ ft, $B = 10$ ft, $t = 16$ in., $\gamma_s = 110$ pcf, $\mu = 0.4$, $\phi = 34°$, $q_a = 3000$ psf, $P_i = 500$ lb/lin. ft.

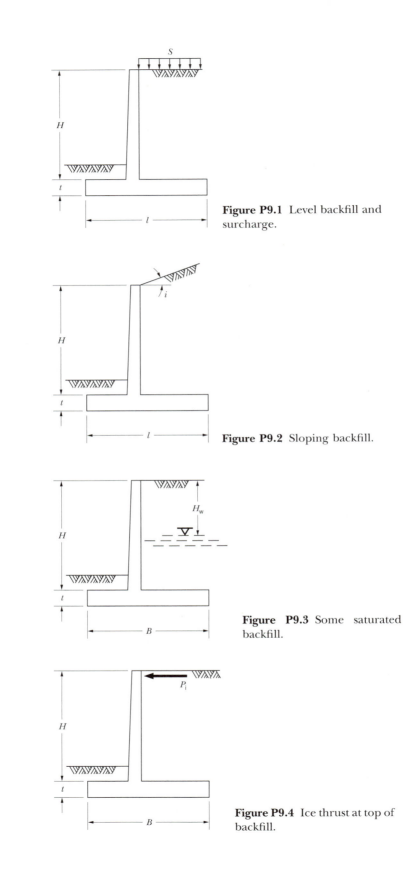

Figure P9.1 Level backfill and surcharge.

Figure P9.2 Sloping backfill.

Figure P9.3 Some saturated backfill.

Figure P9.4 Ice thrust at top of backfill.

9.5   For the wall in Fig. P9.5, $H = 20$ ft, $B = 11$ ft, $t = 16$ in., $\gamma_s = 105$ pcf, $\mu = 0.50$, $\phi = 30°$, $p_a = 4500$ psf, $S = 350$ psf, $l_1 = 5$ ft.

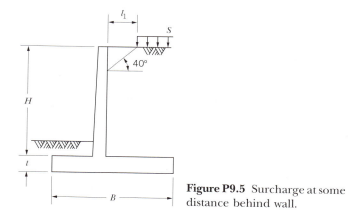

Figure P9.5 Surcharge at some distance behind wall.

Analysis of Internal Forces and Moments

Determine the bending moment and shear forces (M and V) for the stem, toe, and heel near the junction point of base and stem for the given conditions.

9.6   Use the retaining wall in Fig. P9.1 and the dimension and conditions of Problem 9.1.

9.7   Use the retaining wall in Fig. P9.2 and the dimension and conditions of Problem 9.2.

9.8   Use the retaining wall in Fig. P9.3 and the dimension and conditions of Problem 9.3.

9.9   Use the retaining wall in Fig. P9.4 and the dimension and conditions of Problem 9.4.

9.10   Use the retaining wall in Fig. P9.5 and the dimension and conditions of Problem 9.5.

Design of Component Parts of Wall

Determine a suitable design for the toe, heel, and stem. The design should indicate the quantity of reinforcing steel required, the cut-off points, and the size of the members. Base your design on the conditions given below and the ACI Code specifications. Assume $f'_c = 4000$ psi and $f_y = 60,000$ psi.

9.11   Using the data of Problem 9.1 as a starting point, design the wall shown in Fig. P9.1. Neglect the uplift-soil reaction.

9.12   Using the data of Problem 9.2 as a starting point, design the retaining wall shown in Fig. 9.2.

9.13   Using the data of Problem 9.3 as a starting point, design the retaining wall shown in Fig. P9.3.

9.14   Using the data of Problem 9.4 as a starting point, design the retaining wall shown in Fig. P9.4.

9.15   Using the data of Problem 9.5 as a starting point, design the retaining wall shown in Fig. P9.5.

Design of L-shaped Walls

The following five problems concern the design of L-shaped retaining walls. The conditions are those given in Problems 9.1 through 9.5 and Figs. P9.1 through P9.5; however, the toe is to be omitted.

9.16   Given: Problem 9.1 and Fig. P9.1. Design an L-shaped wall.

9.17   Given: Problem 9.2 and Fig. P9.2. Design an L-shaped wall.

9.18   Given: Problem 9.3 and Fig. P9.3. Design an L-shaped wall.

9.19   Given: Problem 9.4 and Fig. P9.4. Design an L-shaped wall.

9.20   Given: Problem 9.5 and Fig. P9.5. Design an L-shaped wall.

Design of Counterfort Walls

9.21   Design a counterfort retaining wall for the following conditions: $\gamma_s = 116$ pcf, $\phi = 32°$, $f'_c = 4$ ksi, $f_y = 60$ ksi, $H = 38$ ft, $D = 4.5$ ft, $\mu = 0.5$, $\alpha = 15°$, $q_a = 4.5$ ksf.

9.22   Design a reinforced-earth retaining wall for the following data: height = 2 m; granular backfill; $\phi = 32°$; $\delta = 22°$; $\gamma = 18.3$ kN/m³; factor of safety = 1.2. Assume steel construction.

BIBLIOGRAPHY

1.   Akinmusuru, J. O., and J. A. Akinbolade, "Stability of Loaded Footings on Reinforced Soil," *ASCE J. Geotech. Eng. Div.,* vol. 107, June 1981.

2.   Al-Hussaini, M., and E. B. Perry, "Field Experiment of Reinforced Earth Wall," *ASCE J. Geotech. Eng. Div.,* vol. 104, Mar. 1978.

3.   Binquet, J., and K. L. Lee, "Bearing Capacity Tests on Reinforced Earth Slabs," *ASCE J. Geotech. Eng. Div.,* vol. 101, Dec. 1975.

4.   Chang, J. C., and R. A. Forsyth, "Design and Field Behavior of Reinforced Earth Wall," *ASCE J. Geotech. Eng. Div.,* vol. 103, July 1977.

5.   Chang, J. C., and R. A. Forsyth, "Finite Element Analysis of Reinforced Earth Wall," *ASCE J. Geotech. Eng. Div.,* vol. 103, July 1977.

6.   Danilevsky, A., "Safety Factor of Dams and Retaining Walls," *ASCE J. Geotech. Eng. Div.,* vol. 108, Jan. 1982.

7.   Frydman, S., and I. Keisser, "Earth Pressure on Retaining Walls Near Rock Faces," *ASCE J. Geotech. Eng. Div.,* vol. 113, June 1987.

8.   Juran, I., and B. Christopher, "Laboratory Model Study on Geosynthetic Reinforced Retaining Walls," *ASCE J. Geotech. Eng. Div.,* vol. 115, July 1989.

9.   Lee, K. L., B. D. Adams, and J. J. Vagneron, "Reinforced Earth Retaining Walls," *ASCE J. Soil Mech. Found. Div.,* vol. 99, Oct. 1973.

10.   Richardson, G. N., and K. L. Lee, "Seismic Design of Reinforced Earth Walls," *ASCE J. Geotech. Eng. Div.,* vol. 101, Feb. 1975.

11.   Rowe, R. K., ''Reinforced Embankments: Analysis and Design,'' *ASCE J. Geotech. Eng. Div.*, vol. 110, Feb. 1984.

12.   Sherif, M. A., I. Ishibashi, and C. D. Lee, ''Earth Pressures Against Rigid Retaining Walls,'' *ASCE J. Geotech. Eng. Div.*, vol. 108, May 1982.

13.   Sherif, M. A., Y-S. Fang, and R. I. Sherif, ''K_a and K_0 Behind Rotating and Non-Yielding Walls,'' *ASCE J. Geotech. Eng. Div.*, vol. 110, Jan. 1984.

CHAPTER 10

Sheet-Pile Walls; Braced Cuts

10.1 INTRODUCTION

It was mentioned in Section 8.1 that the lateral pressure on a retaining structure depends on several factors: (1) the physical properties of the soil, (2) the time-dependent nature of the soil, (3) the imposed loading, (4) the interaction between soil and retaining structure at the interface, and (5) the general characteristics of the deformation in the soil–structure composite.

With flexible earth-retaining structures, the last of the above factors assumes particular importance. As a flexible wall deflects, changes occur in the magnitudes and distributions of earth pressures against the wall from those predicted by the methodologies discussed in the previous chapter for rigid walls. Thus, magnitudes and locations of pressure resultants P_a and P_p are different from those predicted for rigid walls. Included in the group of flexible retaining structures are sheet-pile installations and braced excavations. These will be discussed in this chapter. Figures 10.1 through 10.4 depict a range of uses for sheet piles. Figure 10.5 shows the typical sequence of steps for installing sheet piles.

Figure 10.1 Cellular construction with steel sheet piling is used extensively for deepwater port facilities (e.g., wharves, graving docks, breakwaters). Illustrated is a cellular marginal wharf under construction owned by the Philadelphia Port Corporation. (Courtesy of Bethlehem Steel Company.)

10.2 CANTILEVER SHEET-PILE WALLS

A cantilever sheet-pile wall is constructed by driving sheet piling to a depth sufficient to develop a cantilever beam-type reaction to resist the active pressures on the wall. That is, the embedment length must be adequate to resist both lateral forces as well as a bending moment.

Lateral deflections in cantilever sheet-pile walls may be rather large for high walls since the piling is relatively flexible and the moment varies as the cube of the height. Also, erosion, scour, or soil compression in front of the wall may further compound this deflection. Hence, wall lengths or heights are generally limited to about 5 m. Also, their use is primarily short-term rather than for the long-term, more permanent type of installation.

The embedment depth varies with different soils, perhaps less than the cantilever length for very dense soils to twice this length for very loose soils. Likewise, the pressure distribution varies with different soils and different water conditions. The actual stress distribution is rather hypothetical, perhaps as depicted in Fig. 10.6b. For computation purposes, however, the distribution is generally simplified as shown in Fig. 10.6c. The lateral deflection is estimated in Fig. 10.6a.

Figure 10.2 The 65-ft single-span bridge (Banks Road Bridge, Tomkins County, New York) is supported directly on sheet piling abutments, which are capped with a steel channel and a steel distribution beam, thereby eliminating a traditional reinforced concrete cap. The abutment walls and sheet piling wing walls were tied back to anchors with cable. (Courtesy of Bethlehem Steel Company.)

Cantilever Sheet Piling in Granular Soils

The following analysis will be based on a simplified pressure distribution shown in Fig. 10.7, assuming a homogeneous granular soil. For a layered system (density and/or material deviations), the appropriate values for γ and ϕ should be used in the calculations. Also, if the ground surface is other than level, either the wedge theory or that of Coulomb should be used to approximate the lateral forces. Otherwise, the design concept is identical.

For granular soils it is reasonable to assume the water table to be at the same level on each side of the wall. Thus, the pressure distributions (including surcharge effects, etc.) can be established from average values of K_a and K_p. If a safety factor

Figure 10.3 Segment of a 3000-ft-long sheet-pile wall along the Buffalo River. (Courtesy of Bethlehem Steel Company.)

is to be incorporated, either reduce the K_p value (by perhaps 30–50%) or increase the depth of penetration (by perhaps 20–40%). This will give a safety factor of approximately 1.5–2.0.

With the distribution established, the depth to zero pressure, distance a, can be calculated by similar triangles as

$$a = \frac{p_a}{\gamma_b(K_p - K_a)}$$

By summing forces in the horizontal direction, we can relate distances z as follows (from Fig. 10.7): that is, $\Sigma F_x = 0$,

$$R_a + R_p' - R_p = 0 \tag{a}$$

But

$$[R_p' - R_p] = (p_p + p_p')\frac{z}{2} - p_p\frac{Y}{2}$$

Substituting in Eq. (a),

$$R_a + (p_p + p_p')\frac{z}{2} - p_p\frac{Y}{2} = 0$$

Solving for z, we obtain

$$z = \frac{p_p Y - 2R_a}{p_p + p_p'} \tag{b}$$

From $\Sigma M_{\text{BOTTOM}} = 0$, we get

$$R_{a}(Y + \bar{y}) + (p_{p} + p_{p}')\frac{z}{2}\cdot\frac{z}{3} - p_{p}\left(\frac{Y}{2}\right)\left(\frac{Y}{3}\right) = 0$$

or

$$6R_{a}(Y + \bar{y}) + (p_{p} + p_{p}')z^{2} - p_{p}Y^{2} = 0 \qquad\qquad (c)$$

Substituting for z, from Eq. (b), we get

$$6R_{a}(Y + \bar{y}) + \left(\frac{1}{p_{p} + p_{p}'}\right)(p_{p}^{2}Y^{2} - 4p_{p}YR_{a} + 4R_{a}^{2}) - p_{p}Y^{2} = 0$$

Figure 10.4 An 85-ft × 100-ft (in plan) cofferdam was used to construct Francis Scott Key bridge piers in the Main Channel into Baltimore Harbor (Patapsco River). Typically, sheet piles are driven into the bottom in a cell-type configuration, with the piles braced on the inside to prevent inward collapse as the excavation and dewatering (pumping) and subsequent construction of foundation components (e.g., footings, piers) takes place. (Courtesy of Bethlehem Steel Company.)

(a)

(b)

(c)

Figure 10.5 (a) Sheet piles spread out for crane pick-up (note holes in piles for this purpose) and subsequent installation. (b) Pile-driving hammer is supported by the crane and is sometimes controlled by an operator in the crane cab but usually outside. (c) Group of installed piles. The piles are generally installed individually, with interlocking joints as shown. Thus, the installed pile becomes the "guide" for the next, etc. (Courtesy of Syro Steel Company.)

If we multiply by $(p_p + p'_p)$, we get

$$6(p_p + p'_p)R_a(Y + \bar{y}) + p_p^2 Y^2 - 4p_p Y R_a + 4R_a^2 - p_a^2 Y^2 - p_a p'_p Y^2 = 0$$

Substituting for $p_p = \gamma_b(K_p - K_a)Y = CY$, we have

$$6R_a(CY^2 + CY\bar{y} + p'_p Y + p'_p \bar{y}) - 4CY^2 R_a + 4R_a^2 - CY^3 p'_p = 0$$

Dividing by $-Cp'_p$ and rearranging, we get

$$Y^3 - \left(\frac{2R_a}{p'_p}\right)Y^2 - 6R_a\left(\frac{\bar{y}}{p'_p} + \frac{1}{C}\right)Y - \frac{2R_a}{Cp'_p}(2R_a + 3p'_p \bar{y}) = 0 \qquad (10\text{-}1)$$

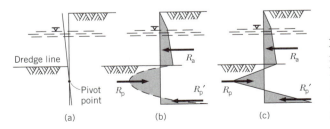

Dredge line

Pivot point

(a)   (b)   (c)

R_a R_p R'_p

Figure 10.6 Sheet-pile deformation and pressures in granular soil. (a) Deflection curve. (b) Likely pressure distribution. (c) Assumed pressure distribution.

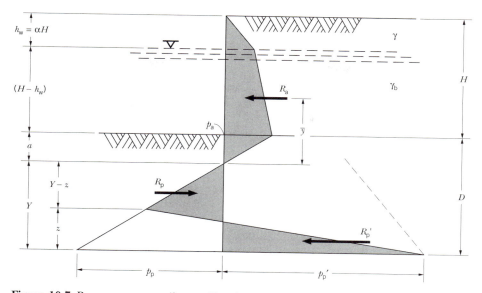

Figure 10.7 Pressures on cantilever piling in granular soils.

where
$$C = \gamma_b(K_p - K_a)$$
$$p'_p = \gamma h_w K_p + \gamma_b K_p(H + D - h_w) - \gamma_b K_a(Y + a)$$

$$a = \frac{p_a}{\gamma_b(K_p - K_a)}$$

$$p_a = K_a[\gamma h_w + \gamma_b(H - h_w)] \text{ (Fig. 10.8)}$$
$$R_a = R_1 + R_2 + R_3 + R_4$$
$$R_a = \begin{cases} R_1 = \frac{1}{2}K_a\gamma h_w^2 \\ R_2 = \gamma h_w K_a(H - h_w) \\ R_3 = \frac{1}{2}K_a\gamma_b(H - h_w)^2 \\ \\ R_4 = \dfrac{p_a^2}{2\gamma_b(K_o - K_a)} \end{cases}$$

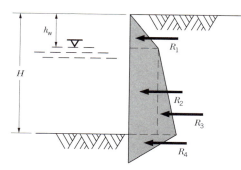

Figure 10.8 Active pressure on the wall.

The solution to Eq. 10-1 is usually a trial-and-error one, obtained by first assuming a value for D (Fig. 10.7), then determining p_p, p_p', a, z, R_a, etc., then solving for Y and subsequently D. Estimates for D are given in Table 10.1.

Table 10.1 Approximate Values for D

Standard Penetration Resistance, N (blows/0.3 m)	Relative Density, D_r	Depth, D
0–4	Very loose	$2.0H$
5–10	Loose	$1.5H$
11–30	Medium dense	$1.25H$
31–50	Dense	$1.0H$
Over 50	Very dense	$0.75H$

Maximum moment occurs where the shear is zero (Fig. 10.9).

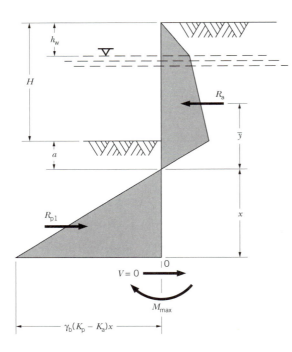

Figure 10.9 Forces on wall above point of zero shear.

From ΣM_0 (at $V = 0$), we have

$$M_{max} = R_a(\bar{y} + x) - R_{p_1}\left(\frac{x}{3}\right) = R_a(\bar{y} + x) - R_a\left(\frac{x}{3}\right)$$

or

$$M_{max} = R_a(\bar{y} + \tfrac{2}{3}x)$$

Substituting for x, we have

$$M_{max} = R_a \left\{ \bar{y} + \frac{2}{3} \left[\frac{2R_a}{\gamma_b(K_p - K_a)} \right]^{1/2} \right\} \tag{10-2}$$

$$R_{p_1} = \tfrac{1}{2}\gamma_b(K_p - K_a)\,x^2$$

For equilibrium, from $\Sigma F_{HORIZ.} = 0$

$$R_{p_1} = R_a$$

Thus,

$$x = \left[\frac{2R_a}{\gamma_b(K_p - K_a)} \right]^{1/2}$$

Hence, a section modulus can be obtained and, subsequently, a suitable pile section can be selected.

Figures 10.10, 10.11, and 10.12 (25) provide an expedient format for determining the depth D and bending moment M_{max} for three cases of γ_b (usually $\gamma_b = \frac{1}{2}\gamma$ is a reasonable approximation for all practical purposes; this is depicted in Fig. 10.11 and Table 10.2). Example 10.1 illustrates the procedure via these graphs.

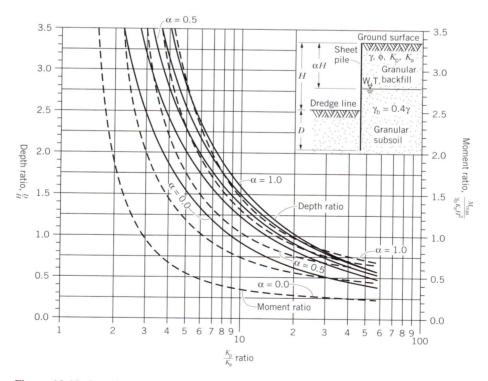

Figure 10.10 Cantilever steel sheet pile, granular subsoil with granular backfill.

$$\gamma_b = 0.4\,\gamma.$$

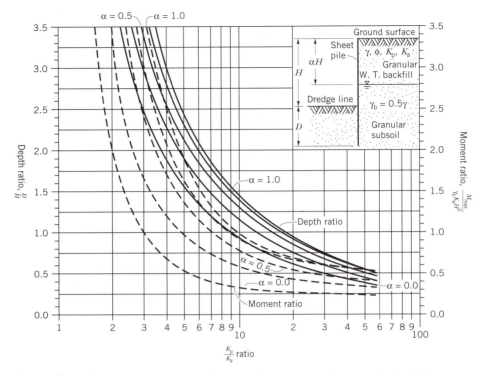

Figure 10.11 Cantilever steel sheet pile, granular subsoil with granular backfill.

$$\gamma_b = 0.5\,\gamma$$

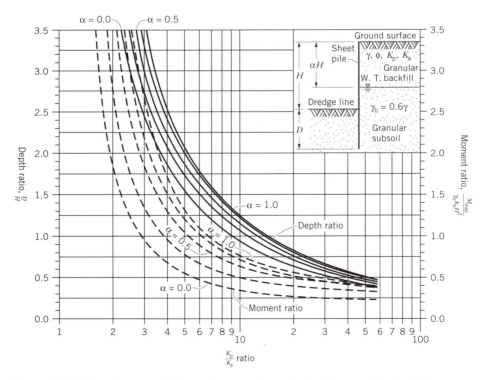

Figure 10.12 Cantilever steel sheet pile, granular subsoil with granular backfill.

$$\gamma_b = 0.6\,\gamma$$

Table 10.2 Values of D/H, for $\gamma_b = 0.5\gamma$. Cantilever Steel Sheet Pile, Granular Subsoil and Backfill (see Fig. 10.11)

		For $\alpha = $ ⌐↓				
ϕ	K_p/K_a	0.0	0.25	0.5	0.75	1.0
10	2.02	4.18	4.64	5.13	5.94	7.37
11	2.17	3.76	4.21	4.57	5.34	6.52
12	2.33	3.35	3.80	4.14	4.78	5.91
13	2.50	3.07	3.51	3.83	4.34	5.34
14	2.68	2.79	3.23	3.44	3.93	4.79
15	2.88	2.63	2.96	3.16	3.63	4.36
16	3.10	2.38	2.71	2.99	3.34	4.05
17	3.34	2.23	2.55	2.74	3.07	3.66
18	3.59	2.09	2.41	2.58	2.80	3.37
19	3.86	1.95	2.27	2.34	2.65	3.20
20	4.16	1.82	2.04	2.20	2.49	2.93
21	4.48	1.69	2.01	2.06	2.35	2.77
22	4.83	1.66	1.88	1.93	2.21	2.52
23	5.21	1.54	1.75	1.90	2.07	2.38
24	5.62	1.42	1.63	1.77	1.94	2.23
25	6.07	1.40	1.61	1.65	1.81	2.09
26	6.56	1.28	1.49	1.52	1.68	1.96
27	7.09	1.26	1.37	1.51	1.56	1.83
28	7.67	1.15	1.36	1.39	1.53	1.70
29	8.31	1.14	1.25	1.37	1.41	1.67
30	9.00	1.03	1.23	1.26	1.40	1.55
31	9.76	1.01	1.12	1.24	1.28	1.43
32	10.59	1.00	1.11	1.13	1.26	1.41
33	11.51	0.90	1.00	1.12	1.15	1.29
34	12.51	0.89	0.99	1.01	1.14	1.27
35	13.62	0.78	0.98	1.00	1.02	1.16
36	14.84	0.77	0.88	0.89	1.01	1.14
37	16.18	0.77	0.87	0.88	0.90	1.03
38	17.67	0.76	0.86	0.87	0.89	1.02
39	19.32	0.65	0.76	0.77	0.89	0.91
40	21.15	0.65	0.75	0.76	0.78	0.90
41	23.18	0.65	0.75	0.76	0.77	0.89
42	25.45	0.64	0.64	0.75	0.76	0.78
43	27.98	0.54	0.64	0.65	0.76	0.77
44	30.80	0.53	0.64	0.64	0.65	0.77
45	33.97	0.53	0.63	0.64	0.65	0.66

Example 10.1

Given Figure 10.13.

Find   (a) Embedment depth D; use $D_{SF} = 1.3D$ via Eq. 10-1

   (b) Maximum moment, M_{max}, via Eq. 10-2

   (c) D via Figure 10.10, 10.11, or 10.12, whichever is applicable

   (d) M_{max} via Figure 10.10, 10.11, or 10.12, whichever is applicable

Procedure
$$p_a = \gamma H K_a = (135 \text{ pcf})(15 \text{ ft})(0.307) = 622 \text{ psf}$$
$$C = \gamma_b'(K_p' - K_a') = (70)(3 - 0.333) = 186.7 \text{ pcf}$$
$$a = p_a/c = 622/186.7 = 3.33 \text{ ft}$$
$$R_a = \tfrac{1}{2}p_a H + \tfrac{1}{2}p_a a = \tfrac{1}{2}(622)(15) + \tfrac{1}{2}(622)(3.33) = 5701 \text{ lb/ft of wall}$$

Find $\bar{y}$ by ΣM_0:

$$R_a \bar{y} = \tfrac{1}{2}p_a H\left(a + \frac{H}{3}\right) + \tfrac{1}{2}p_a a\left(\frac{2a}{3}\right)$$

$$R_a \bar{y} = \tfrac{1}{2}(622)(15)(3.33 + \tfrac{15}{3}) + \tfrac{1}{2}(622)(3.33)\left(\frac{2 \times 3.33}{3}\right)$$

$$\bar{y} = 7.22 \text{ ft}$$

$$p_p' = \gamma H K_p + \gamma_b(Y + a)K_p - \gamma_b(Y + a)K_a; \qquad \text{note that } h_w = H$$

p_p' is a function of Y; it is one of the terms in Eq. 10-1. Thus, the procedure for the solution of Eq. 10-1 is to assume a value for Y and solve p_p' and Eq. 10-1 via a trial-and-error approach, as follows:

From Eq. 10-1

$$Y^3 - \left(\frac{2R_a}{p_p'}\right)Y^2 - 6R_a\left(\frac{\bar{y}}{p_p'} + \frac{1}{C}\right)Y - \frac{2R_a}{Cp_p'}(2R_a + 3p_p'\bar{y}) = 0$$

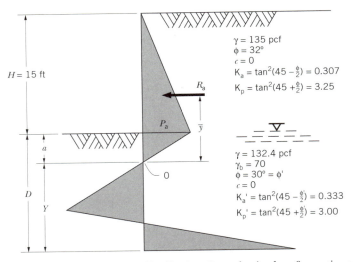

$\gamma = 135$ pcf
$\phi = 32°$
$c = 0$
$K_a = \tan^2(45 - \frac{\phi}{2}) = 0.307$
$K_p = \tan^2(45 + \frac{\phi}{2}) = 3.25$

$\gamma = 132.4$ pcf
$\gamma_b = 70$
$\phi = 30° = \phi'$
$c = 0$
$K_a' = \tan^2(45 - \frac{\phi'}{2}) = 0.333$
$K_p' = \tan^2(45 + \frac{\phi'}{2}) = 3.00$

$H = 15$ ft

R_a

P_a

$\bar{y}$

a

0

D

Y

Figure 10.13 Pressure distribution for cohesionless formations.

Try various Y values, with results as tabulated:

Y	12 ft	15 ft	18 ft	17.5 ft	17.4 ft
p_p'	9443	10,003	10,563	10,470	10,451
Eq. 10.1	-2355	-1391	374	17	-50

Example: say $Y = 17.5$ ft

$$p_p' = \gamma H K_p + \gamma_b'(Y + a)K_p' - \gamma_b'(Y + a)K_a'$$
$$= (135)(15)(3.25) + (70)(17.5 + 3.33)(3.00) - 70(17.5 + 3.33)(0.333)$$
$$p_p' = 10{,}470 \text{ (see tabulated value)}$$

and Eq. 10-1

$$(17.5)^3 - \left[\frac{2(5701)}{10{,}470}\right](17.5)^2 - 6(5701)\left[\frac{7.22}{10{,}470} + \frac{1}{186.7}\right](17.5)$$

$$- \frac{2(5701)}{186.7(10{,}470)}\left[2(5701) + 3(10{,}470)(7.2)\right] = 0$$

$$5359 - 334 - 3619 - 1389 = 17 \approx 0$$
$$\therefore \text{ Use } Y = 17.5 \text{ ft}; \ D = Y + a = 17.5 + 3.33 = 20.88 \text{ ft}$$

(a)
$$D_{SF} = 1.3(20.88) = 27.08 \text{ ft; use } D = 27 \text{ ft}$$

For M_{max}:

$$M_{max} = R_a(\bar{y} + \tfrac{2}{3}x)$$

$$x = \left[\frac{2R_a}{\gamma_b(K_p' - K_a')}\right]^{1/2} = \left[\frac{2(5701)}{70(3 - 0.333)}\right]^{1/2} = 7.82 \text{ ft}$$

Hence,

$$M_{max} = (5701)[7.22 + \tfrac{2}{3}(7.82)]$$
(b)
$$M_{max} = 70.882 \text{ ft-lb/ft of wall}$$

Now to find D and M_{max} via Figs. 10.10, 10.11, or 10.12

$$\frac{\gamma_b}{\gamma} = 0.52 \text{ (Figures 10.11 and 10.12)}$$

$$\alpha = 1$$
$$K_p/K_a = 3/0.307 = 9.77 \approx \text{ say } 10$$

From Fig. 10.11 ($\gamma_b = 0.5\gamma$)   From Fig. 10.12 ($\gamma_b = 0.6\gamma$)
$D/H \approx 1.4$ $D/H \approx 1.25$

Via interpolation (assume straight line)
$$D/H \approx 1.37$$
Hence,

$$D = 1.37(15) = 20.55 \text{ ft}$$

(c) $D_{SF} = 1.3(20.55) = 26.71$ ft (compared with 27 ft via Eq. 10-1)
For M_{max}:
From Fig. 10.11

$$(\gamma_b = 0.5\gamma) \Rightarrow \frac{M_{max}}{\gamma_b K_a H^3} = 1.02$$

From Fig. 10.12

$$(\gamma_b = 0.6\gamma) \Rightarrow \frac{M_{max}}{\gamma_b K_a H^3} = 0.76$$

Via interpolation

$$\frac{M_{max}}{\gamma_b K_a H^3} = 0.968$$

Hence,

(d) $\qquad M_{max} = (0.968)(70)(0.307)(15)^3 = 70{,}208$ ft-lb/ft of wall

(compare with 70,882 via Eq. 10-2.)

Cantilever Sheet Piling in Cohesive Soils

The pressures normally assumed to act on a sheet-pile wall embedded in a cohesive soil are shown in Fig. 10.14 for granular and clay backfill. However, it should be noted that changes such as strength, consolidation, shrinkage, or water in cracks, which are normally time related and rather indeterminate, may result in appreciable

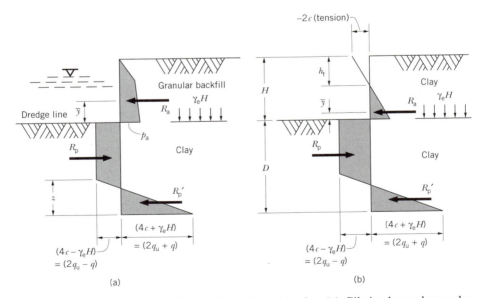

Figure 10.14 Pressures on cantilever piles anchored in clay. (a) Pile in clay and granular backfill. (b) Pile in clay ($\phi = 0$).

changes in the magnitude and location of the pressure resultants acting on the wall.

Immediately after installation, it is common practice to calculate pressures assuming that the clay derives all its strength from cohesion and none from internal friction ($c = \frac{1}{2}$ unconfined compressive strength; $\phi = 0$). $\gamma_e H$ represents the effective pressures at dredge line.

The procedure for the analysis, once the pressure distribution has been established, is basically the same as that previously described. As before, we determine z from the sum of the forces in the x direction, $\Sigma F_x = 0$. Hence,

$$R_a + [R_p' - R_p] = 0 \qquad \text{(a)}$$

But

$$[R_p' - R_p] = (4c - q + 4c + q)\frac{z}{2} - (4c - q)D$$

or

$$[R_p' - R_p] = 4cz - (4c - q)D$$

Substituting in Eq. (a),

$$R_a + 4cz - (4c - q)D = 0$$

Solving for z,

$$z = \frac{(4c - q)D - R_a}{4c} \qquad \text{(b)}$$

and by summing moments about the bottom of the pile, we get

$$R_a(D + \bar{y}) - (4c - q)D\left(\frac{D}{2}\right) + 8c\left(\frac{z}{2}\right)\frac{z}{3} = 0$$

or

$$D^2(4c - q) - 2R_a(D + \bar{y}) - \frac{8cz^2}{3} = 0$$

and substituting for z, we get

$$D^2(4c - q) - 2R_a(D + \bar{y}) - \left(\frac{8}{3}c\right)\left(\frac{1}{4c}\right)^2 [(4c - q)D - R_a]^2 = 0 \qquad \text{(10-3)}$$

A trial-and-error solution is the normal procedure. For a safety factor greater than 1, one may increase the length by 20–40%.

M_{max} occurs where the shear, V, is zero. Thus, from Fig. 10.15,

$$M_{max} = R_a(x + \bar{y}) - (4c - q)(x)\left(\frac{x}{2}\right) \qquad \text{(c)}$$

For equilibrium, (i.e., $\Sigma F_x = 0$)

$$R_a = (4c - q)x$$

or

$$x = \left(\frac{R_a}{4c - q}\right)$$

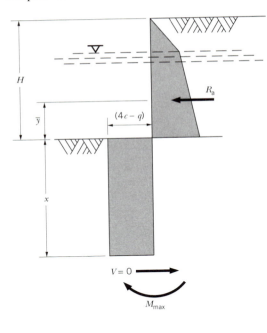

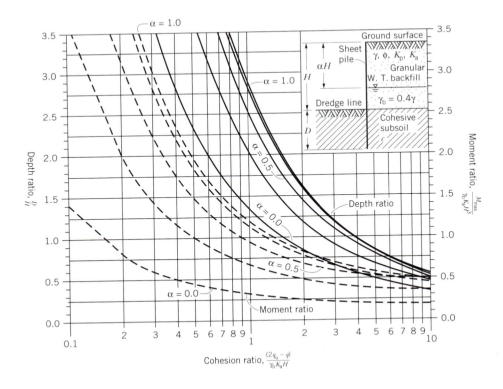

Figure 10.15 Forces on wall above point of zero shear.

Figure 10.16 Cantilever steel sheet pile, cohesive subsoil with granular backfill.

$$\gamma_b = 0.4\,\gamma.$$

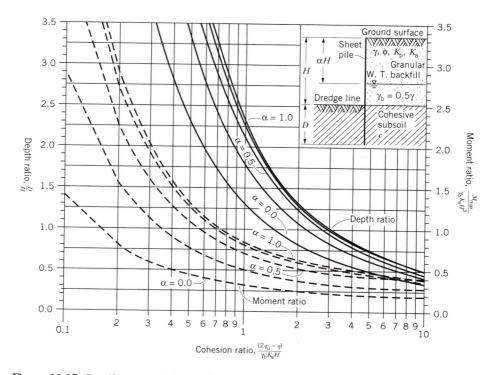

Figure 10.17 Cantilever steel sheet pile, cohesive subsoil with granular backfill.

$$\gamma_b = 0.5 \, \gamma.$$

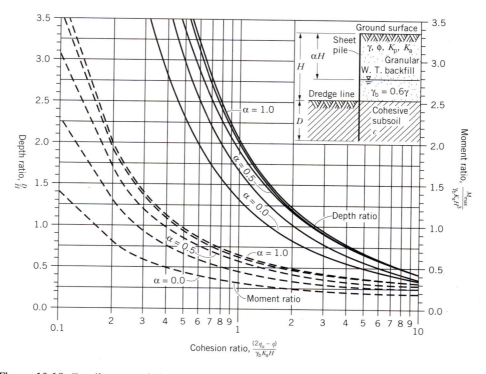

Figure 10.18 Cantilever steel sheet pile, cohesive subsoil with granular backfill.

$$\gamma_b = 0.6 \, \gamma.$$

Substituting in Eq. (c)

$$M_{max} = R_a\left[\left(\frac{R_a}{4c-q}+\bar{y}\right)-\tfrac{1}{2}\frac{R_a}{4c-q}\right] \qquad (10\text{-}4)$$

Figures 10.16, 10.17, and 10.18 (24) provide a relationship for D and M_{max} in terms of c (i.e., $q_u = 2c$), for various values of α. Example 10.2 illustrates the use of such graphs.

Example 10.2

Given Figure 10.19.

Find (a) Embedment depth, D, via Eq. 10-3; use $D_{SF} = 1.3D$
  (b) Maximum moment, M_{max}, via Eq. 10-4
  (c) D via Fig. 10.18 (since $\gamma_b/\gamma \approx 0.6$)
  (d) M_{max} via Fig. 10.18

Procedure

$$R_a = \overbrace{\tfrac{1}{2}(205)6.6}^{R_1} + \overbrace{205(9.9)}^{R_2} + \overbrace{\tfrac{1}{2}(387-205)(9.9)}^{R_3}$$

$$R_a = 677 + 2030 + 901 = 3608 \text{ lb/ft of wall}$$

$$R_a\bar{y} = R_1 y_1 + R_2 y_2 + R_3 y_3$$

$$\bar{y} = \frac{677\left(9.9+\dfrac{6.6}{3}\right) + 2030\left(\dfrac{9.9}{2}\right) + 901\left(\dfrac{9.9}{3}\right)}{3608} = 5.88 \text{ ft}$$

$$4c - q = 4(982) - 1260 = 2668 \text{ psf}$$

$$4c + q = 4(982) + 1260 = 5188 \text{ psf}$$

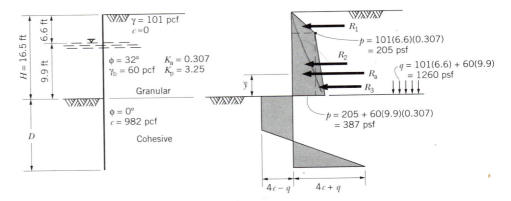

Figure 10.19 Sheet piles with cohesive embedment and granular backfill.

Via Eq. 10-3

$$D^2(4c - q) - 2DR_a - \frac{R_a(12c\bar{y} + R_a)}{2c + q} = 0$$

$$D^2(2668) - 2D(3608) - \frac{3608[12(982)(5.88) + 3608]}{2(982) + 1260} = 0$$

$$2668\,D^2 - 7216D - 81{,}581 = 0$$

$$D = \frac{7216 \pm \sqrt{(7216)^2 + 4(2668)(81{,}581)}}{2(2668)}$$

(a) $D = 7.04$ ft

For M_{max}, from Eq. 10-4 we have

$$M_{max} = R_a \left[\frac{R_a}{4c - q} + \bar{y} - \tfrac{1}{2}\frac{R_a}{4c - q} \right]$$

$$= 3608 \left[\left(\frac{3608}{2668} + 5.88 \right) - \tfrac{1}{2}\frac{3608}{2668} \right]$$

$$= 3608[(7.230) - 0.676]$$

(b) $M_{max} = 23{,}646$ ft-lb/ft of wall

To use Fig. 10.18, we note

$$\alpha = \frac{6.6}{16.5} = 0.4$$

$$\frac{4c - q}{\gamma_b K_a H} = \frac{2q_u - q}{\gamma_b K_a H} = \frac{2668}{(60)(0.307)(16.5)} = 8.78$$

From Fig. 10.18

$$\frac{D}{H} \cong 0.43$$

(c) Hence,

$$D = (0.43)(16.5) = 7.09 \text{ ft}$$

For M_{max}, from Fig. 10.18,

$$\frac{M_{max}}{\gamma_b K_a H^3} = 0.28$$

Hence

$$M_{max} = (0.28)(60)(0.307)(16.5)^3$$

(d)

$$M_{max} = 23{,}168 \text{ ft-lb/ft of wall}$$

10.3 ANCHORED BULKHEADS

Anchored sheet-pile walls or *anchored bulkheads* are a type of retaining wall found in waterfront construction, which is used to form wharves or piers for loading and unloading ships or barges. Briefly, the construction of such walls is accomplished by first driving the sheet piling into the soil to a designated depth, and then attaching a tie-rod support near or at the upper end of the pile and anchoring it to concrete blocks (*deadmen*), brace piles, sheet piles, or other means of support located at a point safely away from any potential slip surfaces. Figure 10.20 shows typical wales and tie rods installed as part of a tie-back system. Frequently, the water side of the pile is dredged in order to acquire greater depth, while the backfill side is built up so as to obtain a level backfill surface. Depending on the soil conditions, one level of tie usually suffices for heights of around 10 m. More than one anchor may be necessary for higher walls in order to reduce the tie force, or to decrease the bending moment or deflection of the sheet piles. Figure 10.21 depicts a retaining wall constructed with steel soldier piles and wooden lagging, with tie back anchors at more than one level.

Figure 10.22 shows the typical forces acting on an anchored sheet-pile wall. The active thrust is resisted by the tie-rod and by the passive resistance. Hence, we need adequate embedment depth in order to develop sufficient passive resistance to resist the active thrust. Likewise, the tie-rod force must be sufficient not only to resist the unbalanced forces, but also to minimize lateral movement of the walls. It is customary to assume that the point where the tie-rod is attached to the wall does not move laterally.

Pressures on and stability of anchored sheet piling are affected by factors such as the depth of piling penetration, the relative stiffness of the piling, the soil properties (e.g., compressibility, cohesive or cohesionless), the amount of anchor yield (e.g., elongation of the rods, yielding in the deadmen or brace-pile support), fluctuating water levels on one or both sides of the pile (e.g., tidal variations), or variable surcharge loads.

Figure 10.20 Wales and tie-rods provide tie-back support for the sheet pile wall, New York Avenue into Washington, D.C. (Courtesy of Bethlehem Steel Company.)

Figure 10.21 Retaining structure consisting of soldier piles driven partially into the base of excavation, at specified spacing and supported by tie-backs and wales at one or more levels. Lagging spans the space between the soldier piles (I-beams) as depicted by the photo. Also shown is the tie-back installation. (Courtesy of A. B. Chance Company.)

The effect of the depth of embedment on the bending moment is illustrated in Fig. 10.22. For relatively short embedment depths, as indicated in Fig. 10.22a, the toe of the pile is not restricted to any significant degree against rotation. The bending moment near the toe, therefore, is negligible. The corresponding moment diagram is indicated to the right of that diagram. The dashed line depicts the shape of the elastic lines for this condition. On the other hand, as the depth of penetration increases, the soil provides some resistance to the toe of the pile in the form of resistance to both translation and rotation. This results in a bending moment as shown in Fig. 10-22b. Subsequent to calculations, the embedment lengths D_1 and D_2 are determined. Thus, with the available information on shear, moment, and allowable stresses, a selection is made of the pile size or section modulus. However, because of the many unknowns and variables associated with anchored bulkheads, a safety factor is introduced in the form of added embedment length. That is, the

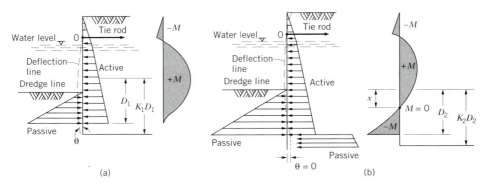

Figure 10.22 Effect of depth of penetration on pressure distribution, bending moment, and elastic line. (a) Relatively small depth of penetration; slope of sheet pile at toe is not zero. (b) Greater depth of penetration; slope of sheet pile at toe is approximately zero, similar to that for a fixed-end condition.

increase in the length of embedment may range anywhere from 20 to 50% and more beyond the length calculated.

10.4 FREE-EARTH-SUPPORT METHOD

Granular Soil

Figure 10.23 shows the active and passive forces as well as the supports provided by the tie-rod acting on an anchored sheet-pile wall in sand. The soil pressures may be computed by the Rankine or Coulomb theory. Furthermore, any bending moment induced by the soil on the toe of the pile is assumed negligible for this analysis. Also, point 0, at the level of the tie, is assumed to be stationary.

For convenience, the active thrust is divided into segments designated by A_1, A_2, and A_3 as shown in Fig. 10.23a. Hence, employing the equations of equilibrium, we can establish the expression necessary to solve for the sole unknown, the embedment depth D. That is, from ΣF_x the force in the tie-rod is

$$T = \mathbf{R}_a - \mathbf{R}_p$$

Similarly, the expressions for the active thrust $\mathbf{R}_a$ and the passive resistance $\mathbf{R}_p$ are given by expressions (a) and (b), respectively:

$$\mathbf{R}_a = \frac{\gamma K_a}{2}(a + b)^2 + \gamma K_a(a + b)(H_w + D) + \frac{\gamma_b K_a}{2}(H_w + D)^2 \qquad \text{(a)}$$

and

$$\mathbf{R}_p = \frac{\gamma_b K_p}{2}D^2 \qquad \text{(b)}$$

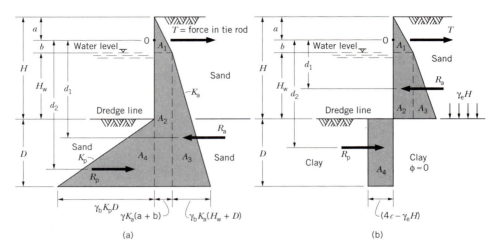

Figure 10.23 Design of anchored sheet piling by free-earth-support method. (a) Totally granular soil. (b) Granular backfill and cohesive soil below dredge line. $\gamma_e H$ = weight of backfill and surcharge. If $p_p \geq 0$, $4c - \gamma_e H \geq 0$, or $c \geq \gamma_e H/4$; unconfined compressive strength $q_u \geq \gamma_e H/2$.

By summing moments about point 0 (level of tie), we get

$$d_1 \mathbf{R}_a = -\frac{\gamma K_a}{2}(a+b)^2 \left(\frac{a+b}{3} - b\right) + \gamma K_a(a+b)(H_w + D)\left(\frac{H_w + D}{2} + b\right)$$

$$+ \frac{\gamma_b K_a}{2}(H_w + D)^2 [\tfrac{2}{3}(H_w + D) + b] \tag{c}$$

and

$$d_2 \mathbf{R}_p = \frac{\gamma_b K_p}{2} D^2 (H_w + b + \tfrac{2}{3}D) \tag{d}$$

From equilibrium we know that the moments about point 0 must equal to zero; thus this is satisfied by

$$d_1 \mathbf{R}_a = d_2 \mathbf{R}_p$$

Substituting the equivalent terms from expressions (c) and (d) one gets an expression of the following form:

$$C_1 D^3 + C_2 D^2 + C_3 D + C_4 = 0 \tag{10-5}$$

The solution to Eq. 10-5 is frequently carried out by a process of trial and error. One notes that the only unknown is D. Hence the solution to Eq. 10-5 yields the embedment depth D. For design purposes it is advisable to increase the depth 20–40 or 50% to provide a safety factor against possible passive failure. One commonly used factor (or increase) is given by

$$\text{design depth } D_d \approx (\sqrt{2})D$$

Note that D is only one of the three items that need to be determined; the other two are the tie-rod tension and the bending moment. The force in the tie rod is necessary for the design of the tie rod itself, wales, and anchorage (see Section 10.7). The bending moment is necessary for the selection of a pile section. Example 10.3 gives the sequence of steps followed.

Figures 10.24, 10.25, and 10.26, (24) provide rather expedient means for estimating (frequently interpolations, typical of graphs, may make the results less accurate than calculations via equations) the embedment depth, D, tie rod force, T, and maximum bending moment, $M_{\max}$. Note: The tie-rod depth is at $H/4$, a frequently (but not always) used ratio.

Cohesive Soils

Figure 10.27 illustrates a case where the sheet piling is embedded in clay and backfilled with a granular material. The active thrust is determined in accordance with the procedure described above, by either Coulomb's or Rankine's theory. The passive resistance, however, is assumed to be a rectangular block with the intensity of pressure as indicated by Fig. 10.23b. The procedure for determining the embedment length D is simply a matter of satisfying equilibrium, very much similar to that described previously, and illustrated in Example 10.3 (for granular soils). In a similar manner, the tie-rod force, T, and maximum bending moment, $M_{\max}$, are determined and, subsequently, one selects the sheet-pile section and designs the wales, tie rod, and anchorage.

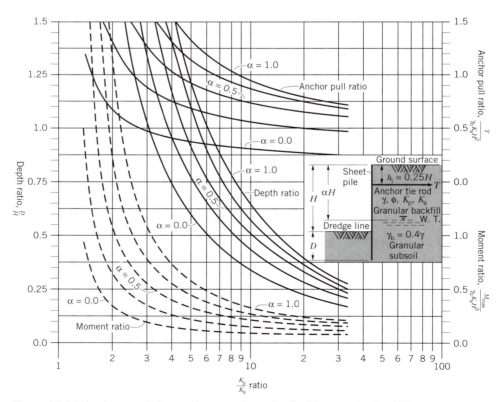

Figure 10.24 Anchor steel sheet pile, granular subsoil with granular backfill.

$$\gamma_b = 0.4\,\gamma.$$

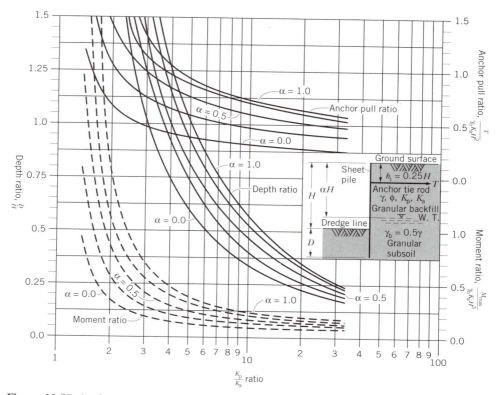

Figure 10.25 Anchor steel sheet pile, granular subsoil with granular backfill. $\gamma_b = 0.5\,\gamma$.

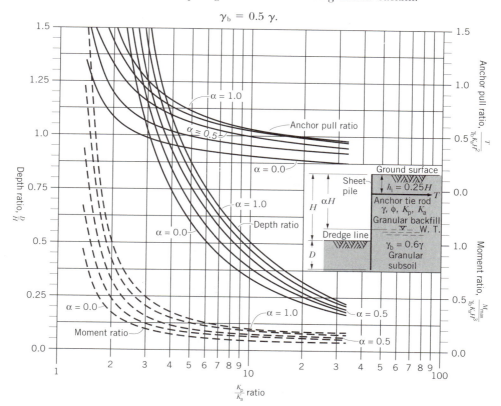

Figure 10.26 Anchor steel sheet pile, granular subsoil with granular backfill. $\gamma_b = 0.6\,\gamma$.

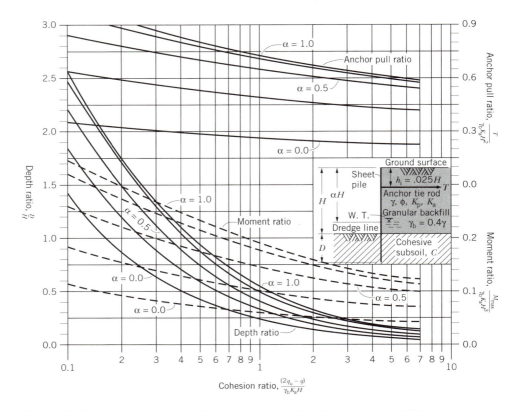

Figure 10.27 Anchor steel sheet pile, cohesive subsoil with granular backfill.

$$\gamma_b = 0.4\,\gamma.$$

One may estimate the values for D, T, and M_{max} via Figs. 10.27, 10.28, and 10.29 (24), provided the conditions fit the graph stipulation (e.g., γ, depth of tie of $H/4$).

10.5 ROWE'S MOMENT-REDUCTION METHOD

Because of the flexibility of steel sheet piling, the soil pressure for such installations differs significantly from the classical hydrostatic distribution. Generally, the bending moment decreases with increasing flexibility of the piling. Hence, the bending moment calculated via the free-earth-support method results in a design usually regarded as too conservative. Rowe (19, 21) proposed a method for reducing the moment in anchored sheet piling, thereby yielding what appears to be a more realistic design.

The significant factors related to Rowe's moment-reduction theory are:

1. The relative density of the soil.
2. The relative flexibility of the piling, expressed in terms of a flexibility number (units are in fps)

$$\rho = H^4/EI$$

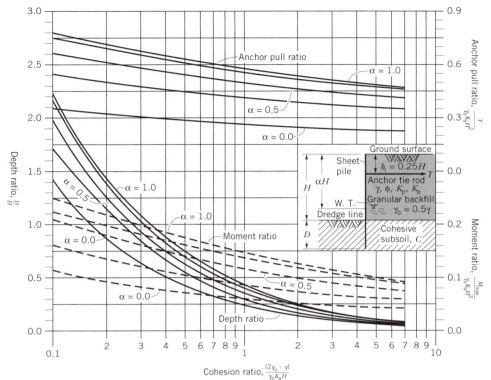

Figure 10.28 Anchor steel sheet pile, cohesive subsoil with granular backfill.

$$\gamma_b = 0.5\,\gamma.$$

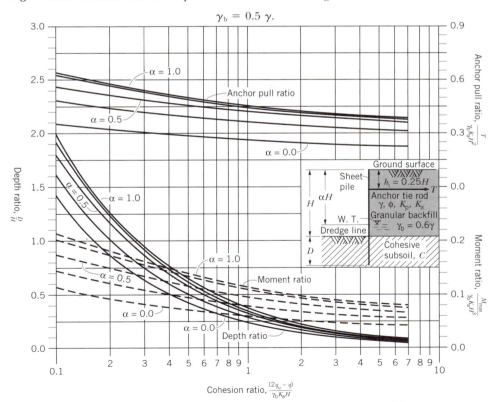

Figure 10.29 Anchor steel sheet pile, cohesive subsoil with granular backfill.

$$\gamma_b = 0.6\,\gamma.$$

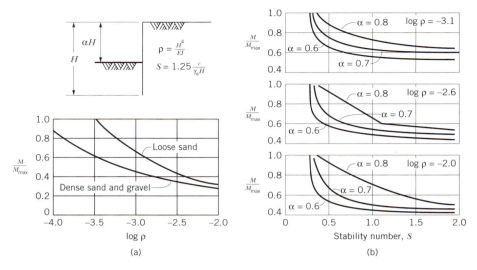

Figure 10.30 Moment-reduction curves after Rowe (19, 22). (a) Sheet piling in sand. (b) Sheet piling in clay.

where H = total length of piling (ft)
        E = modulus of elasticity of section (psi)
        I = moment of inertia of section per foot of wall (in.^{4})

3. For cohesive soils a stability number was defined as

$$S = 1.25c/\gamma_e h$$

4. The relative height of piling α.

Figure 10.30 gives a series of curves developed by Rowe that relate the ratio of reduced design moment M to M_{max} determined by the free-earth-support method to log ρ for sand or stability number S for clays. (Again, note that H here represents the *total* length of the pile.) An interpolation is to be made for sands that range between loose and dense and for α values that are different from those shown. For known values of H and calculated values of ρ and S, the ratios of M/M_{max} can be determined from the corresponding charts in Fig. 10.30. The actual value of M per foot for a given pile can be determined from $\sigma = M/S$; Table 10.3 provides section properties for some common sheet piling. This value must be equal to or slightly larger than the value obtained via the chart. Example 10.3 may be useful in illustrating the steps.

Example 10.3

Given   A sheet-pile wall as shown in Fig. 10.31.

   Find   (a) Total embedment depth of the sheet pile.
            (b) Force in the tie rod per meter of wall, using the free-earth-support method.

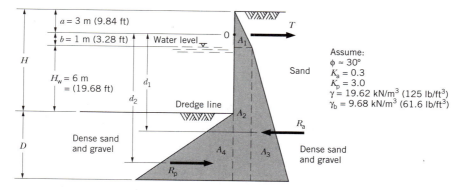

Figure 10.31 Case of granular embedment and granular backfill.

(c) Select a suitable piling section based on Rowe's moment-reduction method.

(d) Compare values for D, M_{max}, and T using Fig. 10.25.

Procedure   (a)

$$\overbrace{\phantom{\frac{\gamma K_a}{2}(a+b)^2}}^{A_1} \quad \overbrace{}^{A_2} \quad \overbrace{\phantom{\frac{\gamma_b K_a}{2}(H_w+D)^2}}^{A_3}$$

$$\mathbf{R}_a = \frac{\gamma K_a}{2}(a+b)^2 + \gamma K_a(a+b)(H_w+D) + \frac{\gamma_b K_a}{2}(H_w+D)^2$$

$$\mathbf{R}_a = \frac{19.62 \times 0.3}{2}(3+1)^2 + 19.62 \times 0.3(3+1)(6+D)$$

$$+ \frac{9.68 \times 0.3}{2}(6+D)^2$$

$$\mathbf{R}_a = 47.09 + 23.54(6+D) + 1.45(6+D)^2$$

$$\overbrace{\phantom{\tfrac{1}{2}\gamma_b K_p D^2}}^{A_4}$$

$$\mathbf{R}_p = \tfrac{1}{2}\gamma_b K_p D^2 = \tfrac{1}{2} \times 9.68 \times 3 \times D^2 = 14.52\,D^2$$

From the sum of the moment about point 0 in Fig. 10.31, ΣM_0 (level of tie rod),

$$\mathbf{R}_a d_1 = -\frac{\gamma K_a}{2}(a+b)^2\left(\frac{a+b}{3}-b\right)$$

$$+ \gamma K_a(a+b)(H_w+D)\left(\frac{H_w+D}{2}+b\right)$$

$$+ \frac{\gamma_b K_a}{2}(H_w+D)^2[\tfrac{2}{3}(H_w+D)+b]$$

$$= -47.09(\tfrac{4}{3}-1) + 23.54(6+D)\left(\frac{6+D}{2}+1\right)$$

$$+ 1.45(6+D)^2[\tfrac{2}{3}(6+D)+1]$$

Table 10.3 Properties of Some Sheet Piling Sections (Manufactured by U.S. Steel Corp.)

Profile	Designation	Driving Distance per Pile (in.)	Weight		Web Thickness (in.)	Section Modulus		Area	Moment of Inertia	
			Per Foot Pile (lb)	Per Square Foot of Wall (lb)		Per Pile (in.³)	Per Foot of Wall (in.³)	Per Pile (in.²)	Per Pile (in.⁴)	Per Foot of Wall (in.⁴)
	Interlock with each other									
	PSX32	$16\frac{1}{2}$	44.0	32.0	$\frac{29}{64}$	3.3	2.4	12.94	5.1	3.7
	PS32*	15	40.0	32.0	$\frac{1}{2}$	2.4	1.9	11.77	3.6	2.9
	PS28	15	35.0	28.0	$\frac{3}{8}$	2.4	1.9	10.30	3.5	2.8
	Interlock with each other									
	PSA28*	16	37.3	28.0	$\frac{1}{2}$	3.3	2.5	10.98	6.0	4.5
	PSA23	16	30.7	23.0	$\frac{3}{8}$	3.2	2.4	8.99	5.5	4.1
	PDA27	16	36.0	27.0	$\frac{3}{8}$	14.3	10.7	10.59	53.0	39.8
	PMA22	$19\frac{5}{8}$	36.0	22.0	$\frac{3}{8}$	8.8	5.4	10.59	22.4	13.7

Interlock with each other
and with PSA23 or PSA28

Section									
PZ38	18	57.0	38.0	3/8	70.2	46.8	16.77	421.2	280.8
PZ32	21	56.0	32.0	3/8	67.0	38.3	16.47	385.7	220.4

Interlocks with itself and
with PSA23 or PSA28

Section									
PZ27	18	40.5	27.0	3/8	45.3	30.2	11.91	276.3	184.2

Diagram dimensions: $\frac{1}{2}$ in., $\frac{3}{8}$ in., 12 in., 11 $\frac{1}{2}$ in., 12 in.

Suggested Allowable Design Stresses—Sheet Piling

Steel Brand or Grade	Minimum Yield Point (psi)	Allowable Design Stress* (psi*)
ASTM A328	38,500	25,000
ASTM A572 GR 50 (USS EX-TEN 50)	50,000	32,000
ASTM A690 (USS MARINER GRADE)	50,000	32,000

* Based on 65% of minimum yield point. Some increase for temporary overstresses generally permissible.

Simplifying and combining terms,

$$R_a d_1 = 810.7 + 286.6D + 30.6D^2 + 0.97D^3$$
$$R_p d_2 = 14.5D^2(H_w + b + \tfrac{2}{3}D) = 14.5D^2(6 + 1 + \tfrac{2}{3}D)$$
$$R_p d_2 = 101.6D^2 + 9.7D^3$$
$$R_a d_1 - R_p d_2 = 0$$
$$D^3 + 8.15D^2 - 32.9D - 96.7 = 0$$

Solving by trial and error,

$$D = 4.35, \text{ say } 4.4 \text{ m}$$

Design: $D_d \approx D\sqrt{2} \approx 4.4\sqrt{2} \approx 6.2$ m

---

Answer | Hence, the total length L of the sheet pile is
$$L = 10 + 6.2 = 162 \text{ m } (53.1 \text{ ft})$$

(b) For $D = 4.4$ m, from R_a and R_b above,

$$R_a = 47.09 + 23.54(6 + 4.4) + 1.45(6 + 4.4)^2 = 449 \text{ kN}$$
$$R_p = 14.5(4.4)^2 = 281.1 \text{ kN}$$

we find:

$$T = R_a - R_p = 449 - 281.1 = 167.9$$

Answer | $$T = 167.9 \text{ kN/m } (11.5 \text{ kips/ft})$$

(c) Figure 10.32 depicts the forces on the wall for a distance x below the water level, where the maximum moment M_{max} occurs. From equilibrium, $\Sigma F_x = 0$,

$$1.05x^2 + 23.54x + 47.1 - 167.9 = 0$$

or

$$x^2 + 22.5x = 115.38$$

Solving for x,

$$x = 4.3 \text{ m} \qquad \text{(below water table)}$$

Thus,

$$M_{max} = (4.31 + 1)167.9 = 891.55$$
$$- \left(4.31 + \frac{4}{3}\right)47.1 = -265.73$$

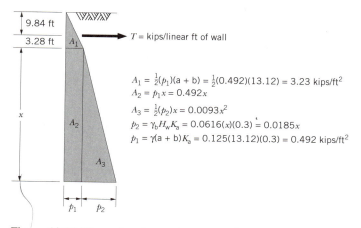

$A_1 = \frac{1}{2}(p_1)(a+b) = \frac{1}{2}(0.492)(13.12) = 3.23 \text{ kips/ft}^2$

$A_2 = p_1 x = 0.492x$

$A_3 = \frac{1}{2}(p_2)x = 0.0093x^2$

$p_2 = \gamma_b H_w K_a = 0.0616(x)(0.3) = 0.0185x$

$p_1 = \gamma(a+b)K_a = 0.125(13.12)(0.3) = 0.492 \text{ kips/ft}^2$

Figure 10.32 Tie and active pressures on sheet piles.

$$-(4.31)\left(\frac{4.31}{2}\right)23.5 = -218.68$$

$$-(4.31)^2\left(\frac{4.31}{3}\right)1.05 = -27.94$$

$$M_{\max} = 379.2 \text{ kN-m/m of wall}$$
$$= 1023 \text{ kip-in./ft of wall}$$

The total length of the pile is 16.2 m = 53.1 ft = H. Then,

$$\rho = \frac{H^4}{EI} = \frac{(53.1)^4}{30 \times 10^6 I} = \frac{0.265}{I}$$

The pile section properties are given in Tables 10.3 and 10.4 for the three selected piles.

Table 10.4 Pile Section Properties

Properties	Pile Section		
	PZ38	PZ32	PZ27
I (per ft)	280.8	220.4	184.2
$\rho = H^4/EI$	9.47×10^{-4}	1.21×10^{-4}	1.44×10^{-4}
$\log \rho$	-3.02	-2.92	-2.84
Section modulus S	46.8	38.3	30.2
Moment capacity $= \sigma S = \frac{25}{12}S$, kip-ft	97.5	79.8	62.9
$M/M_{\max}$	0.89	0.73	0.58

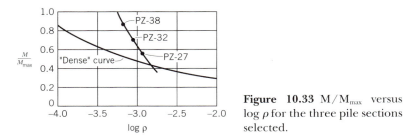

Figure 10.33 M/M_{max} versus $\log \rho$ for the three pile sections selected.

From Fig. 10.30a, the moment-reduction curve corresponding to "dense" sand is reproduced in Fig. 10.33. The "dense" classification is part of the data given.

For the three sections selected, the ratio of M/M_{max} was plotted against the corresponding $\log \rho$ values. Any of the section plotting above the "dense" curve would be satisfactory regarding stress; any below would not. In our case section $PZ27$ is adequate. Note that the allowable design stress used was 25,000 psi.

Answer

Section $PZ27$ is adequate.

Note: Figure 10.25 shows the location of the tie rod at $H/4$; in our example problem the location is at $3H/10$. Thus, the graph answers may quite likely be close but not expected to be exactly the same as those calculated above. Hence, from Fig. 10.25, for $\gamma_b = 0.5$, $K_p/K_a = 10$, and $\alpha = 0.4$, we have

$$\frac{D}{H} \cong 0.43$$

$$\Rightarrow D = (0.43)(10) = 4.3 \text{ m (calculated } D = 4.35)$$

$$\frac{M_{max}}{\gamma_b K_a H^3} \cong 0.13$$

$$\Rightarrow M_{max} = (0.13)(9.68)(0.3)(10)^3$$
$$M_{max} = 377 \text{ kN-m/meter of wall} \qquad \text{(calculated } M = 379)$$

$$\frac{T}{\gamma_b K_a H^2} \cong 0.58$$

$$\Rightarrow T = (0.58)(9.68)(0.3)(10)^2$$
$$= 168.4 \text{ kN/m} \qquad \text{(calculated } T = 167.9)$$

10.6 FIXED-EARTH-SUPPORT METHOD

This method assumes that the toe of the pile is restrained from rotating, as indicated in Fig. 10.34a. The deflection of the sheet piling is indicated by the dashed line. The corresponding moment diagram for this case is shown in Fig. 10.34b. Point C

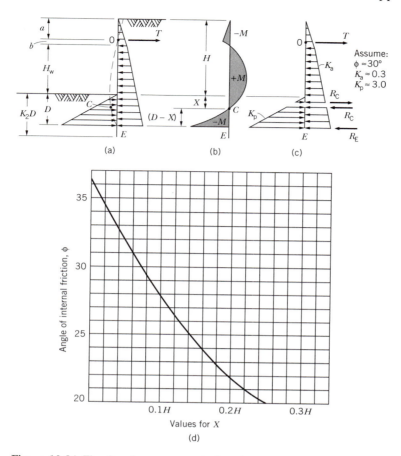

Figure 10.34 Fixed-end-support equivalent beam method in sand.

is a point of contraflexure. At this point the pile could be assumed to act as a hinge (zero bending moment). Hence, the portion of the piling above point C can be treated as a beam that resists the net earth pressures via the force T and the shear R_C, as indicated in Fig. 10.34c. This is known as the *equivalent-beam method*, first proposed by Blum (1). Blum established a theoretical relationship between the angle of internal friction and the distance from the point of contraflexure to the dredge line X, as shown in Fig. 10.34d.

The lateral pressures, both active and passive, are determined by either the Rankine or the Coulomb formula, as before. For a given angle of internal friction ϕ the value for X can be calculated as a function of H from Fig. 10.34d. Hence by summing moments about 0 (anchor level), the shear R_C is determined. With R_C known, the summation of moments about point E yields a relationship where the only unknown is D. The depth D determined from this expression is generally increased from 20 to 40% or more (K_2 is 1.2–1.4D, etc.). The force T in the tie can likewise be determined by summing moments about the hinge at point C.

With the value of the embedment depth calculated, one may proceed to determine the shear and the moment, as before, and subsequently to select the appropriate size for the sheet piling. Example 10.4 illustrates the sequence of steps summarized above.

Example 10.4

Given   The sheet-pile wall of Example 10.3.

Find   Using the fixed-earth-support method, determine: (a) the force T in the tie rod and (b) the embedment D of sheet piling.

Procedure   (a) The active and passive pressures at the toe of the pile are:

$$p_a = \gamma K_a(a + b) + \gamma_b K_a(H_w + D)$$
$$p_a = (19.62)0.3(4) + (9.68)0.3(6 + D)$$
$$p_a = 41 + 2.9D$$

and

$$p_p = \gamma_b K_p D = 9.68 \times 3 \times D = 29D$$

For $H = 10$ m, $X = 0.068H = 0.068(10) = 0.68$ m. The active forces above point C, namely, $A_1\ A_2$, A_3, and A_4, are (Fig. 10.35):

$$A_1 = \tfrac{1}{2}\gamma K_a(a + b)^2 = \frac{19.62}{2}(0.3)(4)^2 = 47.09; \bar{y} = 6.68 + \tfrac{4}{3} = 8.01$$

$$A_2 = \gamma K_a(a + b)(H_w + X) = 19.62(0.3)(4)(6.68) = 157.27; \qquad \bar{y} = 3.34$$

$$A_3 = \tfrac{1}{2}\gamma_b K_a(H_w + X)^2 = \frac{9.68}{2}(0.3)(6.68)^2 = 64.79; \qquad \bar{y} = 2.23$$

$$A_4 = \tfrac{1}{2}(19.75)(X) = \frac{19.75}{2}(0.68) = 6.72; \qquad \bar{y} = \frac{0.5}{3}$$

$$\Sigma M_{C(\text{top})}: T \times 7.68 + 6.72 \times \frac{0.68}{3} = 47.09(8.01) + 157.27(3.34)$$

$$+ 64.79(2.23)$$

$$7.68T = 377.19 + 525.28 + 144.48 - 1.52$$

Answer

$$T = 136.12 \text{ kN/lin m}$$

(Note: by the free-earth-support method, Example 10.3, $T = 167.9$ kN/m.

(b) $$\Sigma F_X: T + R_C = \Sigma A = 47.09 + 157.27 + 64.79 - 6.72$$
$$136.12 + R_C = 262.43$$
$$R_C = 126.31$$

$$R_{p_1} = \left(\frac{29D + 19.75}{2}\right)(D - X);$$

$$\bar{y}_{p_1} = \left(\frac{D-X}{3}\right)\left(\frac{2 \times 19.75 + 29D}{29D + 19.75}\right)$$

$$R_{a_1} = \left(\frac{41 + 2.9D + 42.94}{2}\right)(D-X);$$

$$\bar{y}_{a_1} = \left(\frac{D-X}{3}\right)\left(\frac{2 \times 42.94 + 41 + 2.9D}{41 + 2.9D + 42.94}\right)$$

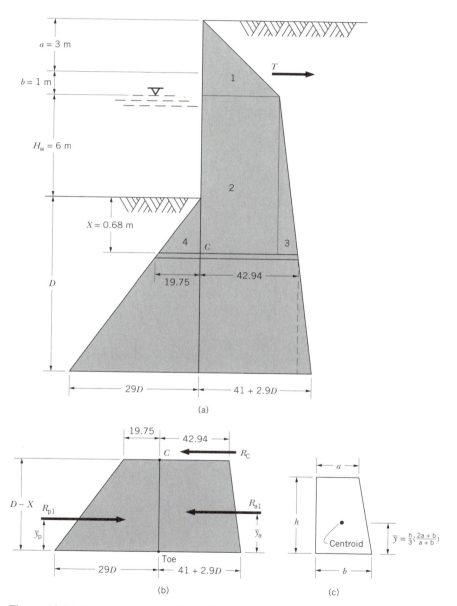

Figure 10.35 Forces acting on the wall above and below the point of contraflexure, C.

Thus, $\sum M_{\text{toe}}: R_C(D - X) = R_{p_1}\bar{y}_p - R_{a_1}\bar{y}_a$, or

$$R_C = \frac{D - X}{6}(26.1D - 87.4)$$

Substituting $R_C = 126.31$ $X = 0.68$, and simplifying

$$26.1D^2 - 105.13D - 698.44 = 0$$

Solving for D,

Answer

> $$D = 7.57 \text{ m}; \ D\sqrt{2} = 10.7 \text{ m}$$
> (Note: via Example 10.3, $D = 4.35$ and $D\sqrt{2} = 6.2$ m)

10.7 WALES, TIE RODS, AND ANCHORAGES

The tie-back support provided for sheet-pile installations is illustrated by the general arrangement shown in Fig. 10.36. A wale is placed in a horizontal position in front

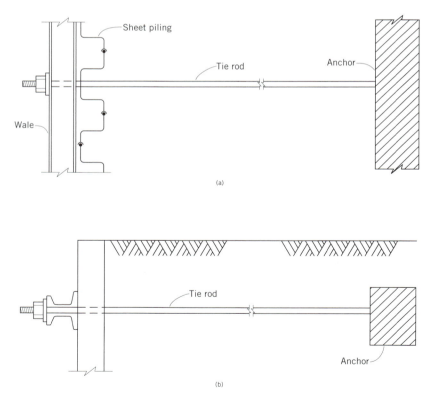

Figure 10.36 General arrangement of wale, tie rod, and anchor. (a) Plan view. (b) Elevation view.

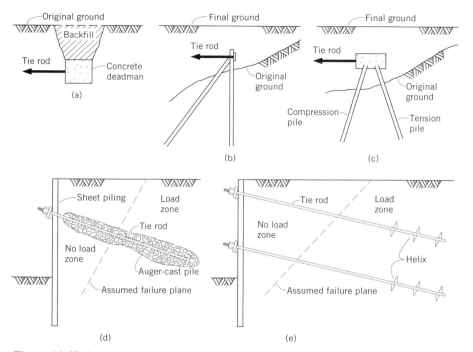

Figure 10.37 Some common types of anchors. (a) Deadman anchor, either precast or cast-in-place concrete. (b) and (c) Braced piles. (d) Tie rod anchored in cast-in-place pile. (e) Helical anchors.

of the piling or attached to the backfill face of the piling if a flush-front face is required. The latter is generally the case for harbor installations. Wales are usually fabricated of two-channel sections placed back to back and separated sufficiently to insert the tie rod through as shown (also see Fig. 10.20).

The tie rod may be a cable or a steel bar threaded to permit vertical alignment and tension adjustments. Paint, asphaltic material, or concrete encasement are materials commonly used to provide corrosion protection to the tie rod; in Fig. 10.37a the rod is protected by concrete. Bearing piles are frequently used to support tie rods that rest on soft and compressible material in order to minimize sagging due to settlement. The spacing of the tie rods depends on the forces assumed by each rod. The design forces, particularly in cohesive soils, are usually assumed to be 20–30% larger than those computed via the previously discussed methods, with allowable stresses approaching 80–90% of the yield stress of the steel.

The wales may be conservatively designed as simply supported beams with spans equal to the distances between tie rods or, more realistically, as continuous beams. Splices in channel sections of wales should be staggered to avoid possible weak points. Governing concerns in wale evaluations are bending and web crippling. The latter problem is frequently overcome by web reinforcement of the section.

Figure 10.37 shows a variety of anchor schemes used in the tie-back system. Deadman-type anchors shown in Fig. 10.37a are constructed either by pouring a concrete beam in place, or by embedding a precast beam. Care should be taken

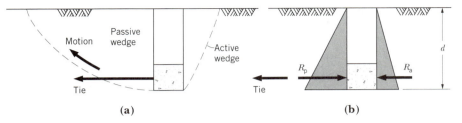

Figure 10.38 Forces related to deadman-type anchor. (a) Reaction elements on deadman. (b) Active and passive resultants on deadman.

to provide reliable lateral resistance by placing the precast beam directly against the original soil and/or to tamp fill material in front of the beam. The deadman is placed at a sufficient distance from the sheet-pile wall to permit full development of the passive resistance. Hence, the ultimate capacity per unit length of the deadman is approximately

$$T_u = \mathbf{R}_p - \mathbf{R}_a \tag{10-6}$$

where $\mathbf{R}_p$ and $\mathbf{R}_a$ may be regarded as the passive and active resultant forces. Their magnitude is estimated on the assumption that the earth wedge in front of the deadman slides upward and forward, as shown in Fig. 10.38a. An approximate allowable capacity in granular soils per unit length may be obtained from Eq. 10-7:

$$T_a = \frac{1}{F} (A \gamma K_p d^2) \tag{10-7}$$

where T_a = allowable capacity per unit length of deadman
F = factor of safety
A = coefficient, usually assumed 0.6
γ = unit weight of soil
K_p = coefficient of passive pressure
d = depth to bottom of deadman

The allowable capacity per unit length in cohesive soils ($\phi \approx 0$) is sometimes approximated by Eq. 10-8:

$$T_a = \frac{d^2}{F} \left(A \gamma K_p + \frac{c}{2} \right) \tag{10-8}$$

where c = soil cohesion.

The anchors shown in Figs. 10.37b, 10.37c, and 10.37d develop their capacity from the piles. Installation and capacity of piles are topics discussed in Chapter 11.

10.8 BRACED CUTS

With the possible exception of excavations in rock formations, steep cuts of any significant depth in soils are laterally supported by either retaining walls or braced sheeting. Generally, retaining walls are used as supports for a more permanent construction. On the other hand, braced sheetings are commonly used for temporary supports, perhaps during construction such as for tunneling and subways, deep trenches, deep building foundations, and so on. Also, the braced supports are appreciably more flexible than most permanent supports such as concrete retaining walls.

Figure 10.39 shows two commonly used techniques for lateral bracing. In Fig. 10.39a steel plates or wood sheeting, placed either horizontally or vertically, is supported directly by cross members or *struts*. This arrangement may be expanded to include structural members, such as *soldier beams* or *wales*, placed either vertically or horizontally at some designated spacing, as the need dictates. These, in turn, are also supported by the horizontal struts. This method of bracing is generalized as *lagging*.

Also common is the form of bracing shown in Fig. 10.39b. In this case steel sheet piling is driven into the ground, generally to a depth greater than the excavation in order to develop bottom anchorage. As the excavation progresses, wales are placed horizontally along the length of the excavation at intermittent depths. They, in turn, are supported by the horizontal struts placed at designated spacings.

The pressure exerted against bracing is different from that usually assumed for retaining walls. The cause for this difference lies in the character of "wall" movements and the subsequent pattern of deformation of the soil associated with the two different types of support. As the top strut is wedged against the wale, it will permit relatively little movement at that point. With advancing excavation below the wale, however, some horizontal movement of the soil will be permitted by the relatively flexible sheeting until the next strut is fixed in position. In other words, the soil in the vicinity of the first strut is rather immobilized, but the lower stratum slips down and toward the excavation until the second strut is fixed in position. Figure 10.40 depicts (although exaggerated) deflections typical of braced excava-

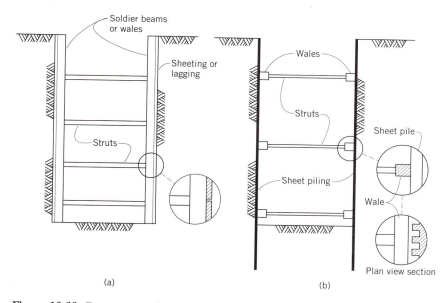

Figure 10.39 Common methods of bracing cuts. (a) Lagging. (b) Sheet piles.

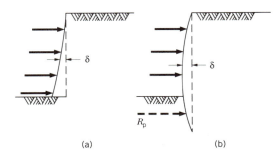

Figure 10.40 Deflections of braced excavations. (a) Lagging. (b) Sheet piling.

tions. In the process, however, the soil below has a tendency to drag on the soil above it. This continues with progressing excavation until the final strut is in place. The state of stress associated with this type of deformation is closely related to a condition generalized as *arching-action*. Furthermore, the distribution of pressure is greatly affected by the deformation conditions, the nature of the soil, and the manner of installing the braces.

At the present time the pressure distribution on braced sheeting is not totally understood. Indeed, almost all of the design for braced sheeting is based on empirical methods derived as a result of field tests (13, 21, 22, 25). Figure 10.41a provides a qualitative comparison of the type of pressure distribution obtained from field observations and measurements, and the theoretical distribution given by the Rankine theory (dashed line). Field data indicate that the resultant pressure is 10–15% higher than that predicted by the Rankine distribution. Furthermore, the point at which the resultant force acts is higher than that for the Rankine pressure. It acts near the midpoint of the depth, compared to the third of the height from the bottom for the Rankine distribution.

Figures 10.41b, 10.41c, and 10.41d show the pressure distribution recommended for the design for various types of soils. Although other distributions have been recommended for this purpose, the ones shown appear to give acceptable results

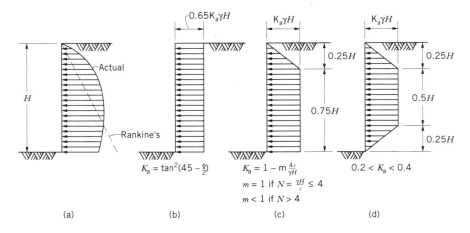

Figure 10.41 Pressure distributions on braced excavations. (a) General. (b), (c), and (d) Distributions used for design. [After Terzaghi and Peck, 1967 (26).]

for most design problems. On the other hand, judgment should not be discarded for any individual excavation considered.

Example 10.5

Given   An excavation as shown in Fig. 10.42.

Find   Forces in struts per meter depth normal to page.

Procedure   Refer to Fig. 10.41c:

$$N = \frac{\gamma H}{c} = \frac{17.8 \times 7}{25} = 4.98 > 4$$

N is only slightly above the value of 4 recommended. Thus, let us assume $m \approx 0.8$. (Values of m measured ranged from 1 in Chicago to 0.4 in Oslo, Norway.) Reliable values of m might better be based on actual measurements of strut loads. Thus,

$$K_a = 1 - m\frac{4c}{\gamma H} = 1 - 0.8\left(\frac{4 \times 25}{17.8 \times 7}\right) = 0.36$$

and

$$K_a \gamma H = 0.36 \times 17.8 \times 7 = 44.86 \text{ kN/m}^2$$

The pressure distribution is as shown in Fig. 10.43. For calculating the strut forces, each strut is assumed to carry the corresponding load based on a simple span.

$$P_1 = 0.5(1.75)(44.86) = 39.25 \text{ kN}$$

P_1 acts at 1.17 m from top or 1.33 m from ②

$$P_2 = 0.75(44.86) = 33.64 \text{ kN}$$

P_2 acts at 0.375 m from ②

$$P_3 = 1.5(44.86) = 67.28 \text{ kN}$$

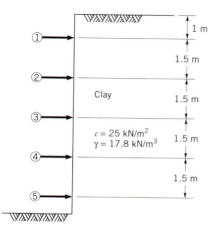

Figure 10.42 Forces in struts.

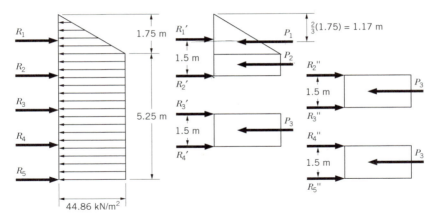

Figure 10.43 Determination of strut forces in braced cuts.

This force is of the same magnitude for the three rectangles shown.

$\sum M_②$:

$$R_1 = \frac{39.25 \times 1.33 + 33.64 \times 0.375}{1.5} = 43.21 \text{ kN}$$

$\sum F_x$:

$$R_2' = P_2 + P_1 - R_1 = 39.25 + 33.64 - 43.21 = 29.68$$
$$R_2'' = \tfrac{1}{2} P_3 = 33.64$$

Hence,

$$R_2 = R_2' + R_2'' = 63.32 \text{ kN}$$

Since $R_3' = R_3'' = R_4' = \cdots = R_2''$,

$$R_3 = 2R_3' = 2R_2'' = 67.28 \text{ kN}$$
$$R_4 = 67.28 \text{ kN}$$
$$R_5 = \tfrac{1}{2} P_3 = 33.64 \text{ kN}$$

Answer

$$R_1 = 43.21 \text{ kN}$$
$$R_2 = 63.32 \text{ kN}$$
$$R_3 = 67.28 \text{ kN}$$
$$R_4 = 67.28 \text{ kN}$$
$$R_5 = 33.64 \text{ kN}$$

Problems

Reference is made to Fig. P10.1 for Problems 10.1 through 10.9. Using free-earth-support method and comparing against graphs, determine for these problems:

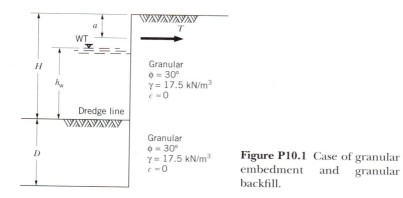

Granular
$\phi = 30°$
$\gamma = 17.5$ kN/m^3
$c \approx 0$

Granular
$\phi = 30°$
$\gamma = 17.5$ kN/m^3
$c \approx 0$

Figure P10.1 Case of granular embedment and granular backfill.

(a) The force in the tie rod per meter of wall.
(b) The total length of the sheet pile by the method stipulated and data given for the corresponding problems.

10.1	$H = 8$ m; $h_w = 4$ m; $a = 2.0$ m.
10.2	$H = 8$ m; $h_w = 2$ m; $a = 2.0$ m.
10.3	$H = 8$ m; $h_w = 0$; $a = 2.0$ m.
10.4	$H = 8$ m; $h_w = 4$ m; $a = 2.0$ m.
10.5	$H = 8$ m; $h_w = 2$ m; $a \doteq 2.0$ m.
10.6	$H = 8$ m; $h_w = 0$; $a = 2.0$ m.
10.7	$H = 10$ m; $h_w = 4$ m; $a = 2.5$ m.
10.8	$H = 10$ m; $h_w = 2$ m; $a = 2.5$ m.
10.9	$H = 10$ m; $h_w = 0$; $a = 2.5$ m.

Reference is made to Fig. P10.10 for Problems 10.10 through 10.12. Using fixed-earth-support method and comparing against graphs, determine for these problems:

(a) The force in the tie rod per meter of wall.
(b) The total length of the sheet pile by the method stipulated and for the data corresponding to the following problems.

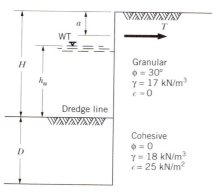

Granular
$\phi = 30°$
$\gamma = 17$ kN/m^3
$c \approx 0$

Cohesive
$\phi = 0$
$\gamma = 18$ kN/m^3
$c = 25$ kN/m^2

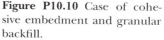

Figure P10.10 Case of cohesive embedment and granular backfill.

10.10 $H = 8$ m; $h_w = 4$ m; $a = 2.0$ m.

10.11 $H = 8$ m; $h_w = 2$ m; $a = 2.0$ m.

10.12 $H = 8$ m; $h_w = 0$; $a = 2.0$ m.

For Problems 10.13 through 10.17 determine the forces in the struts shown in Fig. P10.13 for the data given:

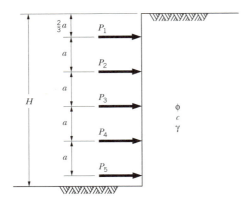

Figure P10.13 Forces in a braced excavation.

10.13 $H = 7$ m; $\gamma = 18$ kN/m³; $\phi = 32°$; $c = 0$; sand.

10.14 $H = 7$ m; $\gamma = 18$ kN/m³; $\phi = 0°$; $c = 30$ kN/m²; soft clay.

10.15 $H = 7$ m; $\gamma = 18$ kN/m³; $\phi = 0°$; $c = 30$ kN/m²; stiff clay.

10.16 $H = 6$ m; $\gamma = 17.5$ kN/m³; $c = 0$; $\phi = 32°$; sand.

10.17 $H = 6$ m; $\gamma = 17.5$ kN/m³; $c = 25$ kN/m²; $\phi = 0$; soft clay.

10.18   For the data of Problem 10.2 select a suitable piling section based on Rowe's moment-reduction method.

BIBLIOGRAPHY

1. Blum, H., *Einspannungsverhältnisse bei Bohlwerken,* W. Ernst and Sohn, Berlin, Germany, 1931.

2. Boghart, A., "Use of Stable Program in Tied-Back Wall Design," *ASCE J. Geotech. Eng. Div.,* vol. 115, Apr. 1989.

3. Broms, B., and H. Stille, "Failure of Anchored Sheet Pile Walls," *ASCE J. Geotech. Eng. Div.,* vol. 102, Mar. 1976.

4. Browzin, B. S., "Anchored Bulkheads: Horizontal and Sloping Anchors," *ASCE J. Geotech. Eng. Div.,* vol. 107, May 1981.

5. Clough, G. W., and Y. Tsui, "Performance of Tied-Back Walls in Clay," *ASCE J. Geotech. Eng. Div.,* vol. 100, Dec. 1974.

6. Clough, G. W., and L. A. Hansen, "Clay Anisotropy and Braced Wall Behavior," *ASCE J. Geotech. Eng. Div.,* vol. 107, July 1981.

7. Clough, G. W., and M. W. Reed, "Measured Behavior of Braced Wall in Very Soft Clay," *ASCE J. Geotech. Eng. Div.,* vol. 110, Jan. 1984.

8. Davis, A. G., and C. Plumelle, "Full-Scale Tests on Ground Anchors in Fine Sand," *ASCE J. Geotech. Eng. Div.,* vol. 108, Mar. 1982.

9. Golder, H. Q., et al., "Predicted Performance of Braced Excavation," *ASCE J. Soil Mech. Found. Div.,* vol. 96, May 1970.

10. Haliburton, T. A., "Numerical Analysis of Flexible Retaining Structures," *ASCE J. Geotech. Eng. Div.,* vol. 94, Nov. 1968.

11. Hanna, T. H., and I. I. Kurdi, "Studies on Anchored Flexible Retaining Walls in Sand," *ASCE J. Geotech. Eng. Div.,* vol. 100, Oct. 1974.

12. Kay, J. N., and M. I. Qamar. "Evaluation of Tie-Back Anchor Response," *J. Geotech. Eng. Div.,* vol. 104, Jan. 1978.

13. Lambe, T. W., "Braced Excavations," *4th Pan Am. Soil Mech. and Found. Eng. Conf.,* ASCE, 1970.

14. Lambe, T. W., et al., "Measured Performance of Braced Excavation," *ASCE J. Soil Mech. Found. Div.,* vol. 96, no. SM3, May 1970.

15. Mana, A. I., and G. W. Clough, "Prediction of Movements for Braced Cuts in Clay," *ASCE J. Geotech. Eng. Div.,* vol. 107, June 1981.

16. Nataraj, M. S., and P. G. Hoadley, "Design of Anchored Bulkheads in Sand," *ASCE J. Geotech. Eng. Div.,* vol. 110, Apr. 1984.

17. O'Rourke, T. D., "Ground Movements Caused by Braced Excavations," *ASCE J. Geotech. Eng. Div.,* vol. 107, Sept. 1981.

18. Poulos, H. G., and M. F. Randolph, "Pile Group Analysis: A Study of Two Methods," *ASCE J. Geotech Eng. Div.,* vol. 109, Mar. 1983.

19. Rowe, P. W., "Anchored Sheet Pile Walls," *Proc. Inst. Civ. Eng.,* vol. 1, pt. 1, 1952.

20. Rowe, P. W., "A Stress–Strain Theory for Cohesionless Soil with Applications to Earth Pressures at Rest and Moving Walls," *Geotechnique,* vol. 4, June 1954.

21. Rowe, P. W., "Sheet Pile Walls in Clay," *Proc. Inst. Civ. Eng.,* vol. 7, July 1957.

22. Rowe, P. W., "Anchored Sheet-Pile Walls," *Proc. Inst. Civ. Eng.,* 1962.

23. Swatek, E. P., Jr., et al., "Performance of Bracing for Deep Chicago Excavation," *5th Pan Am. Soil Mech. and Found. Eng. Conf.,* ASCE, vol. 1, pt. 2, 1972.

24. Schroeder, W. L., and P. Roumillac, "Anchored Bulkheads with Sloping Dredge Lines," *ASCE J. Geotech. Eng. Div.,* vol. 109, June 1983.

25. Stock, J. F. "Development of Design Curves for Cantilever and Anchored Steel Sheet Piles," Master of Science Thesis, 1992, Youngstown State University.

26. Terzaghi, K., and R. B. Peck, *Soil Mechanics in Engineering Practice,* 2nd ed., Wiley, New York, 1967.

27. Tsinker, G. P., "Anchored Sheet Pile Bulkheads: Design Practice," *ASCE J. Geotech. Eng. Div.,* vol. 109, Aug. 1983.

28. Turabi, D. A., and A. Balla, "Distribution of Earth Pressure on Sheet-Pile Walls," *ASCE J. Geotech. Eng. Div.,* vol. 94, Nov. 1968.

29. Ulrich Jr., E. J., "Internally Braced Cuts in Overconsolidated Soils," *ASCE J. Geotech. Eng. Div.,* vol. 115, Apr. 1989.

30. Ulrich Jr., E. J., "Tieback Supported Cuts in Overconsolidated Soils," *ASCE J. Geotech. Eng. Div.,* vol. 115, Apr. 1989.

31. Vaid, Y. P., and R. A. Campanella, "Time-Dependent Behavior Earth Wall," *ASCE J. Geotech. Eng. Div.,* vol. 103, July 1977.

32. Vidal, H., "The Principle of Reinforced Earth," *Highw. Res. Rec.* 282, 1969.

CHAPTER 11

Deep Foundations:
Piles, Drilled Piers,
and Caissons

11.1 INTRODUCTION

Most building sites possess adequate bearing capacity to permit a spread-footing
design, especially for light structures. Certain weak strata are sometimes improved
by a form of pretreatment such as compaction, chemicals, or preloading, as described in Chapter 5. Then there are times when, because of economic or physical
restrictions, or perhaps unusually heavy loads, sites with poor bearing capacity
cannot be improved by such procedures. In such cases the building loads are
transmitted to a deeper, more suitable soil or rock formation via piles, drilled piers,
or caissons. Foundations that encompass piles, drilled piers, or caissons are generally
classified as deep foundations. It is such foundations that are covered in this chapter;
shallow foundations were covered in Chapters 6 and 7.

   The functional features of piles, drilled piers, or caissons are rather similar: Each
is typically subjected to an axial compressive load, although some lateral forces are
usually inevitable; also, sometimes piles and caissons serve as anchors for special
installations (e.g., guy lines, certain underpinning designs). The difference between
piles, drilled piers, and caissons lies primarily in their physical size and method of
installation.

   In general, a deep foundation is more expensive than an ordinary spread footing
design. On the other hand, it is frequently more economical and, generally, much

more reliable than procedures associated with soil stabilization. Of course, specific cost estimates should be done for the various approaches based on specific sites and building requirements, and only subsequent to proper evaluation of the subsurface conditions via a detailed program of testing and analysis. Indeed, a subsurface evaluation should always be viewed as a most important prerequisite in the context of foundation selection.

Piles

Piles are specially installed, relatively slender columns used to transmit the structural loads to a lower, firmer soil or rock formation. Their diameter (or equivalent dimension) is generally 750 mm or less; this is a commonly used dimension to delineate piles from driller piers or caissons. The piles may be of **concrete**, **steel**, **timber**, or a **composite** of a steel shell filled with concrete. Still another type is one commonly referred to as a compaction pile; it is installed by either injection, through vibrations or pounding into the soil stratum, a column of sand and gravel, or lean concrete.

Some of the considerations related to the selection of the type of piles are (1) corrosive properties of the stratum, (2) fluctuations in the water table (and possibly related dry-rotting effects associated with such fluctuations), (3) ease of installation, (4) length requirement, (5) availability of material, (6) installation equipment, (7) restrictions on driving noise and vibrations, and (8) costs.

The load transfer via piling may be realized from skin friction, from end-bearing, or from a combination of both. If a very large percentage of the total load is resisted by the side or skin friction, the pile is known as a **friction pile**. On the other hand, if the pile rests on a very firm stratum (say rock), such that almost all of the load is transmitted directly to the stratum via tip resistance, the pile is known as an **end-bearing pile**. Almost invariably, however, each of the two types derives some of its load capacity from a combination of end-bearing and skin friction. The percentage load assumed by either friction or end-bearing is less discernible in compaction piles. All types discussed in this chapter fall in the general category of bearing piles. Their primary function is to resist vertical loads; they are not to be confused with sheet piles, whose primary function typically is to resist lateral forces.

Piles may be driven (hammered) into the strata, or they may be installed by a cast-in-place process. Furthermore, they may be driven vertically or on a batter, they may be tapered, or they may be bell-shaped bottom. Figure 11.1 shows a slightly tapered steel shell being installed dynamically (hammered) via a diesel hammer (upon reaching a designated resistance to penetration, the shell is filled with concrete). Figure 11.2 depicts the installation of battered piles.

The capacity of the pile is typically estimated by the foundation designer based on soil data and somewhat empirical procedures, and on the designer's experience. However, it is common practice and highly advisable, that the capacity be verified by subjecting one or more piles to a load test. Typically, test piles are injected for this purpose and the pile load test carried out prior to the placement of the project piles. In this manner, any adjustments that may need (say length) to be made subsequent to the test results could be made prior to the installation of the piles for the project. Occasionally, for driven piles, the load test is eliminated if careful observation of the pile's resistance to penetration are made during the driving sequence.

a   b

Figure 11.1 (a) A 14-in.-diameter Monotube pile being lifted to be placed in the "leads" and, subsequently, to be driven by a Delmag D-15 diesel hammer. Ohio Department of Transportation test pile project over Tuscarawas River, Tuscarawas County, Ohio. (b) A 14-in. Monotube pile driven by a Conmaco 5125 air hammer, JFK International Airport. (Courtesy Monotube Pile Corporation.)

It is common practice to design and test single piles, even when they are part of a pile group or cluster. It is important to note that, depending on the number of piles, pile spacing, method of installation, soil characteristics, and load distribution, the behavior of a group or composite of piles may not be totally reflective of that of the individual piles. For example, it is likely that the load capacity of the composite is less than that of the sum of the individual piles. The efficiency of a pile group is less than that of an individual pile in soil formations, but may approach 100% efficiency for piles bearing directly on rock. Also, the settlement of the pile group is likely to be greater than that of an individual pile, or corresponding load per pile, in soil formations.

Drilled Piers and Caissons

In a general sense, drilled piers and caissons are larger piles; they are usually 750 mm or more in diameter. The terms *drilled pier* and *caisson* are frequently used interchangeably by engineers. Typically, for drilled piers, a shaft is drilled into the soil, which is then filled with concrete. The shaft may be cased with a metal shell (a casing) in order to maintain the shaft from collapsing and sometimes to facilitate the cleaning of the bottom of the shaft as needed. The casing may be left in place as part of the pier, or it may be gradually withdrawn as the shaft is filled with concrete. The lower part of the shaft may be uncut, or belled out to develop a larger end-bearing area, thereby increasing the capacity of the drilled pier for a given allowable end-bearing pressure. Typically, drilled piers and caissons are designed

Figure 11.2 Shown are 14-in.-diameter Monotube piles being driven by a Delmag D16-32 diesel hammer for Minnesota Department of Transportation bridge near Mall of America (a shopping mall). It is common practice to batter (notice slope of lead and piles in trench) piles to resist horizontal forces. In the background are the anchored sheet piles. (Courtesy of Monotube Pile Corporation.)

as end-bearing members, neglecting the side friction from soil formations, but accounting in a rather empirical fashion the friction from rock when the shaft is drilled some length into the rock formation (socketed into rock). Common reasons for using drilled piers instead of piles are as follows: Pile capacities are insufficient to carry the required load; larger end-bearing area is needed and this is best accomplished with larger diameter shafts or bell-bottoms; driving noises and vibrations are not allowed for the site.

Caissons are used rather extensively for construction of bridge piers and abutments in rivers and lakes and commonly used in connection with construction of waterfront structures such as docks, wharves, and sea walls. Caissons may be open-end or box type, with a shell-type configuration constructed to form a cell. These could be constructed at the site or constructed at one location (closed bottom) and transported to the building site and, subsequently, sunk onto a previously prepared soil base where the structural component is to be built. These shells provide a controlled area from excavating and dewatering, as needed, until the bearing stratum is reached.

Generally, such box-type caisson construction is rather expensive. Sheet-pile cofferdams are usually more economical, particularly for relatively shallow depths of 30 to 35 ft or less.

11.2 TIMBER PILES

One of the oldest types of piles, the wood pile, is still one of the most common. Frequently it is the cheapest. Essentially, wood piles are made from tree trunks with the branches and bark removed. Southern pine and Douglas fir are common types, although hardwoods are sometimes used. The choice is frequently dictated by the availability of the material, and this may vary in different parts of the country.

The installation of this type of pile is by driving. Normally, the pile is driven with the small end down. Typically, the pile has a natural taper, with a top cross section of twice or more that at the bottom end. Depending on the resistance to driving and obstacles such as boulders within the soil stratum, wooden piles are sometimes tied with steel bands at the bottom end and/or at the top end to avoid splitting.

The size of the pile is frequently dictated by the load capacity expected, by the type of soil, and sometimes by building codes. Generally, 20-m maximum lengths are common, although lengths of over 30 m have been used on numerous occasions. Although the piles could withstand hard handling, their lengths are frequently dictated by transportation restrictions and local availability.

As a general rule, wood piles have a long life provided they are not subject to alternate wetting and drying, or to attacks from marine borers. On the other hand, subjected to repetitious cycles of wetting and drying, the pile may experience rapid decay through dry-rotting within months or perhaps one or two years; cedar is generally more resistant to such effects. For this reason wood piles are almost always cut off below the permanent groundwater. Furthermore, most frequently wooden piles are treated with preservatives such as creosote oil, which is impregnated into the wood. This is an effective means of preventing dry-rotting, particularly for that part of the pile that may undergo cycles of wetting and drying (e.g., marine installations), as well as against damage from most animal and plant attacks. As a general rule, the top end of the pile should be protected, generally by lead paint, zinc coating, or some other treatment, before the concrete cap is poured. Needless to say, such protective treatment, as well as additional costs that may be incurred if excavation, dewatering, and sheeting should prove necessary in connection with cutoffs below the surface, are economic considerations that should be part of the selection process.

Overdriving of timber piles may result in splitting, crushing, or shearing of the pile. Buckling of the pile may be still a further effect from overdriving. This is sometimes detected by a sudden reduction in penetration resistance. As a general rule, however, close observations of pile behavior made during the driving operations is the best insurance against such an occurrence. Furthermore, designing for a rather limited capacity, perhaps up to 30 tons, may be further assurance against failure from overdriving.

11.3 STEEL PILES

Steel piles are usually rolled-H, fabricated shapes, or pipe piles. Sometimes sections of sheet piles are used to form a pile unit, usually in the form of a box. Because of the relative strength of steel, steel piles withstand driving pressures well and are usually very reliable end-bearing members, although they are found in frequent use as friction piles as well. Pipe piles are normally, although not necessarily, filled with concrete after driving. Prior to driving, the bottom end of the pipe pile usually is capped with either a flat or a cone-shaped point welded to the pipe. Sometimes the pipe is driven open-ended instead of capped, and the soil is removed after the completion of driving either by augering, or by air or water jetting the soil from within the pipe. In soil formations that contain large boulders or rock, steel tips, as shown in Fig. 11.3, are welded to H-piles to protect them from splitting or tip damage.

Strength, relative ease of splicing, and sometimes economy are some of the advantages cited in the selection of steel piles. Corrosion is a negative aspect of steel piles. Corrosive agents such as salt, acid, moisture, and oxygen are common enemies of steel. Steel piles in disturbed soils appear to corrode more rapidly due to an apparent higher oxygen content in the disturbed soil than in an undisturbed soil. The oxygen content in an undisturbed soil appears to decrease significantly after a few feet from the surface. Furthermore, because of the corrosive effect that salt water has on steel, steel piles have a rather restrictive use for marine installations. Peat deposits, strata containing coal or sulfur, acid clays, and swampy soils are usually of common concern. In general, however, soils with a pH of less than 7 (acidic) should be looked at carefully. Coating such piles with paint or encasing them in a concrete shell may provide some protection against corrosion, although the paint may be damaged during the installation process. For a limited life span of the pile, it is sometimes the practice to compensate for the anticipated corrosion by specifying a greater thickness of the steel section. Sometimes cathodic protection is used to reduce such corrosive effects.

Figure 11.3 A cast steel tip shown welded to HP10 × 42 pile. Such tips are common for strata of large boulder or rock formations in order to minimize splitting or tip damage to the pile. (Courtesy ADF Corporation.)

11.4 CONCRETE PILES

Under comparable circumstances, concrete piles are much more immune to the detrimental effects of the corrosive elements that rust steel piles or to the effects that cause decay of the wooden piles. Concrete is sometimes used as a protective coating for steel piles. Furthermore, concrete is generally more available than steel in most geographic regions, and in many instances at more economically advantageous terms than wood.

Concrete piles may be of a *precast* or a *cast-in-place* type. Although both types may be plain or reinforced, straight shaft or tapered, they are more conspicuous in their differences rather than in their similarities.

Precast Concrete Piles

These are concrete piles that are formed, cast to specified lengths and shape, and cured before they are driven or jetted into the ground. Reinforced with conventional steel or prestressed, these piles are generally cast with a pointed tip, have square or octagonal cross section, and are usually tapered. They are frequently used in marine installations, or where part of the pile may serve as a column above ground. Although especially designed for each project, the sizes of the precast piles, in both diameter and length, are frequently governed by handling stresses. Usually they are limited to less than 25-m length and are generally less than 0.5 m in diameter. The pile capacity is usually limited to about 75 tons.

While cutoffs and splices may be required to adjust for differences in length, these procedures pose a less-than-desirable situation, particularly for splicing. For example, to splice a new section onto one already driven, the abutting ends of both sections must be chipped so as to expose the reinforcing bars of both. They are, subsequently, connected by fresh concreting at this junction—a rather time-consuming effort due to both labor and curing time required for the new joint.

Cast-in-Place Piles

The installation process of cast-in-place piles may consist of driving a steel tubing or casing into the ground and then filling it with concrete (*cased pile*); or concrete may be cast into a driven shell that is subsequently extracted as the concrete is poured (*uncased* or *shell-less pile*). An increasingly popular shell-less type is constructed by casting through the auger, under pressure, without the aid of a steel shell.

Depending on wall thickness and general strength, a steel shell or pipe may be driven with or without the aid of a mandrel. That is, for relatively thin shells, a mandrel or core fits into the shell in order to prevent inward collapse and buckling of the shell during the driving operations. Once the shell has been driven to the desired depth, the mandrel is extracted and the shell filled with concrete. Many pipe-type sections and some relatively thin-fluted sections, however, may frequently be driven without a mandrel. Sometimes the casing is driven open-ended with subsequent removal of the soil within the pipe, usually via jetting. Subsequently, the casing is filled with concrete.

One method of obtaining a shell-less pile consists of extracting the driven shell while the concrete is either hammered down or restricted from moving upward with the pipe by means of a mandrel. Under sufficient force and hammering action, a bulb-shaped mass of concrete could be formed at the lower end of the pile for additional bearing support.

An *auger-cast pile* is a shell-less type formed by using a continuous-flight-auger with a hollow stem. The augers range in size from about 25 to 40 cm in diameter, with the flights welded to a hollow stem, which may range from around $7\frac{1}{2}$ to 10 cm in diameter. The tip of the auger has a small opening, which is plugged up during the downward augering. The hole is augered to a predetermined depth. A concrete grout is then pumped by a positive-displacement pump through high-pressure hoses connected to the stem of the auger. Under pressure, the concrete mortar expels the plug from the tip of the auger and permeates into the stratum and into the cavity below the tip of the auger created as the auger is slowly withdrawn. With the pump continuously pumping, the auger is extracted slowly enough to permit maximum grout penetration into the hole and into the adjoining stratum, as well as to prevent choking of the pile by lateral soil pressures, but fast enough to prevent the grout from coming up the shaft. The soil displaced by the auger is brought up by the auger flights during the auger extraction. Reinforcing, in the form of either individual bars, steel sections, or several bars connected by ties or spirals, similar to concrete-column reinforcement, is pushed into the grout at the end of the casting while the grout is still wet.

Particularly advantageous for sand and gravel strata, the auger-cast piles usually provide excellent friction resistance, are vibration free, and usually are more economical than other piles for comparable purposes. Indeed, because it requires neither splicing nor fixed-length dimensions, this type of pile may be injected to varying lengths throughout a site in a most expeditious manner. Furthermore, particularly for granular soils, it compares most favorably with the other types in terms of load capacity. Figure 11.4 shows some common types of cast-in-place piles.

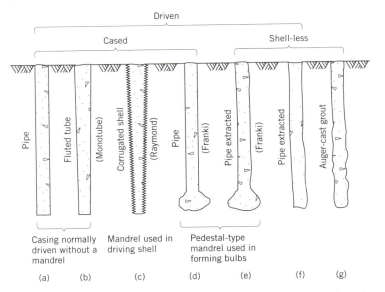

Figure 11.4 Some common types of cast-in-place piles. Their lengths and capacities vary for various types and soil conditions.

11.5 ULTIMATE CAPACITY OF A SINGLE PILE

The pile support is realized from two sources: (1) *end-bearing* and (2) *side friction*. As mentioned previously, although the percentage of the pile load assumed by the friction or by the end-bearing will generally vary widely with different situations, static equilibrium dictates that

$$Q_u = Q_p + Q_s \tag{11-1}$$

where Q_u = total pile load
   Q_p = point resistance (end-bearing)
   Q_s = side-friction resistance

Figure 11.5a shows the forces acting on a pile as described by Eq. 11-1. The magnitude of Q_p can be estimated with acceptable accuracy by using a form of the general bearing capacity formula presented in Chapter 4. The shape and depth correction factors are replaced with compensating terms more representative of the conditions associated with different types of soil, types and lengths of piles, methods of installation, and other factors discussed later in this chapter. However, in general Q_p may be estimated as

$$Q_p = A_p \left(\bar{c} N_c + \gamma L N_q + \frac{\gamma}{2} B N_\gamma \right) \tag{11-2}$$

The total support provided by friction Q_s could be estimated as the product of the surface area (product of perimeter p and length L) of the pile in contact with the soil times the average shear resistance per unit area s_s developed between the soil and the pile, that is, $Q_s = pLs_s$. For round piles, $Q_s = 2\pi RLs_s$. However, since the unit shear resistance s_s may vary widely over the length of the pile, an average unit resistance is perhaps difficult to ascertain with any degree of accuracy. In that case it would be more logical to express the total resistance derived from friction as the sum of unit resistances. These are expressed by Eq. 11-3, where ΔL is the

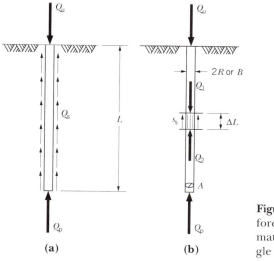

(a)   **(b)**

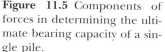

Figure 11.5 Components of forces in determining the ultimate bearing capacity of a single pile.

increment of pile length. For a layered system and/or varied sections, (Fig. 11.5b),

$$Q_s = \sum (\Delta L)\, p_i\, s_{si} \tag{11-3}$$

For uniform soil characteristics (s_s = constant) and constant section (p = constant),

$$Q_s = pLs_s \tag{11-3a}$$

Substituting the equivalent quantities from Eqs. 11-2 and 11-3 into Eq. 11-1, one obtains a general ultimate load bearing capacity for piles as given by Eq. 11-4. That is, for a layered system and/or varying pile sections the ultimate load becomes

$$Q_u = A_p\left(\bar{c}N_c + \gamma L N_q + \tfrac{1}{2}\gamma B N_\gamma\right) + \sum \Delta L p_i s_{si} \tag{11-4}$$

As a special case, for uniform soil strength (s_s = constant) or constant pile section (p = constant) and round piles of radius R, we have

$$Q_u = \pi R^2\left(\bar{c}N_c + \gamma L N_q + \gamma R N_q\right) + 2\pi R L s_s \tag{11-4a}$$

where

$$
\begin{aligned}
Q_u &= \text{ultimate bearing capacity of a single pile} \\
s_{si} &= \text{shaft resistance per unit area at any point along pile} \\
B &= \text{general dimension for pile width} \\
A_p &= \text{cross-sectional area of pile at point (bearing end)} \\
R &= \text{radius of pile at a given increment of length} \\
p_i &= \text{perimeter of pile in contact with soil at any point} \\
L &= \text{total length of embedment of pile} \\
\gamma &= \text{unit weight of soil} \\
\bar{c} &= \text{effective cohesion of soil} \\
N_c, N_q, N_\gamma &= \text{bearing capacity factors}
\end{aligned}
$$

11.6 FRICTION PILES IN COHESIONLESS SOILS

When a pile is driven into cohesionless soil, the density of the stratum is increased due to (1) the volume displacement (volume equal to the volume of the imbedded portion of the pile); and (2) the densification resulting from the driving vibrations. Increased density results for a distance of a few pile diameters. In general, a significant increase for a distance of 2 diameters and a noticeable increase up to 8 diameters are rather common for cohesionless layers.

The ultimate bearing capacity Q_u is again the sum of the forces realized from skin friction and end-bearing, as expressed by Eq. 11-1. The expansion of this equation into the version given by Eq. 11-4 can be used to determine the capacity of piles in a granular formation with some modification. That is, it is readily apparent that for the cohesionless case ($c = 0$) the first term of Eq. 11-4 drops out. Furthermore, the third term (N_γ term) is relatively small when compared with the middle term. Hence, a reasonable approximation for the ultimate bearing capacity of piles in cohesionless material could be given by Eq. 11-5:

$$Q_u = A_p \gamma L N_q + \sum \Delta L p_i s_{si} \tag{11-5}$$

The above equation implies that both end-bearing and total skin resistance would increase with increasing depth for a homogeneous stratum. However, experimental evidence and field observations indicate that the end-bearing capacity reaches some upper limit, and does not increase infinitely with depth. It appears that crushing, compressibility, and general failure in the zone near the pile tip, as well as other factors, impose an upper limit on the ultimate bearing capacity of a given pile. Letting $\bar{\sigma}$ represent the effective overburden pressure at the pile tip, the tip resistance could be represented by

$$Q_p = A_p \bar{\sigma} N_q \leq A_p \sigma_e \qquad \text{(a)}$$

where σ_e represents some limiting value for end-bearing for which the depth D equals or is greater than the critical depth (that is, $D \leq D_c$). Although specific values for σ_e still appear questionable at this stage, values of 25 kN/m² for loose sand and 100 kN/m² for dense sand appear as reasonable approximations for the upper limit.

There is lack of agreement among experts regarding the manner in which to compute the skin resistance, which is represented by the last terms in Eq. 11-4. The disagreement lies primarily in the computation of the unit shear resistance s_s developed between pile and soil. In general, however, the value for s_s could be estimated by Eq. (b):

$$s_s = K_s \bar{\sigma} \tan \delta \qquad \text{(b)}$$

and, therefore, the value for Q_s is given by Eq. (c):

$$Q_s = \sum p_i \Delta L K_s \bar{\sigma} \tan \delta \qquad \text{(c)}$$

where
K_s = average coefficient of earth pressure on the pile shaft (see Table 11.1)
$\bar{\sigma}$ = average effective overburden pressure along pile shaft
δ = angle of skin friction
$\tan \delta$ = coefficient of friction between soil and pile surface

The values of K_s may vary, and are frequently more accurately derived from the results of pile load tests at the building site. These values may range from a value equal to an earth pressure coefficient at rest K_0, to a value equivalent to some passive pressure, approximately 3 to 4 times this value. The lower values are generally associated with bored or jacketed piles into loose granular material, while the upper range applies for driven or pressure-injected piles into relatively dense material. Generally, a value of K_s between 1 and 2 appears reasonable, with Table 11.1 providing some approximate values for K_s for some common types of driven piles.

On the basis of the foregoing discussion, the ultimate bearing capacity of a pile in a cohesionless material could be expressed by Eq. 11-6:

$$Q_u = A_p (\bar{\sigma} N_q) + \sum p_i \Delta L K_s \bar{\sigma} \tan \delta \qquad \text{(11-6)}$$

Table 11.1 Approximate Values for K_s for Some Driven Piles

Pile Type	Range of K_s
Concrete	1.5 ± 10%
Pipe	1.1 ± 10%
H-section	1.6 ± 10%

It is to be noted that experimental evidence indicates that the value of N_q is influenced by such factors as the friction angle ϕ, the compressibility of the soil, the method of pile installation, and, to a lesser extent, the type and shape of the pile. Hence, the values obtained from Fig. 4.11 are indeed approximate. On the other hand, they are deemed acceptable for estimating purposes, particularly in the lower range of values for $\phi = 25°$ or less.

11.7 BEARING CAPACITY BASED ON STANDARD PENETRATION TESTS OF COHESIONLESS SOILS

A widely used and reasonably reliable method for predicting the load capacity of driven piles into cohesionless soils is an empirical correlation of capacity and blow counts from *standard penetration tests* (SPT). For very coarse material, including large gravel and/or boulders, the blow count on the split-spoon sampler may be exceedingly high and not representative of the true characteristics of the soil to support the driven pile.

The correlation between blow counts and pile capacity has not been adopted as a standard by the profession. Quite to the contrary, wide variations in views exist among practitioners regarding such interpretations. Frequently, the engineers will use guidelines based on personal experience for given localities and geographic districts for estimating pile bearing capacities and will verify the design by appropriate tests in the field. Meyerhof (67) suggests a formula that may be used for a *cohesionless* material, particularly in sand deposits, given by Eqs. (a) and (b):

$$q_p = 40N\frac{L}{2R} \leq 400N \quad (kN/m^2) \tag{a}$$

$$f_s = 2N \quad (kN/m^2) \tag{b}$$

where f_s is the average shear resistance per unit surface area and R is the radius of the pile. Hence, on the basis of the above expressions, the total capacity of the pile may be approximated by Eq. 11-7, consistent with the limits for the first term of Eq. (a) above. That is,

$$Q_u = A_p q_p + A_s f_s$$

or

$$Q_u = A_p\left(40N\frac{L}{2R}\right) + 2NA_s \tag{11-7}$$

where A_s is the surface area that develops the friction = (perimeter) (length).

The end-bearing resistance and average skin support of bored piles in cohesionless material are less than that for driven piles of comparable size. In general, the ultimate capacity of a bored pile may be safely estimated as approximately half that of a driven pile. On the other hand, pressure-injected piles installed in cohesionless soil via the auger-cast method (cast under pressure) provide a frictional resistance as well as an overall ultimate capacity that is favorably comparable to (and in some instances higher than) those of many driven piles. This is generally attributed to

Table 11.2 Values of q_p and f_s Used by the Author as a Guide in Estimating Capacities of Auger-cast Piles Injected under Pressure in Cohesionless Soil*

Average Blow Count Readings N	<10	10–15	16–20	21–25	26–35	36–50	>50
q_p (kN/m²)†	0–200	200–300	320–400	420–500	520–700	720–1000	1000
f_s (kN/m²)	0–30	30–45	45–60	60–75	75–105	105–150	175

* For diameters of 25–40 cm (10–16 in.) and $L \geq 6$ m long.
† For N readings at tip.

(1) a densification of the surrounding stratum induced by some volume displacement and (2) relatively deep lateral penetration of the cement grout (perhaps several centimeters) into the cohesionless stratum. Such penetration creates (1) a shear surface anywhere between 10 and 20% larger and (2) an end-bearing area sometimes 25 to 30% larger than for a driven pile of comparable size. Furthermore, the friction angle between soil and pile usually approaches ϕ. In fact, the value of ϕ is frequently increased around the periphery of the pile due to injection of the mortar grout into the surrounding soils. Figure 11.6 shows an auger-cast pile installation where some served as bearing piles and some for a cut-off wall.

Table 11.2 gives a range of values for bearing pressures at the pile tip and side friction f_s with resistance numbers obtained from the standard penetration test. These are perhaps most suitable in sands and small gravels, and less so in cohesive materials; they are least applicable for very coarse gravels and boulders. These results should be verified by testing of a typical pile or piles in accordance with the accepted testing procedures (e.g., ASTM D-1143-74). The essence of such procedures is briefly described in Section 11.10.

11.8 FRICTION PILES IN COHESIVE SOILS

Piles driven into soft saturated clays tend to (1) disturb the clay around the pile, (2) increase the pore-water pressure, (3) increase compressibility in the clay, and (4) remold the clay to varying degrees for a distance of approximately one pile diameter. Temporarily there is an apparent loss of pile capacity. However, the pore-water pressure dissipates rather rapidly, and after some consolidation the shear strength is regained, frequently exceeding the initial value. The increased densities, consolidation effects, and increased horizontal stress after pile driving contribute to an increase in the frictional resistance between soil and pile. This recovery normally takes place within 30 days—ordinarily less than the time in which the total building load is applied to the pile.

The ultimate bearing capacity of a single pile in clay could be estimated by the general expression given by Eq. 11-1. The resistance provided by end-bearing Q_p and that from the friction Q_s could be determined by use of Eqs. 11-2 and 11-3, respectively. As mentioned previously, the N_γ term is relatively small in comparison

a

b

with the other two terms and therefore may be (conservatively) neglected. Hence, the total resistance from end-bearing could be expressed by Eq. 11-8:

$$Q_p = A_p(\bar{c}N_c + \gamma L N_q) \tag{11-8}$$

Based on Fig. 4.11, the value of N_c for a value of $\phi = 0$ is 5.14. Skempton (118) found the value for N_c to be approximately 7.5 for long footings and approximately 9 for circular or square footings when the ratio of footing depth to width exceeded a value of about 10. Values for N_c ranging from 5 for a very sensitive normally consolidated clay to 10 for insensitive, overconsolidated clay are common.

The total resistance from friction Q_s may be estimated from Eq. 11-3. For convenience, this is reproduced as

$$Q_s = \sum (\Delta L) p f_s \tag{11-3}$$

where f_s is the unit skin friction resistance in clay. According to Meyerhof (67), the values for f_s could be approximated as given by Eqs. (a) and (b). For driven piles,

$$f_s = 1.5 c_u \tan \phi \tag{a}$$

For bored piles,

$$f_s = c_u \tan \phi \tag{b}$$

where c_u = average cohesion, undrained condition
ϕ = angle of internal friction of the clay

Table 11.3 shows a relationship between the unit skin friction resistance f_s and the unconfined compressive strength of clay.

Although values for f_s may vary with different sources, particularly for clayey soils, virtually all authorities agree that the values should be used only as guides for estimating the capacity of the pile. As mentioned previously, the actual capacity should be determined via load tests of actual piles in the field.

Based on the foregoing discussion, an expression for estimating the ultimate bearing capacity of a pile installed in a clayey stratum could be given by Eq. 11-9:

$$Q_u = A_p(\bar{c}N_c + \gamma L N_q) + A_s f_s \tag{11-9}$$

◀ **Figure 11.6** The construction of a 16-ft-deep machinery pit in a high water table formation (1.5 ft below surface). (a) The author is experimenting with a 16-in.-diameter auger-cast to serve for bearing support (some are visible at the bottom of the excavation, with 4 #6 bars protruding up from the pile) as well as for water cut-off on the four sides. The side piles are 26 ft long, with approximately 17 ft above the bottom of the excavation, and braced as excavation progressed (the fine sand and silt was too soft for a cantilever support). The injection pressure and pile-overlapping [see lower left-hand portion of part (b)] provided acceptable seal against water infiltration; of course, some dewatering was necessary as would have been the case if steel sheet piles were used. These piles also eliminated the need for forming the outside of the walls. (b) Reinforcing steel for walls being placed subsequent to pouring the bottom of the pit. Note dowel steel protruding up from the base.

Table 11.3 Values of f_s in kN/m² Based on
Unconfined Compression Tests on Clay
[Tomlinson (126)]

Unconfined Compressive Strength of Clay (kN/m²)	f_s (kN/m²)	
	Concrete or Timber	Steel
0–72	0–34	0–34
72–144	34–48	34–48
144–288	48–62	48–57
288	62	57

Example 11.1 _____

Given   (1) An auger-cast pile; diameter 0.41 m; $L = 12$ m. (2) Dry sand and gravel stratum; $\gamma = 19.4$ kN/m³; $\phi \cong 32°$; $c \approx 0$.

Find   The ultimate bearing capacity of the pile.

Procedure   From Fig. 4.11, for $\phi = 32°$, $N_q \cong 23$ and $N_\gamma \cong 21$, and

$$s_{s\,avg} = \tfrac{1}{2}K_s\gamma L \tan\phi = \tfrac{1}{2}(1.5)(19.6)(12)\tan 32° = 109$$

$K_s \approx 1.5$ from Table 11.1. Furthermore, let us assume the shear resistance varies linearly with depth, and $\overline{\phi} \approx \phi$. For the typical auger-cast pile, installed under *pressure*, the above assumptions seem reasonable. Hence, using Eq. 11-4a,

$$Q_u = \pi\left(\frac{0.41}{2}\right)^2\left[0 + 19.4(12)(23) + 19.4\left(\frac{0.41}{2}\right)(21)\right] + 2\pi\left(\frac{0.41}{2}\right)(12)(109)$$

$$Q_u = 0.132[5354.1 + 83.5] + 1686$$

Answer

$$Q_u = 2408 \text{ kN } (541 \text{ kips})$$

Example 11.2 _____

Given   (1) An auger-cast pile; diameter 0.41 m; $L = 12$ m. (Note that this is the same size pile as given in Example 11.1.)

        (2) Sandy silt stratum; $\gamma = 18.4$ kN/m³; $\phi = 16°$; $c = 13$ kN/m²; water table at surface.

Find   The ultimate bearing capacity of the pile.

Procedure   The assumptions made in Example 11.1 will appear valid here as well. Hence, for $\phi = 16°$; $N_q = 4$; $N_\gamma = 3$; $N_c = 12$,

$$s_{avg} = 13 + \tfrac{1}{2}(1.5)(18.4 - 9.81)(12)\tan 16° = 35.5, \text{ say } 36 \text{ kN/m}^2$$

Thus,

$$Q_u = \pi \left(\frac{0.41}{2}\right)^2 \left[13(12) + 8.59(12)(4) + 8.49\left(\frac{0.41}{2}\right)(3)\right] + 2\pi\left(\frac{0.41}{2}\right)(12)(36)$$

$$Q_u = 0.132[156 + 412.3 + 5.3] + 649$$

$$Q_u = 75.72 + 649$$

Answer

$$Q_u = 725 \text{ kN (163 kips)}$$

11.9 PILE CAPACITY BASED ON DRIVING RESISTANCE

In his book *Pile Foundations* Chellis (19) provides a fascinating short summary of implements man has used to drive piles. The primitive pile-driving equipment consisted of a variety of rams, which were worked by hand power or perhaps crudely improvised forms of hammers. During the medieval times, the equipment revealed improved features, which incorporated some mechanical advantages. With the invention of power engines, the energy for the pile-driving equipment is provided solely by power hammers. Figure 11.7 shows a pile-driving hammer and a cushion that is placed between the hammer and pile. There are numerous types and manufacturers of such hammers. In a general category, the following are basic features of several types:

1. *Single-acting hammer.*   Steam or compressed air raises the ram, then permits it to drop freely onto the top of the pile.
2. *Double-acting hammer.*   Here steam or compressed air not only lifts but also pushes down on the ram.
3. *Diesel hammer.*   As the name implies, this hammer is operated by a diesel engine, which raises the ram and then permits it to fall freely.

Dynamic Formulas

Numerous empirical formulas have been developed in an attempt to predict the capacity of a driven pile from its resistance to penetration. The basis of most of these formulas is a transfer of the kinetic energy of the hammer to the pile and to the soil.

The energy E delivered by the hammer is assumed transformed into a force P_u transmitted into the pile, as the pile penetrates a distance s and energy loss ΔE. Symbolically, this could be represented by Eq. (a):

$$E = P_u s + \Delta E \tag{a}$$

If we represent the loss ΔE by $\Delta E = P_u C$, then

$$E = P_u s + P_u C = P_u(s + C) \tag{b}$$

a

b

Figure 11.7 (a) Delmag D-48 hammer driving an 18 ft × 18 in.-diameter concrete pile with a Fredco cap. Port Orange, Florida. (b) Insertion of Hamortex cushion in a Vulcan #1 hammer bowl, Tom's River, New Jersey. (Courtesy of Reedrill Inc.)

where C is an empirical constant. Common values for C are 0.1 in. and 1.0 in. for single-acting steam hammers and for drop hammers, respectively. Hence, solving for P_u, we have,

$$P_u = E/(s + C) \tag{11-10}$$

Equation 11-10 has been modified by many in attempts to account for variables, such as hammer weight and efficiency, pile properties (e.g., weight, length, cross section, modulus of elasticity), and coefficient of restitution for different pile materials. Table 11.4 gives some of these formulas. One notes that not all of the many variables are accounted for in each of these formulas. Hence, it is recommended that, generally, two or more formulas should be used to arrive at a reasonable average of pile capacity; some judgment should be used in selecting the pile capacity to be used for the specific site and structure.

Table 11.4 Some Commonly used Dynamic Formulas

Name	Formula	Recommended Safety Factor
AASHTO	$*P_u = \dfrac{2h(W_r + A_r p)}{s + 0.1}$	6
Canadian National Building Code	$**P_u = \dfrac{e_h E_h}{s + C_1 C_2} \cdot \dfrac{W_r + n^2(0.5 W_p)}{W_r + W_p}$	3
ENR—Modified (Engr. News Record)	$**P_u = \dfrac{1.25 e_h E_h}{s + 0.1} \cdot \dfrac{W_r + n^2 W_p}{W_r + W_p}$	6
Eytelwein	$**P_u = \dfrac{e_h E_h}{s + 0.1(W_p/W_r)}$	6
Navy—McKay	$**P_u = \dfrac{e_h E_h}{s\left(1 + 0.3\dfrac{W_p}{W_r}\right)}$	6

* For double acting steam hammers use A_r = ram cross section. For single acting and gravity rams take $A_r p = 0$.

** Author's recommended values for:

Hammer	e_h	Material	n
Single-acting	0.8	Wood piles	0.25
Double-acting	0.85	Wood cushion on steel	0.32
Drop	0.85	Steel-on-steel anvil	0.5

Symbols

A_r = ram cross section; E = modulus of elasticity of pile; e_h = efficiency of hammer; E_h = rated hammer energy; h = fall of ram; L = length of pile; n = coefficient of restitution; P_u = ultimate capacity of pile; s = distance of pile penetration per hammer blow; W_p = total weight of pile + cap and block + shoe; W_r = weight of ram. $C_1 = \dfrac{3P_u}{2A_p}, \dfrac{\text{kips}}{\text{in.}^2}$; $C_2 = \dfrac{L}{E} + 0.0001, \dfrac{\text{in.}^3}{\text{kips}}$.

The Wave Equation

The shortcomings associated with the pile-driving formulas are manifested in that they do not relate to the soil–pile interaction and the time-related behavior of the pile system. These variables were accounted for by Smith (120) as well as by a number of other investigators via an analogy, termed the *wave equation*, of the pile behavior and a mathematical model. For those interested, Bowles (11) gives a more detailed discussion of the approach.

The analogy consists of assuming the pile system to be comprised of a series of small concentrated masses, as shown in Fig. 11.8. The springs simulate the axial resistance of the pile. Elastic deformation, tamping, and friction constitute the surface restraints. The propagation of the elastic wave through the pile is analogous to that induced by a longitudinal impact on a long rod. Thus, a partial differential

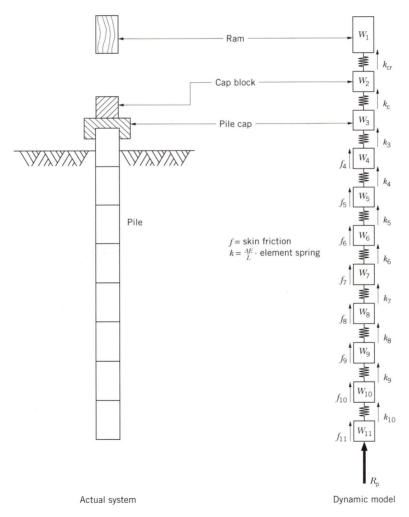

Figure 11.8 Dynamic model used in wave equation analysis.

equation can be written to describe the model. This can be solved numerically, with the aid of digital computers.

The wave equation analysis is a useful tool in determining the pile capacity, in selecting appropriate pile-driving equipment (e.g., hammer weights, caps, lengths), and in establishing guidelines (stresses) in the pile-driving effort. With improved computer capabilities, the method is finding increasing use in the geotechnical engineering field.

Example 11.3

Given

Hammer	*Pile*
DALMAG D-12 (double acting)	HP 10 × 42
$E_h = 16,500$ ft-lb	(wt/ft = 42 lb/ft)
Rate = 40 to 60 bpm	$R = 40$ ft
$W_h = 2750$ lb	$W_p = 40 \times 42 = 1680$ lb
Stroke = 130 in.	

Find   Average P_u when $s = 0.2$ in./blow from the following formulas: Mod. ENR; Eytelwein; Navy-McKay.

Procedure   *Via Mod. ENR Formula*

$$P_u = \frac{1.25\, e_h E_h}{s + 0.1} \cdot \frac{W_r + n^2 W_p}{W_r + W_p}$$

For steel on steel $e_h = 0.85$; $n = 0.5$

$$E_h = 16,500 \text{ ft-lb} = 198,000 \text{ in-lb}$$

Hence,

$$P_u = \frac{1.25(0.85)(198,000)}{0.2 + 0.1} \left[\frac{2750 + (0.5)^2 1680}{2750 + 1680} \right]$$

$$P_u = 501,797 \text{ lb}$$

$$P_a = P_u/6 = 83,636 \text{ lb}$$

Via Eytelwein Formula

$$P_u = \frac{e_h E_h}{s + 0.1(W_p/W_r)}$$

$$P_u = \frac{0.85(198,000)}{0.2 + 0.1\left(\dfrac{1680}{2750}\right)}$$

$$P_u = 644,828 \text{ lb}$$

$$P_a = P_u/6 = 107,471 \text{ lb}$$

Via Navy-McKay Formula

$$P_u = \frac{e_h E_h}{s\left(1 + 0.3\,\dfrac{W_p}{W_r}\right)}$$

$$P_u = \frac{0.85(198{,}000)}{0.2\left[1 + 0.3\left(\dfrac{1680}{2750}\right)\right]} = \frac{168{,}300}{0.237}$$

$$P_u = 710{,}126 \text{ lb}$$

$$P_a = P_u/6 = 118{,}354 \text{ lb}$$

Average

$$P_a = 103{,}154 \text{ lb}$$

It is a widely accepted consensus that virtually all the dynamic pile-driving formulas are but rough approximations of pile-driving resistance. The lack of accuracy of these formulas, however, does not imply the abandonment of their use. Quite to the contrary, they could be useful if properly supplemented by adequate load tests. Indeed, a common procedure is to observe the penetration resistance for a driven pile, then load-test it, and proceed to relate the load capacity to the resistance to penetration by use of one of these formulas.

11.10 PILE LOAD TEST

The most reliable means of determining the load capacity of a pile is to subject it to a static load test. The test procedure consists of applying static load to the pile in increments up to a designated level of load and recording the vertical deflection of the pile. Occasionally, the pile load comes from dead weight (e.g., soil, pig iron) balanced on top of the pile. Most frequently, however, the load is transmitted by means of a hydraulic jack placed between the top of the pile and a beam supported by two or more reaction piles. Figure 11.9 is a schematic view of a typical load test arrangement; Figure 11.10 shows an actual test setup. In this case, four piles are used as reaction piles. Although two reaction piles (one on each end) may provide sufficient reaction resistance, a four-pile reaction setup provides more stability to the physical setup. The vertical deflection of the top of the pile is commonly measured by mechanical gauges attached to a beam, which spans over the test pile and which is totally independent (unattached) of the load setup.

The results from pile tests could be the source of most useful information in at least two general aspects: (1) in determining the ultimate bearing capacity of the pile, and (2) in evaluating the deflection characteristic of the pile. With regard to the ultimate bearing capacity, one may reflect on several additional and related tangible benefits. For example, if the capacity of the pile is different from that desired, the pile length, diameter, and details of installation (e.g., driving criteria, injection pressures, material content) can be adjusted prior to the installation of

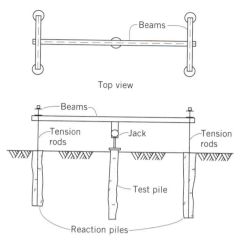

Figure 11.9 Schematic arrangement for pile-load test.

the rest of the piles. Hence, quite apparently, the economic and safety aspects of the design could be more authentically evaluated. Adjustments could likewise be made to account for deflections (settlement), thereby providing a design that would fit within the tolerable deformation requirements. A side, but equally important benefit from the installation and testing of a test pile, prior to the installation of the rest, is general information about the site conditions with regard to potential problems (e.g., boulders, water problems) which may or may not be completely reflected in prior soil subsurface investigations.

While we have reiterated recommendations that pile capacities based on analytical procedures or pile-driving formulas, and so on, should be verified by load tests, one should not assume that load tests are a cure-all, or that they are unquestionably accurate. Indeed, though tests are considered the most reliable manner of determining the pile capacity, the results from such tests may be misleading if the relevant factors are overlooked:

1. A time lapse should be provided between the time of installation and the time of test loading; this may be anywhere from three to four days for a granular stratum to perhaps a month for clayey soils. This is the time normally required for the respective soils to regain the strength lost during the driving operation. In the case of concrete piles a minimum time is also required to develop the material strength.

2. The specific location for the installation and subsequent testing of the pile must be representative of the overall site if the test results are to be representative of the rest of the piles. It is common practice on the part of the engineers to select the most unfavorable conditions of the site (one that is expected to be of the least capacity), which thus results in a conservative installation for the rest of the site.

3. The pile characteristics, such as length, diameter, installation method, must closely resemble those of the piles to be installed later. Obviously, the results of the test pile would be subject to question if this were not the case.

Figure 11.10 (a) Auger-cast pile tested at one of the author's projects (General Extrusion Company, Youngstown, Ohio). The load is applied via a calibrated hydraulic jack, resisted by the steel girder. Four reaction piles provide the downward support to the beam. (b) The vertical movement of the pile (as loaded) is measured by two mechanical dial gauges suspended on two angles placed independently of the girder (i.e., resting on the ground).

The manner in which the load is applied to the pile may be designated by local building codes or by the more common ASTM guidelines (e.g., ASTM D-1143-74). Briefly, most codes provide the following guidelines:

1. The load is to be applied in increments of 20 to 25% of the design loads. These increments are applied either at specified time intervals, or after a specified rate of settlement has been observed, usually less than half a millimeter per hour.

2. The level of load is generally twice the design load (200% of design load). Sometimes the load is increased to failure on the second load cycle, after the pile has successfully resisted the 200% load test during the first cycle.

3. The actual load to be used for design purposes is usually 50% of the load that results in a settlement of a specified magnitude (usually not to exceed 25 mm).

Figure 11.11 illustrates a typical load-settlement relationship. Figure 11.11a shows the loading, in this case in 20% load increments, and the unloading in five steps. Figure 11.11b shows the net settlement for the corresponding loads. For a given point, such as point 1, the net settlement is equal to the gross settlement minus the elastic recovery. Symbolically this represents $(\delta_g - \delta_e)$.

The number of piles tested for any given site is generally left as a judgment factor for the engineer. In turn, the engineer generally will take into account such factors as the overall scope of the project (total cost, foundation cost, number of piles,

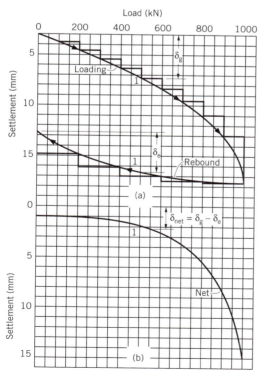

Figure 11.11 Typical load–settlement relationship. (a) Gross settlement. (b) Net settlement.

etc.), the uniformity of the soil characteristics, and perhaps the type of building (height, masonry versus steel, etc.). Inherent in the decision is the general recognition that pile testing is somewhat expensive, and therefore economy is a relevant factor.

11.11 PILE GROUPS

The behavior of a group of piles is different from that of individual piles in a number of ways:

1. In general, the bearing capacity of a group of piles is less than the sum of the individual piles.
2. The settlement of a pile group is larger than that of the individual piles for corresponding levels of load.
3. The efficiency of the pile group is less than that of a single pile.

Yet in spite of these shortcomings, a pile group is a much more common occurrence than a single pile. Single piles lack the overall stability against overturning, a deficiency that is easily overcome by a cluster of piles. Similarly, horizontal thrust in various directions could be more readily resisted by battered piles within the group. Generally, for column supports a minimum of two piles, but more frequently three or more piles, are clustered in a group and connected via a concrete cap to form a unit. For walls, a line of single piles is common.

Load Capacity of Pile Groups

A single pile will transmit its load to the soil in a pressure bulb, as shown in Fig. 11.12a. Depending on the pile spacing, however, the pressure bulbs may overlap. The stresses at points of overlap are obviously larger than those for individual piles at the corresponding elevations. That is, the resulting group pressure bulb is the

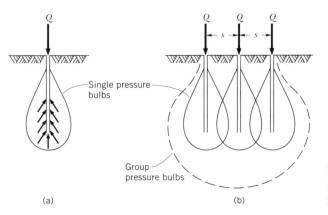

(a)   (b)

Figure 11.12 Pressure bulb for friction pile systems. (a) Single pile. (b) Group pile action.

superposition of the individual effects, as depicted by the group pressure bulb in Fig. 11.12b.

The capacity of the group is less than the sum of the individual pile capacities. Hence tests of individual pile capacities do not provide a totally accurate assessment of the group capacity. By increasing the spacing, thereby minimizing overlapping, the effect of superposition is less, or perhaps it is totally eliminated if spacing is large enough. However, large spacings require a correspondingly large (wide) and thick (and thereby more expensive) concrete cap. Most building codes stipulate minimum spacing. Considerations regarding stratum (e.g., clayey, granular), type of pile (e.g., shape, method of installation), settlement, strength criteria, judgment and experience on the part of the engineer are valuable and frequently weighed factors in the determination of spacing. As a guide, however, a center-to-center spacing of 2 diameters ($2D$) is almost always a minimum requirement, with $3D$ to $3.5D$ being quite common.

The stresses within the soil stratum in which a pile group is located are difficult to estimate for a number of reasons, such as:

1. The distribution of friction stresses along the pile length is rather indeterminate; and so is the superposition of stresses resulting from overlapping of pressure bulbs.

2. Consolidation, fluctuation in the water table, and other time-related variables are not easily assessed.

3. The interaction of a pile cap with the soil is difficult to evaluate with any degree of accuracy.

Although analytical solutions for such stresses were proposed by a number of investigators [e.g., Mindlin (72), Poulos and Davis (89), and others], a widely used empirical method consists of treating the pile group as an "equivalent" footing whose length and width are equal to the length and width of the pile group. The support is provided by the end-bearing of this "footing" and the friction along the sides of this long (deep) footing. This is shown in Fig. 11.13.

The general bearing capacity formula for deep foundations may be used to provide an approximate value for the group capacity. The approach for estimating soil stresses from pile groups is described in the following section.

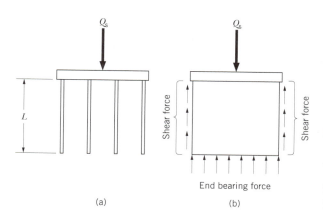

Figure 11.13 Estimate of bearing capacity of pile groups via equivalent deep footing. (a) Pile group. (b) Equivalent deep footing. Q_g = capacity of pile group.

Settlement of Pile Groups

The settlement of pile groups may be viewed as the result of four separate causes:

1. The axial deformation of the pile
2. The deformation of the soil at the pile–soil interface
3. The compressive deformation of the soil between the piles
4. The compressive deformation in the stratum below the tips of the piles

The settlements reflected by items 1 and 2 are relatively small and are, therefore, usually neglected. The settlement described by item 3 is difficult to evaluate with reliable accuracy, and therefore it is also ignored. The settlement reflected by item 4, however, can be estimated by treating the pile group as an equivalent footing (Fig. 11.14) and computing the stresses as described in Section 2.7, for a square, rectangular or a circular shape, as the case may be. When the piles are end-bearing, the base of the equivalent footing is at the pile tip, with the corresponding stress isobars as illustrated in Fig. 11.14a. When the piles are essentially friction type, the base of the equivalent footing is assumed to act at a depth of approximately two-thirds of the pile length, as shown in Fig. 11.14b. The settlement in both instances, however, is computed only for any compressible stratum below the pile tips.

Pile Group Efficiency

The efficiency of a pile group is the ratio of the capacity of the group to the sum of the capacities of the individual piles. Symbolically, this may be expressed by

$$E_g = \frac{Q_g}{nQ_i} \times 100\%$$

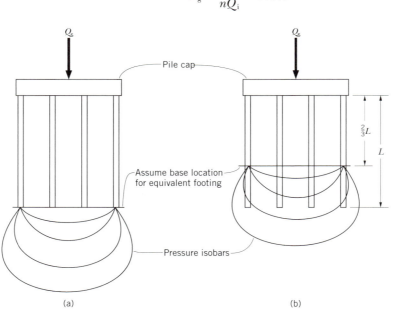

Figure 11.14 Method for determining pressures under pile groups. (a) Group of end-bearing piles in dense sands or sand–gravel deposit. (b) Group of friction piles.

where E_g = efficiency of pile group
Q_g = capacity of group
Q_i = capacity of individual pile
n = number of piles in group

If we assumed the pile group to be an equivalent footing, as shown in Fig. 11.14, the capacity of a pile group of n piles could be estimated as follows: Let Fig. 11.15 represent an arbitrary pile arrangement with a rows and b columns and an equal spacing s. The capacity of the group Q_g is equal to (group perimeter) $\times$ (pile length) $\times$ (friction factor). The capacity of a single pile Q_i is equal to (single-pile perimeter) $\times$ (pile length) $\times$ (friction factor). Symbolically, therefore, the efficiency expression could be written as:

$$E_g = \frac{Q_g}{nQ_i} = \frac{[2s(a+b-2) + 4D]Lf_s}{n(\pi D)Lf}$$

$$E_g = \frac{2s(a+b-2) + 4D}{\pi abD} \times 100 \qquad (11\text{-}11)$$

Equation 11-11 indicates that as the spacing is increased, the efficiency of the group increases. As mentioned previously, however, practical considerations (e.g., footing size, costs) generally limit the spacing to approximately $3D$.

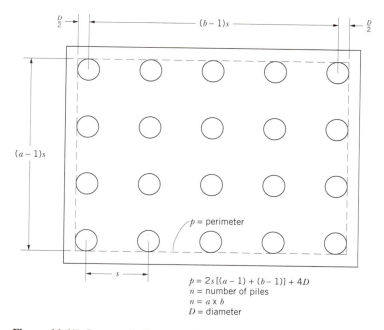

$p = 2s[(a-1) + (b-1)] + 4D$
n = number of piles
$n = a \times b$
D = diameter

Figure 11.15 Group of piles, equal spacing s.

Example 11.4

Given   A pile group with characteristics as shown above in Figs. 11.16a and b.

Find   (a) σ_z at middepth of the clay layer, via an equivalent footing approach depicted in Fig. 11.14b.

        (b) Efficiency of the pile group.

Procedure   (a) The depth at the midsection of the clay layer is 8.5 m below the bottom of the hypothetical footing, as shown in Fig. 11.17c. An estimate of σ_z can be made using Boussinesq's values (Section 2.7) or by a 30° (or 2:1) "slope" method. Via Boussinesq, Table 2.1 for $z = 8.5$ (see Fig. 11.17c) and $m = 1.65/8.5$ and $n = 1.15/8.5$, we have:

$$m = \frac{1.65}{8.5} = 0.195$$

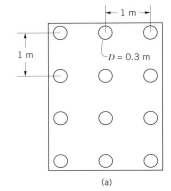

(a)

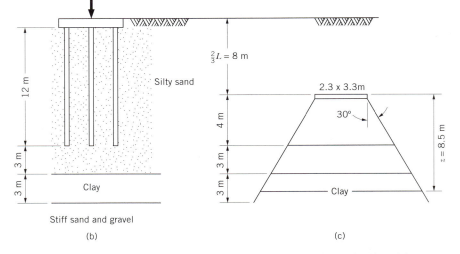

(b)   (c)

Figure 11.16 (a) Top view of pile group. (b) Soil profile and piles. (c) σ_z at middepth of clay.

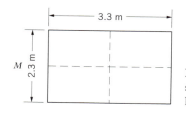

Figure 11.17 Estimate of pressure σ_z at mid-depth of clay layer, via Boussinesq.

$$n = \frac{1.15}{8.5} = 0.135$$

$$f(m, n) = 0.0122$$

$$q = \frac{30,000 \text{ kN}}{2.3 \times 3.3} = 3952 \text{ kN/m}^2$$

$$\sigma_z = 3952(4)(0.0122)$$

Answer

$$\sigma_z = 193 \text{ kN/m}^2 \ (4 \text{ kips/ft}^2)$$

Based on the 30° slope,

$$q_z = \frac{30,000}{(2.3 + 1.16 \times 8.5)(3.3 + 1.16 \times 8.5)}$$

Answer

$$q_z = 188 \text{ kN/m}^2$$

This compares well with the 193 kN/m² via Boussinesq.
(b) The efficiency of a pile group is

$$E_g = \frac{2s(a + b - 2) + 4D}{\pi abD} \times 100$$

or

$$E_g = \frac{2(1)(3 + 4 - 2) + 1.2}{\pi(12)(0.3)} \times 100$$

Answer

$$E_g = 94\%$$

11.12 DRILLED PIERS

The name *drilled piers* is used to describe a cast-in-place pile constructed by drilling a hole, generally 750 mm or larger in diameter, to a designated depth, then filling it with concrete. The pier shaft may be straight (uniform diameter) or may be enlarged ("belled") at the bottom. It may be reinforced or unreinforced. A drilled pier is also sometimes referred to as a drilled shaft or drilled caisson. Figure 11.18 depicts several types of drilled piers. Figure 11.19 depicts the installation of a bell-bottom pier.

The choice of drilled piers over piles is generally based on the relative cost for the two selections, but practical considerations or limitations may also play a pivotal role in the selections. The following are a few points that may play a key role in selecting drilled piers over piles:

1. Fewer piers are needed than piles. Since the load capacity of a drilled pier is typically much larger than that of a pile, fewer piers are needed for a given building.

2. Pile caps needed for pile groups are not needed for drilled piers. Because of their size, piers generally accommodate the column connections, via dowels, anchors, and bearing plates directly to the pier, thereby eliminating the need for the pile cap.

3. One can inspect the bottom of the drilled hole. For the smaller-diameter holes, it is common to make visual inspections from the surface (e.g., via flashlights, mirrors, shavings extracted from the bottom by clean-out buckets); for larger holes it is sometimes desirable to lower a person into the hole for more accurate observations of the bottom.

4. Drilled piers can be designed to resist larger lateral loads and bending moments.

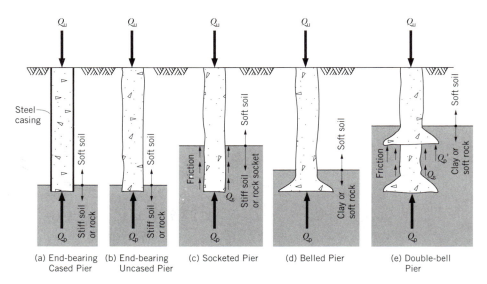

Figure 11.18 Types of drilled piers.

Figure 11.19 (a) Texoma Model 900 on crawlers drilling 30-in.-diameter, 30-ft-deep, belled-bottom piers. (b) Completed drilled shafts. The bell, at the bottom of the shaft, is used to enlarge the bearing area at the pier tip, thereby increasing the pier's capacity for a given allowable bearing pressure (see Fig. 11.27b). (Courtesy Reedrill Inc.)

5. The construction of drilled piers is virtually vibration-free with minimal noise. Driven piles generally create substantial and potentially damaging vibrations to adjoining buildings, and are noisy.

6. Drilled piers are generally more conducive to resisting upward forces. Particularly if "belled," drilled piers could be designed to resist substantial upward pull.

7. Drilled piers could be constructed in relatively difficult strata. Formations that have substantial numbers of boulders, dense sands, or seams of rock (even including some shales) may make it rather difficult for pile installation. Such formations prove to be relatively small obstacles for drilled piers. Also, if one were to penetrate into sound rock formations, such procedures are totally possible with the drilled piers but virtually impossible with piles.

Of course, soil borings are virtually a prerequisite for a reasonable assessment for the selection of drilled piers (versus piles) as well as for the bearing capacity. Indeed, it is frequently specified that the bottom of caissons bearing on rock be cored, usually for 5 ft, in order to assess the rock characteristics (strength, degree of fracture, clay seams, rock quality, etc.). Depending on the importance of the project, the number of such corings may range anywhere from a few to the total number of piers. Also, it is common practice that such assessment be made in the context of available data and in conjunction with the structural engineer or architect.

11.13 THE CONSTRUCTION OF DRILLED PIERS

Much of the early drilled pier construction was done by digging the shaft by hand, with the sides of the excavated hole being lined by boards (laggings), which were braced to provide the lateral support against shaft collapse. Current methods are almost universally done by power drilling, particularly in this country. Figure 11.20 shows two drilling rigs, one to drill a vertical shaft, the other battered holes for a tie-back installation. Figure 11.21 shows typical drill heads, sometimes referred to as cutting bits, for drilling into soil or soft rock; core barrels, shown in Fig. 11.27a, are common tools used to drill into rock. When needed, drill-mud or steel casing (pipe) is inserted into the hole to guard against wall collapse; sometimes sheet piles are substituted for a pipe-type casing for particularly large piers.

Current construction techniques normally fall in one of the three predominant methods of construction: (1) the **dry method**, (2) the **casing method**, and (3) the **slurry method**. As part of the twelve Terzaghi lectures, professor Limon C. Reese presented a schematic description of the basic steps used in constructing drilled shafts via the three methods (107). Figures 11.22 through 11.24 depict the key steps in the construction of drilled shafts, as presented by Professor Reese.

The Dry Method

The dry method of construction can be used in soils that will not cave in during excavation, usually cohesive soils above the water table. There are three stages of construction: (1) The shaft is excavated to the desired depth, (2) the reinforcing

Figure 11.20 (a) Texoma Model Taurus X L truck mounted rig drilling 8-ft-diameter shafts for wind towers in Mohave, California. (b) Texoma Model 800 on crawlers drilling ''tie-back'' holes on wall of foundation excavation. (Courtesy of Reedrill Inc.)

Figure 11.21 Drill heads commonly used to auger through soil or relatively soft rock for drilled pier installation.

steel cage is lowered into the excavation, and (3) the hole is filled with concrete. Figure 11.22 depicts the schematic of the three stages. It shows a straight (uniform diameter) shaft, but the hole may be belled or underreamed.

Concrete may be placed by free fall providing the concrete does not strike the sides of the hole or those of the steel cage. This requires that the concrete be directed vertically through a hopper, which is centered over the hole, or it may be placed with a tremie or pumped into the hole. Concrete that is pumped must meet mix design criteria for adequate pumpability as well as for the drilled shaft construction. It is important that the bottom of the hole be reasonably clean before the concrete is placed. Usually loose material can be removed with a clean-out or muck-bucket.

Belling buckets or underreamers come in a variety of sizes and shapes. The cutting blades are retracted for inserting the bucket into the previously drilled hole. The standard bell configurations are 45° or 60°, measured with the horizontal; hemispherical underreamers are also sometimes used. Rebar cages are fabricated in the field; typically, they consist of longitudinal bars with spiral ties. The cage is

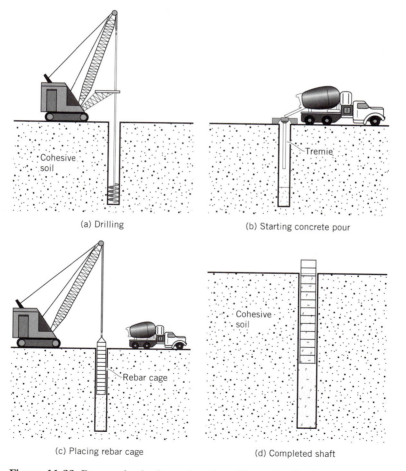

(a) Drilling

(b) Starting concrete pour

(c) Placing rebar cage

(d) Completed shaft

Figure 11.22 Dry method of construction. [From Dr. L. C. Reese (107).]

generally lowered and placed into the hole without the use of guides, but occasionally guides are used to center the cage. The diameter of the cage should be such to leave ample space for free flow of concrete between the walls of the drilled hole and the cage.

Most drilling through soil or relatively soft rock is done using open helix augers. Augers come in a variety of shapes and sizes and have either a knife-blade cutting edge for drilling through soil or hard-surface teeth for stiff soils or soft rock.

The Casing Method

The casing method is used for sites with caving soils, such as for strata consisting of cohesionless soils below the water table. The steel casing is installed and pushed into a stratum of impermeable soil in order to make a seal. The outside of the casing is slightly smaller than the inside of the hole. The casing method cannot be used if the seal is impossible to obtain or if there is no impermeable formation into which the lower portion of the hole can be drilled.

When the caving soil is encountered, a mud or slurry is introduced in the hole to maintain the excavation until a relatively impermeable soil is encountered. Hence, the excavation proceeds until an impermeable formation is encountered, as shown in Fig. 11.23. A kelly is typically used to provide the rotational torque and

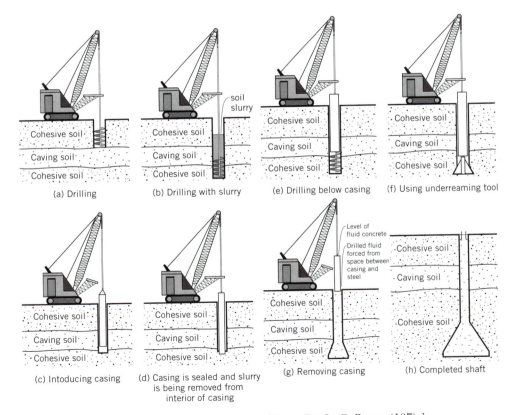

Figure 11.23 Casing method of construction. [From Dr. L. C. Reese (107).]

vertical force needed to push the casing into the impermeable strata and develop the seal.

Once the casing is sealed and the slurry is removed, drilling can continue. Once the drilling is complete, the cage is installed and concrete can be placed by a tremie. The casing should not be removed until a sufficient head of concrete is present to force the slurry trapped behind the casing out of the hole. A high slump concrete is required to prevent friction between the casing and the concrete from lifting the concrete.

A drilling bucket can be used to dig into some soil formation as well as to remove drilling mud or slurry from the bottom of the hole. The cuttings are forced into the barrel by way of the rotating motion of the barrel. When the barrel is full, the bucket is removed from the hole and the contents dumped by opening the bottom door. Generally, drill cuttings are removed from drilling augers by merely spinning the auger at the surface.

When a drilled hole becomes tight, the drill rig is used to twist the casing into the ground. A special tool is sometimes used, called the "twister," which is attached to the kelly. The arms of the twister fit into slots cut into the top of the casing and the casing is screwed down into the ground.

Concrete should be placed as soon as possible after the hole is completed, and should have good flow characteristics. It is imporant that the placement of the concrete results in complete filling of the shaft and without segregation of the concrete particles. Also, it is important that no concrete be dragged upward from friction between the casing and the concrete when the casing is pulled up. This can be generally avoided with concrete with sufficient slump, say 6 in. or 8 in. The casing should be pulled out vertically and as smoothly as possible to break the seal in the cohesive soil; a short jerk by the crane is usually sufficient.

The Slurry Method

When drilling, if soil will tend to squeeze, cave, or slough into the hole, a mud slurry can be used to keep the hole open. When the caving material is encountered, slurry is introduced into the hole. In some cases, the material being drilled is on-site clay of suitable characteristics for mud slurry by merely adding water and mixing. Usually bentonite, a member of the montmorillonite family of clays, is added to form a suitable slurry. The slurry not only provides stability of the excavation, but may prevent suspended particles from settling out, may allow clean displacement by concrete, and may allow for easy pumping.

When the excavation is complete, the slurry at the bottom of the hole may have to be removed using a muck-bucket (see Fig. 11.25 and Fig. 11.27) if it contains too much granular material that has settled out of the slurry. Next, the rebar cage is placed through the slurry. Finally, the concrete is placed via a tremie, displacing the lower-density slurry upward. Figure 11.24 depicts the general procedure for this method.

Engineering properties of a mud slurry that determine its performance are density, viscosity, shear strength, and pH. The control of slurry pH is important in some cases since groundwater or soil may react with a drilling mud and alter its viscosity. For large slurry jobs, a slurry processing plant is used to mix slurry, remove sand and silt, and recirculate the slurry with the pumps. Figure 11.25 shows some of the tools used in this undertaking.

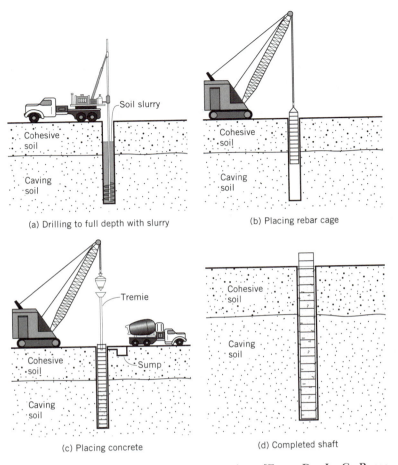

(a) Drilling to full depth with slurry

(b) Placing rebar cage

(c) Placing concrete

(d) Completed shaft

Figure 11.24 Slurry method of construction. [From Dr. L. C. Reese (107).]

The slurry is removed as the concrete is placed via a tremie, as shown in Fig. 11.26. The tremie is always submerged in the concrete so that the least amount of concrete surface area is contaminated with this mud from the shaft drilling, and the lighter-weight slurry is pushed up. The slurry is then typically pumped into tanks or nearby sumps for a use or ultimate disposal once a project is complete. Figures 11.27 and 11.28 depict some of the tools that are used for specific drilling needs and rock formations.

11.14 DESIGN CONSIDERATIONS FOR DRILLED PIERS IN COMPRESSION

As for piles, the load capacity for the drilled piers is derived from the end-bearing as well as friction. Symbolically, the ultimate capacity of the pier is expressed by Eq. 11-1 reproduced here:

$$Q_u = Q_p + Q_s \tag{11-1}$$

Figure 11.25 Drilling buckets can be used to dig as well as remove drilling mud (muck-bucket). The cuttings are forced into the barrel. When the barrel is full, the bucket is lifted from the hole and cuttings are dumped by opening the bottom door. (a) Bucket being lifted out of the hole; (b) the drilling bucket's bottom; (c) the bottom door open. (Courtesy Association of Drilled Shaft Contractors.)

Figure 11.26 Concrete being poured into the tremie, which funnels the concrete into the drilled shaft through the steel cage shown. (Courtesy Association of Drilled Shaft Contractors.)

Figure 11.27 (a) On the left is a "cleaned-out bucket" used to remove loose rock and soil from bottom of shafts. On the right is a core barrel used to cut a cylindrical rock core, which is subsequently extracted to develop a pier shaft. (b) A belling tool used to enlarge the bottom of the shaft, thereby increasing the bearing area and pier capacity. The tool is suitable for reasonably cohesive formations and soft rock formations (the belling shape collapses in granular formations). (c) Casing (steel tube) is used to maintain the shaft from collapsing in granular formations prior to pouring of concrete. Once concrete is poured, the casing is usually removed and reused.

This is the same expression as used for piles. As for piles, the determination for the quantities Q_u and Q_s are somewhat difficult to establish, perhaps even more so for drilled piers. An important aspect in this regard is the scrutiny and techniques employed in constructing drilled piers.

It is important that a thorough subsurface investigation be conducted in order to obtain representative design parameters and in order that the appropriate construction procedure be chosen. For example, the frictional resistance may be significantly altered if the slurry is not sufficiently removed, or when a granular formation around the bottom of the shaft is loosened by potential suction if the slurry is displaced via a jerky motion of the augering tool.

It is important that, wherever practical, load tests be conducted for confirmation of the pier's capacity. The design parameters and assumptions may be off; similarly, soil properties may also differ. Judgment and experience are always desirable factors in both the selection of construction techniques and design of drilled piers.

The design of drilled piers is still in the developmental stage. Reese, Touma, and O'Neill (106) presented the results of a series of tests they conducted on instru-

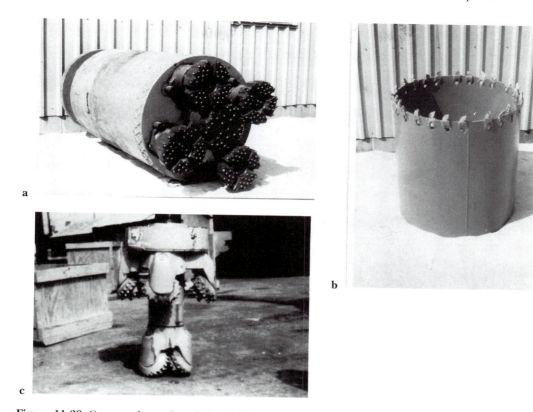

Figure 11.28 Commonly used tools for drilling in rock. (a) and (b) hardened tool heads chew up the rock; (c) core-barrel cuts a cylindrical core from the rock. (Courtesy of International Association of Foundation Drilling.)

mented piers. Figure 11-29 shows some plotted test data from one of their instrumented drilled pier tests. In Fig. 11.29a one notes that a significant amount of the total load is assumed by the sides (frictional resistance) and only a rather insignificant percentage by the bottom, for the corresponding settlement. Also, one notes that a downward movement of less than $\frac{1}{4}$ in. was needed to develop the full load transfer along the sides, while a downward movement of more than 1 in. is required to develop the maximum base resistance.

It is interesting to note that more than half (63%) of the total load was derived from side resistance. Indeed, all 11 tests of drilled piers reported by Reese, Touma, and O'Neill carried a significant portion of the applied load in side resistance, with 9 of the 11 piers exceeding 50%, and 7 of the 11 exceeding 60%. Hence, since only a small downward movement is required to develop the ultimate side resistance (e.g., $\frac{1}{4}$ in.) and since the large percentage of the total load is carried by the sides even at failure, the observation is made by Reese et al. that many drilled piers at working loads will transfer almost all the applied load to the soil in skin friction.

Reese et al. suggested an interim criterion for the design of drilled piers in both clay and sand. They are as follows:

Piers in Clay   The proposed ultimate axial load capacity of a drilled pier in clay may be computed by Eq. 11-1:

$$Q_u = Q_p + Q_s \qquad (11\text{-}1)$$

where Q_u = the ultimate axial load capacity of the pier
   Q_p = the ultimate capacity at the base of the pier
   Q_s = the ultimate capacity of the sides

Their recommendations for Q_p and Q_s in clay is as follows:

$$Q_p = N_c c A_p \qquad (a)$$
$$Q_s = \alpha_{avg} s_u A_s \qquad (b)$$

where A_s = the peripheral area of the stem
   A_p = the area of the base
   c = average undrained cohesion of the soil for a depth of two base diameters beneath the base; the shear strength may be substituted for cohesion for soils having undrained angle of internal friction of 10° or less
   α_{avg} = the ratio of the peak mobilized shear strength to that of the soil, averaged over the peripheral area of the sides

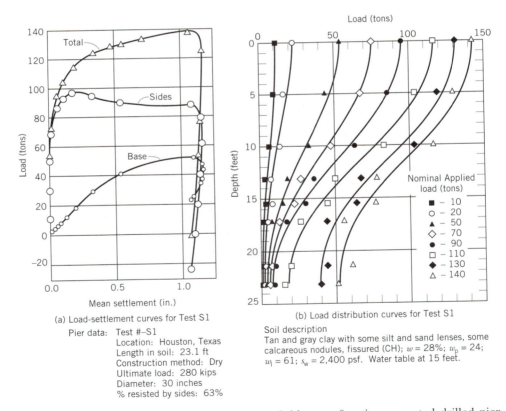

(a) Load-settlement curves for Test S1

Pier data: Test #–S1
   Location: Houston, Texas
   Length in soil: 23.1 ft
   Construction method: Dry
   Ultimate load: 280 kips
   Diameter: 30 inches
   % resisted by sides: 63%

(b) Load distribution curves for Test S1

Soil description
Tan and gray clay with some silt and sand lenses, some calcareous nodules, fissured (CH); $w = 28\%$; $w_p = 24$; $w_l = 61$; $s_w = 2{,}400$ psf. Water table at 15 feet.

Figure 11.29 Some plotted test data, based on field tests of an instrumented drilled pier reported by Reese, Touma, and O'Neill (106).

Figure 11.30 (a) Pier reinforcement, commonly referred to as a cage, is preassembled on the ground and then lowered into the shaft, followed by the concrete, etc. The concrete truck (transit mixer) is waiting to discharge the concrete into the shaft. (b) Some of the first-level concrete columns show the column reinforcement exposed to subsequent lapping with reinforcing bars from the next level. (Parking deck for which the author was the geotechnical consultant.)

α_{avg} = 0.5 for piers installed dry or using lightweight drilling slurry and displacing the slurry with fluid concrete

  = 0.3 for piers installed with some drilling mud that may be entrapped between sides of pier and soil

  = 0.3 for belled piers installed dry in soil about the same stiffness below base as on sides

  = 0.15 for belled piers installed with drilling mud with some possible mud trapped between pier and soil

  = 0 for straight-sided or belled piers with base resting on significantly stiffer soil than soil around the stem

s_u = undrained shear strength of the soil along the sides

N_c = the bearing capacity factor = 9

Hence, based on Reese et al.'s work, the ultimate bearing capacity could be determined by Eq. 11-12:

$$Q_u = N_c c A_p + \alpha_{\text{avg}} s_u A_s \qquad (11\text{-}12)$$

One notes from Fig. 11.29 that the load transfer near the ground surface and near the bottom of the shaft is significantly less than for the interim length; notice the rather vertical orientation of the curves near the bottom and near the top of the pier. Indeed, Reese et al. indicated that a visible gap was observed between the soil and concrete near the surface. Correspondingly, Reese et al. recommended that the top 5 ft and bottom 5 ft of the straight pier be excluded from considerations when calculating the axial load capacity of a drilled pier in clay. Similarly, they recommend that the bottom 5 ft of the stem of a pier above the bell should likewise be excluded: i.e.

- exclude the top 5 ft from consideration in equations
- exclude the periphery of a bell from consideration
- exclude the bottom 5 ft of a straight pier
- exclude the 5 ft of the stem above the bell

Reese et al. also note that the above-mentioned criteria do not encompass provisions for dealing with problems of soils that will change in volume with time or with changes in moisture (e.g., clay that may produce downdrag or uplift, expansive clays). In the final analysis, Reese et al. proposed that the design procedures be used with caution and strongly recommend that load tests be conducted whenever possible to verify the design.

Drilled Piers in Sand   The ultimate axial load capacity of a drilled pier in sand likewise may be computed by

$$Q_u = Q_p + Q_s$$

Based on Reese et al., the base resistance Q_p and side resistance Q_s may be computed as

$$Q_p = \frac{A}{K} q_t \qquad (a)$$

$$Q_s = \alpha_{\text{avg}} C \int_0^H \overline{p}_z \tan \overline{\phi}_z \, dz \qquad (b)$$

$$Q_u = \frac{A}{K} q_i + \alpha_{avg} C \int_0^H \bar{p}_z \tan \bar{\phi}_z \, dz \qquad (11\text{-}13)$$

where
A_b = cross-section of base (ft^2)
K = reduction factor = $0.6B$ (B = pier diameter)
C = circumference of pier (ft)
H = total depth of embedment (ft)
$\bar{p}_z$ = effective overburden pressure (psf) at middepth
$\bar{\phi}_z$ = effective friction angle
q_t = base resistance to downward movement of 5% of diameter
  = 0 for loose sand (relative density designation)
  = 32,000 psf for medium density (relative density designation)
  = 80,000 psf for dense formations (relative density designation)
α_{avg} = a factor that allows correlation with experimental results
α_{avg} = 0.7 for $H < 25$ ft
  = 0.6 for H = 25 ft to 40 ft
  = 0.5 for $H > 40$ ft
dz = differential element of length (ft)

Of course, the allowable (design value) is obtained by dividing the ultimate pier capacity by an appropriate safety factor. In view of the amount of relative downward movement needed to develop side and base resistance, for clay, Reese et al. proposed

$$Q_{design} = Q_s + \frac{Q_p}{3} \qquad (11\text{-}14)$$

In a case of drilled piers in sand, a factor of safety of 2 is normally recommended for computing the working load.

Example 11.5

Given

Soil Description	*Avg. Index Props.*
0–6 ft fine sand, some gravel	w_{avg} = 15%; ϕ = 33°
6–30 ft silty clay	w = 19%; LL = 43; PL = 21; s_u = 3.1 ksf
30–87 ft stiff clay	w = 22%; LL = 55; PL = 24; s_u = 4.5 ksf
water table at 3 ft	

Find A suitable drilled pier design for an axial design load Q_d = 580 kips, assuming a S.F. = 2 for side shear

Solution Assume 4000 psi concrete; $f_c = 0.25 f_c' = 1000$ psi
Thus,

$$\frac{\pi}{4} (D)^2 f_c = Q_d$$

Hence,

$$D = \sqrt{\frac{580,000}{\frac{\pi}{4} (1000)}} = 27.18 \text{ in.}$$

Assume $D = 2.5$ ft $= 30$ in.; $C = \pi D = 7.85$ ft. Also, assume bearing will be in stiff clay, and neglect frictional support from 6 ft of sand. Hence, if $L = 48$ ft, effective $L' = 48$ ft $- 6$ ft $= 42$ ft. Use a casing for about 8 ft, then proceed with a dry installation. Hence $\alpha_{avg} = 0.5$. Accounting for the two different values for s_u, we have

$$Q_s = \alpha_{avg} s_u A_s = (0.5)(7.85)[(3100 \times 24 \text{ ft}) + (4500 \times 18 \text{ ft})]$$

or

$$Q_s = 609{,}945 \text{ lb}$$

For a S.F. $= 2$ for "side" shear, the allowable Q_{as} is

$$Q_{as} = 304{,}972 \text{ lb}$$

For maximum end-bearing development, $\delta \approx 1$ in. (Reese et al.); $Q_{ap} = N_c c A_p$; $N_c \approx 9$.
The average $c \approx [(3100 \times 24 \text{ ft}) + (4500 \times 18 \text{ ft})]/42 = 3700$ psf. Then,

$$\frac{9(3700)(4.91)}{3} \overset{?}{=} 580{,}000 - 304{,}972$$

or

$$54{,}501 \text{ lb} < 275{,}028 \text{ lb} \qquad \text{N.G.}$$

At this point we may increase L, or go with a bell. Let's make this decision after we see how much more length we need for a straight-shaft option. The addition length, L'', needed is

$$L'' = \frac{275{,}028 - 54{,}501}{(\alpha)(A_s)(s_u)} = \frac{220{,}527}{(0.5)(7.85)(4500)}$$

$$L'' = 12.5 \text{ ft}; \qquad \text{for S.F.} = 2, \ L'' = 25 \text{ ft}$$

Hence, the total length of the shaft is

$$L_{tot} = 48 \text{ ft} + 25 \text{ ft} + 5 \text{ ft neglected above base} = 78 \text{ ft}$$

The upper 6 ft of pier friction was already disregarded. From author's experience, it is more economical to drill a straight shaft to this depth than to bell into the stiff clay, even at 32 ft $\pm$. (The "change in tooling" is more costly).
Now to check for total settlement, δ_T.

$$\delta_T = \delta_s + \delta_{1 \text{ in.}}$$

where $\delta_{1 \text{ in.}}$ = amount of travel of base required to develop full point resistance (Reese et al.)

δ_s = axial shortening = $\dfrac{(P_{avg})(L_T)}{AE}$

From Code

$$E_c = 57{,}000 \sqrt{f_c'} = 57{,}000 \sqrt{4000} = 3.6 \times 10^6 \text{ psi}$$
$$L = 78 \text{ ft} = 936 \text{ in.}$$

$$A_a = \frac{\pi}{4}(30)^2 = 706.5 \text{ in.}^2$$

The compressive force in the concrete pier varies from 580 kips at the top to 54.5 kips at the bottom. Hence, assuming an average axial force of $\left(\dfrac{580 + 54.5}{2}\right) = 317.25$ kips, the axial shortening δ_s is approximately

$$\delta_s = \frac{(317{,}250)\,(936)}{(706.5)\,(3.6 \times 10^6)} = 0.12 \text{ in.}$$

and

$$\delta_{tot} = 0.12 + 1 = 1.12 \text{ in.}$$

Assume 1% steel rebars;

$$A_s = (0.01)\,\frac{\pi}{4}\,(30)^2 = 7.06 \text{ in.}^2 \;(9 \text{ \#8}, \, A_s = 7.06 \text{ in.}^2)$$

Hence, the following is the summary:
$D = 30$ in.; $L = 78$ ft; $f'_c = 4000$ psi
$A_s = 9$ #8 bars; $\delta = 1.135$ in.; dry method with casing for upper 8 ft. Check shaft diameter (hole) for 30 in. dimension prior to pouring concrete.

11.15 DRILLED PIERS SOCKETED IN ROCK

When faced with a choice of bearing on the rock's surface or extending the base of the pier down to some point into the rock, the foundation engineer is likely to select the latter for the following reasons:

- The rock quality and bearing capacity usually improve with depth,
- The pier capacity is increased from the shaft friction,
- Higher pier capacity means fewer piers and thus a likely cost savings (i.e., the savings from fewer piers typically more than offsets the cost for additional drilling into rock).

A rock-socketed pier derives its support from three sources: (1) base (end) bearing, (2) shaft friction, and (3) friction from the soil above the rock. These are depicted in Fig. 11.31. Symbolically, the ultimate capacity of the pier is expressed by Eq. 11-15:

$$F = F_B + F_\tau + F_s \tag{11-15}$$

where F = ultimate load capacity of pier
F_B = load generated by the base
F_τ = load mobilized by socket friction
F_s = load provided by soil friction

The problem of designing rock-socketed piers has been around for a long time, but it did not receive significant attention until about the early 1970s when a number of investigators (23, 38, 39, 41, 51, 57, 101, 109, 110, 140, 141) began to focus on it. Indeed, their efforts enhanced greatly our understanding of the behavior of rock-socketed piers; some excerpts of their work are presented below. However, at this point, we still lack much information on pier–soil–rock interaction. Most

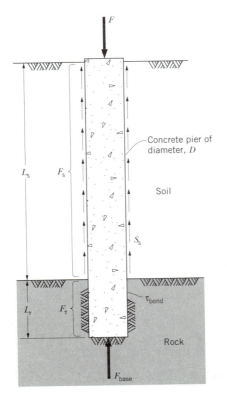

Figure 11.31 Forces acting on a rock-socketed pier.

of the available information is related to the behavior of rock-socketed piers during short-time testing or during construction; it is very scant on long-range behavior of such piers. Correspondingly, we are lacking time-tested design parameters and, therefore, lack the means to ascertain conclusively the behavior of rock-socketed piers. Hence, our current-day design approaches are empirical; the design should be validated by load tests when one is able to justify the cost.

Over time, engineers designed drilled piers under various assumption that:

- all the pier capacity is generated by the wall friction (end-bearing is ignored)
- ignore the socket friction contribution and assume that all the resistance is derived from end-bearing
- designate a specified amount at load to be carried as end-bearing and the remainder by socket friction
- designate the load to be carried by socket friction, and the remainder by end-bearing
- estimate a balance between end-bearing and bond, acting in concert with each other

Ample evidence exists that a substantial amount of resistance is developed by the socket friction, but also that some load is carried by the base (see Figs. 11.31, 11.32, 11.33 and Table 11.5). Furthermore, with time, there is a load transfer from the socket walls to its base. Ladanyi (57) depicts this phenomenon via Fig. 11.32

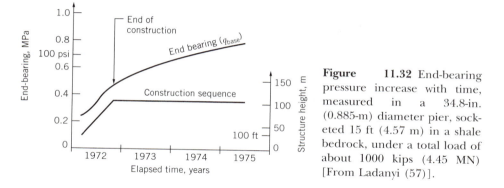

Figure 11.32 End-bearing pressure increase with time, measured in a 34.8-in. (0.885-m) diameter pier, socketed 15 ft (4.57 m) in a shale bedrock, under a total load of about 1000 kips (4.45 MN) [From Ladanyi (57)].

from one instrumented drilled pier where measurements were made for nearly four years. Figure 11.33a shows a similar load transfer phenomenon from the soil formation to the rock socket for a 26-month time span. A like trend is depicted by Table 11.5 for a construction-time span, as reported by Radhakrishnan and Leung (101).

Glos and Briggs (39) also found that end-bearing support provides a very small percentage of the total load. For the two instrumented piers (Fig. 11.34a) socketed in soft rock, the average end-bearing resistance appears to be less than 10% of the total load, as depicted by Figure 11.34c. In Fig. 11.34b one notes that a large percentage of the socket friction is mobilized in the upper part of the socket, dissipating rapidly to a negligible amount past the 4 ft to 5 ft depth (approximately L/D of 2 to 3)—a phenomenon that also was noted by others (51, 57, 101).

It is a widely held consensus that for design of rock-socketed piers the resistance provided by the soil overburden is small (and therefore it could be neglected) and that the base resistance, although relatively small in comparison with skin resistance, should be accounted for in the design. It is also widely recognized that dominant design parameters include the rock strength, the inner-face roughness, as well as the rock-mass modulus and the concrete modulus. It is further widely recognized that the unconfined compressive strengths of rock, as represented by lab tests, are not representative of in-situ strengths; generally, the in-situ strengths are perhaps

Table 11.5 Relative Load Transfer in Overburden and Rock between Load Testing and In-service Loading Condition for Pile 67*

Stratum (1)	*Five-Story* Test Loads (%) (2)	Service Loads (%) (3)	*10-Story* Test Loads (%) (4)	Service Loads (%) (5)	*Full Dead Loads* Test Loads (%) (6)	Service Loads (%) (7)
Soil (0–3.8 m)	53	29	42	27	42	24
Rock (3.8–6.8 m)	41	62	50	62	50	62
End-bearing	6	9	8	11	8	14

* Reported by Radhakrishnan and Leung (101).

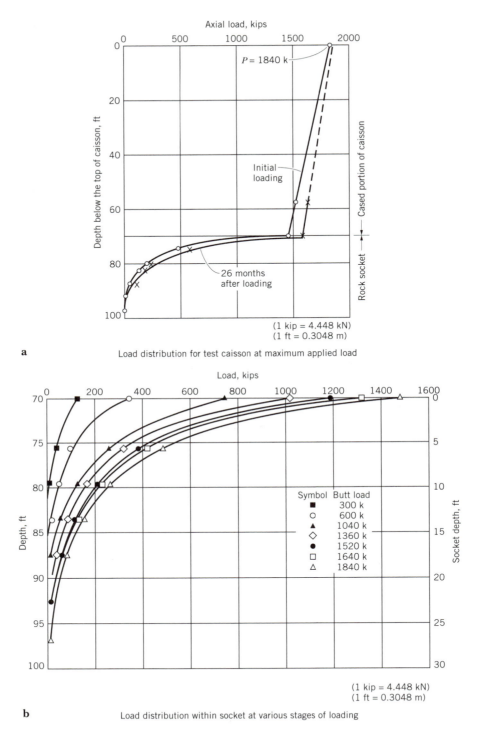

(1 kip = 4.448 kN)
(1 ft = 0.3048 m)

a Load distribution for test caisson at maximum applied load

(1 kip = 4.448 kN)
(1 ft = 0.3048 m)

b Load distribution within socket at various stages of loading

Figure 11.33 Load transfer phenomena for a 24-in.-diameter rock-socketed pier in sound mica schist, as reported by Koutsoftas (51). (a) A transfer from an initial skin resistance along the casing (above the socket) of about 350 kips (i.e., 1840 − 1490) to about 265 kips after 26 months; the transfer of load in the socket resulted in an increase on socket resistance from 1490 kips initially to about 1575 kips. (b) One notes that the large portion of the load is resisted by the upper portion of the socket, with the load rapidly dissipating with depth; the base load resistance is small for the entire 26 months recorded.

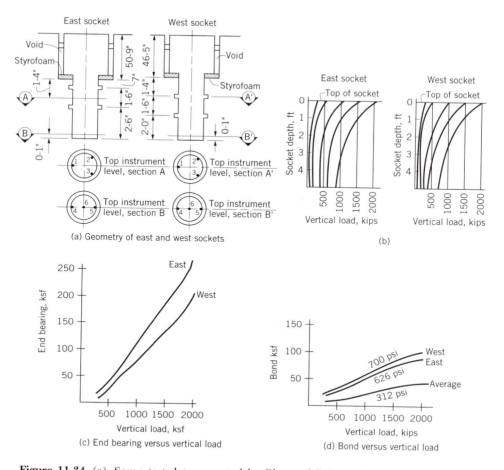

Figure 11.34 (a) Some test data reported by Glos and Briggs (39) on piers socketed in soft rock. (b) Measured decreases in load from side friction vs. socket depth. (c) Measured end-bearing forces for the two piers. (d) The top curves represent the bond mobilized above the top set of gauges; the lower curve represents the average bond over the entire socket length.

4 to 10 time larger than those depicted by lab tests. Furthermore, noted by a number of investigators (38, 87, 101, 109) is that the shear resistance in the socket behaves in a plastic manner, and that no drop-off in shear resistance occurs. See Figure 11.35.

We know that not all rocks are alike and, therefore, not all design parameters can be uniform. Indeed, joints, fissures, soft seams, voids, open joints, soil-filled seams, and so on only add to the complexity of the problem. Also, bentonite, mud, water and other substances may inhibit good bonding, as might the type and speed of drilling equipment. That is, sockets may be constructed by hand, flight augers, buckets, rock rollers, percussion drills, core barrels, and so on—all resulting in different wall roughness. For example, a slow drill will provide a relatively smooth surface in hard rock, whereas a faster drill does the opposite. In this context, it is

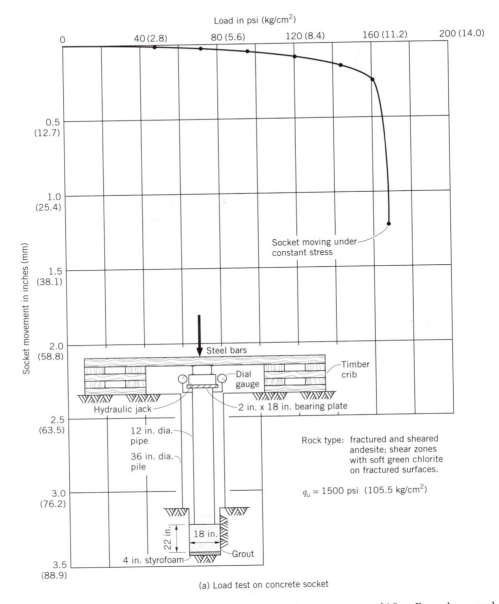

Figure 11.35 Socket friction development with socket movement. [After Rosenberg and Journeaux (109).]

prudent to assess the rock quality by adequate geotechnical investigation and, where economically feasible, to run in-situ tests for both the rock quality and the constructed pier. In this regard, one may deem it advisable to probe with a core drill past (deeper than) the anticipated bottom of the pier to assess the rock quality. Dilatometers (which exert radial all-round pressure), bore-hole jacks (which exert a unidirectional force diameterically opposite on the walls), or a Menard pressure meter, shown in Figure 11.36, may be feasible tools for testing the socket strengths.

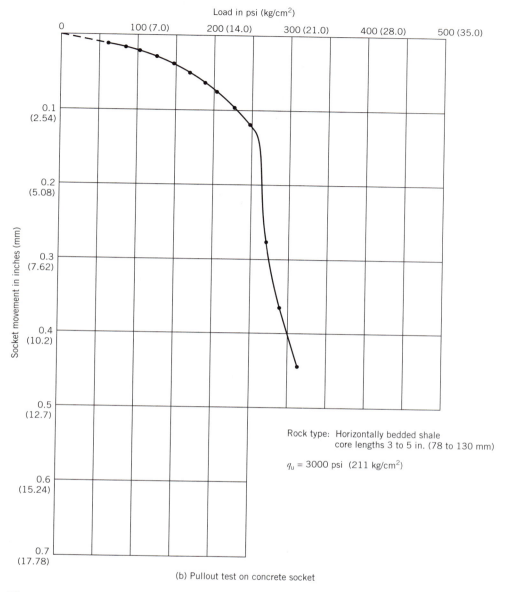

(b) Pullout test on concrete socket

Figure 11.35 (*continued*)

Figure 11.37 provides an estimate for the percentage of the base support to the total, based on theory, assuming no slip occurs along the socket length [after Rowe and Armitage (110)]. However, Rowe and Armitage did propose a procedure for design of a socketed pier allowing for slip.

Many building codes give allowable values from shear and normal stresses for a rock socket. However, while convenient, they are increasingly viewed as overconservative. For this reason, a number of investigators have recommended design proce-

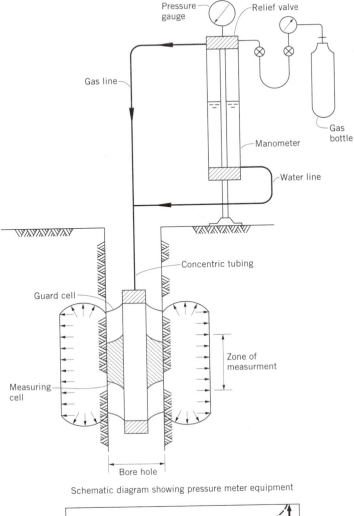

Schematic diagram showing pressure meter equipment

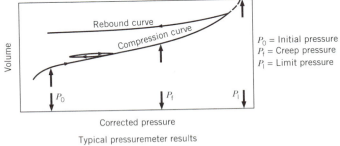

Typical pressuremeter results

Figure 11.36 Typical Menard Pressuremeter, equipment and results. [After Freeman et al. (38).]

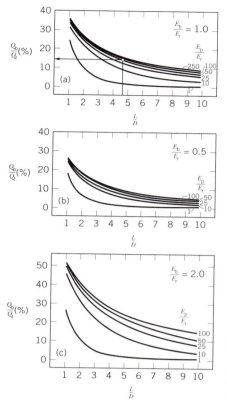

Figure 11.37 Elastic load distribution curves for a complete socketed pier: (a) homogenous, $E_b/E_t = 1.0$; (b) $E_b/E_t = 0.5$; (c) $E_b/E_t = 2.0$.

dures that are viewed as more consistent with test data (39, 57, 87, 101, 109, 110, 126). Ladanyi (57) outlines a step-by-step procedure based on the design concept proposed by Rosenberg and Journeaux (109). It is presented below:

Socket Design Procedure (after Ladanyi)

(a) Select the socket diameter, $2a$, and the design bond stress, $\tau_{bond,adm}$.

(b) Neglect the end-bearing and find the maximum socket length, L_{max}, from

$$L_{max} = \frac{F}{2\pi a \tau_{bond,adm}} \tag{1}$$

(c) Take a value of $L < L_{max}$ and find n from Fig. 11.38 for the given values of L/a and E_r/E_c

(d) Calculate and check:

$$q_{base} = \frac{nF}{\pi a^2} \leq q_{base,adm} \tag{2}$$

and

$$\tau_{bond} = \frac{(1 - n)F}{2\pi a^2 (L/a)} \leq \tau_{bond,adm} \tag{3}$$

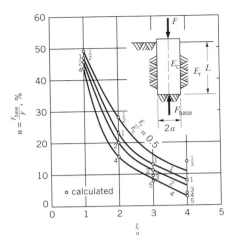

Figure 11.38 End-bearing ratio, $n = F_{base}/F$, as a function of the embedment ratio L/a and the modulus ratio E_r/E_c for an elastic behavior of rock and concrete. [From Ladanyi, deduced from Osterberg and Gill (86).]

(e) Repeat the calculation until

$$\tau_{bond} \approx \tau_{bond,adm} \quad \text{and} \quad q_{base} \leq q_{base,adm}$$

Example 11.6

Given Total pier load = 1800 kips (8 MN); medium strength rock; 4500 psi (31 MPa) concrete; $E_r/E_c = 1$.

Find A suitable socket length for a 4-ft-diameter pier (i.e., $a = 24$ in.)

Procedure Assume allowable stresses for wall friction (bond) and base as $\tau_{bond,adm} = 160$ psi (1.13 MPa) and $\sigma_{base,adm} = 500$ psi (3.21 MPa).

From Eq. 1

$$L_{max} = \frac{1,800,000}{2\pi(24)(160)} = 74.6 \text{ in.}$$

$$L_{max}/a = 3.11$$

For, say, an $L/a = 3 < 3.11$ we get, from Fig. 11.38, $n \cong 0.14$. Thus, from Eq. (2)

$$q_{base} = \frac{nF}{\pi a^2} = \frac{0.14 \times 1,800,000}{\pi(24)^2} = 139 \text{ psi} < 500 \text{ psi} \qquad \text{O.K.}$$

From Eq. (3)

$$\tau_{bond} = \frac{(1-n)F}{2\pi a^2(L/a)} = 142.6 \text{ psi} < 160 \text{ psi} \qquad \text{O.K.}$$

Thus, from

$$\frac{L}{a} = 3, \qquad L = 3(24) = 72 \text{ in. } (1.83 \text{ m})$$

11.16 THE STRUCTURAL DESIGN OF DRILLED PIERS

Piers that are not subject to bending are sometimes designed with no reinforcement. Piers are supported by the surrounding soil; thus slenderness effects are usually not a concern. However, a small amount of reinforcement, perhaps 1% of the gross area of the pier, is desirable for at least the upper 10 ft of a shaft.

The allowable concrete stress f_c recommended by most building codes is $0.25\ f_c$. For steel, the usual suggested allowable stress is $f_s = 0.4 f_y$. Hence, the allowable axial load on the pier is

$$Q_a = A_c f_c + A_s f_s \tag{11-16}$$

An alternative evaluation is to treat the pier as a column, accounting for the factored axial load P_u.

11.17 UNDERPINNING

Among the more common causes for the need to underpin an existing structure are

- excessive settlement
- danger of collapse
- provide additional foundation capacity
- adjacent excavation to a level below the existing foundation
- tunneling (e.g., subway construction) under the foundations

Slabs and lightly loaded footings may be lifted by injecting a cement grout under pressure; this process is known as mudjacking. Sometimes a gel (e.g., sodium silicate/calcium chloride) is injected in a loose sandy formation to increase the shear strength of the soil and thereby enhance bearing capacity or reduce settlement. Another procedure sometimes used to densify a stratum and increase its bearing capacity is the vibroflotation approach, described in Chapter 5. Indeed, the method and details for enhancing the types of supports to existing foundations may vary widely with designers and specific projects. In this section, we reflect on some of the more common methods for underpinning.

Figure 11.39 shows an underpinning scheme applicable to both isolated footings as well as wall footings. Briefly, auger-cast piles are installed on each side of the footing. Auger-cast piles are vibration-free; vibrations from pile driving may cause damage to the existing building. A concrete beam is poured under the footing, supported by the piles, as indicated in the figure. To facilitate this construction, sufficient excavation must be carried out so as to permit tunneling under the footing and subsequent pouring of the beam. Generally, vibrating the concrete is advisable in order to provide good contact with the cleaned bottom of the footing.

For isolated footings, if the percentage of undersoil to be removed for the construction of the beams is relatively small, no temporary column support is needed. On the other hand, if a rather large percentage of the footing support is removed (generally 50% or more), then temporary support for the column may be required. Obviously, the capacity of the four piles must equal the desired capacity of the column load.

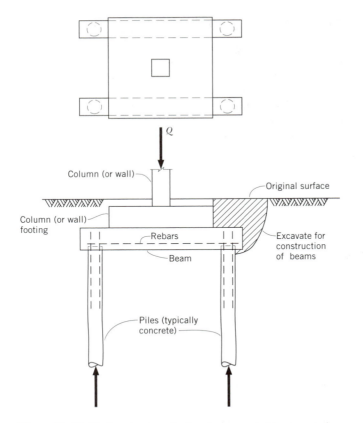

Figure 11.39 Underpinning via simply supported beams under
the foundation.

For wall footings, the spacing of such beams depends on the desired support capacity. Generally, one would expect a series of such beams, with the spacing determined as a function of the wall load and the capacity of these beams. The beams are poured in such a way that the undisturbed soil supporting the wall footing during this construction is not overstressed. This is frequently done by staggering the pours (e.g., beams 1, 3, 5, poured on one day and 2, 4, 6, poured perhaps two or three days later, after the concrete is cured).

With the current technology, auger-cast piles* could be installed with relatively low headroom clearance (8 to 10 ft of headroom). Thus, such installations are now possible. Even for interior installations they are more costly and relatively tedious since more hand work is required for an interior than those where headroom is not a problem.

---

* As described in Section 11.4, auger-cast piles are installed by augering to a predetermined depth, then pumping a cement–sand grout (no coarse aggregate) of a very wet (high slump) consistency. One makes up for the loss of strength due to high water content by increasing the cement content to normally about 10 bags per cubic yard. The steel reinforcement is "pushed" down into the grout-filled pile shortly after pumping is complete. Even H-sections are pushed into such a pile for perhaps up to two 20-ft lengths.

Figure 11.40 shows an underpinning system that is particularly appealing when the interior piling installation is unacceptable or impractical. The concept is applicable to both isolated footings as well as wall footings. Typically, the sequence of steps consists of first installing the piles, followed by appropriate excavation needed for tunneling under the footing and subsequent construction of the beam.

One notes some differences in footing and support conditions in this scheme from the previous installation. The cantilever portion of the beam (the overhanging length) will, of course, deflect, thereby possibly inducing some tilting of the footing and bending into the column. Also, the one pile is in compression while the "outer" pile is in tension. Correspondingly, each pile must be designed accordingly. Of particular note, the tension pile must have reinforcement for its entire length, since one typically does not depend on the tensile stresses in the concrete. Likewise, the force from the tension pile into the beam must be transmitted by adequate anchors or hooks to the upper beam steel (negative bending moment), as shown in the figure. Furthermore, the direction of the frictional soil resistance of one pile is opposite the other.

The spacing of these beams depends on the pile capacities, both compression and tension. The spacing between piles is generally dictated by soil properties; as a rule of thumb, pile spacings are about 5 pile diameters, center to center. Needless

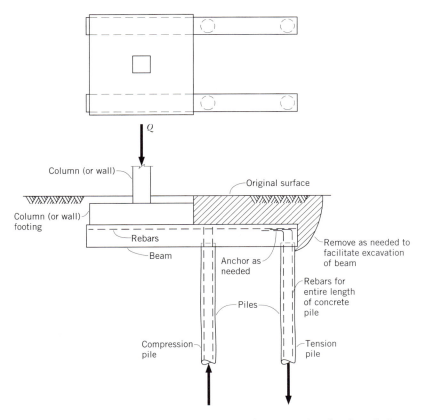

Figure 11.40 Underpinning via overhanging beams under the foundation.

to say, the pile movements (downward for the compression pile and upward for the tension pile) must be accounted for and assessed in the context of the expected behavior of the underpinned footing and column or wall.

Figure 11.41 shows a scheme that is occasionally used when side space is limited and interior piles are not a viable option. Briefly, a vertical H-pile is installed first. Then excavation is done to permit fabrication of the knee-brace joint. As the figure indicates, the horizontal beam is installed directly under the footing. It is welded to the vertical beam and to a compression brace, as shown.

One notes that the vertical member is subjected to a bending moment, which is resisted by the passive soil resultants on each side of the pile. Correspondingly, the vertical beam will deflect horizontally, thereby creating a rotation of the support system and subsequently a possible rotation of the footing itself. Quite obviously, if the soil is soft and compressible, the passive resistance may be of questionable effectiveness; correspondingly, the amount of rotation may be unacceptably large.

Instead of driving (pounding) the vertical beam into the ground, a more common procedure consists of installing an auger-cast pile with a relatively fluid cement grout and the lowering the structural shape (the H-section) into this pile. This envelopment of the H-section has the advantage of (1) eliminating vibrations (and possible further damage to the building), (2) providing some protection against

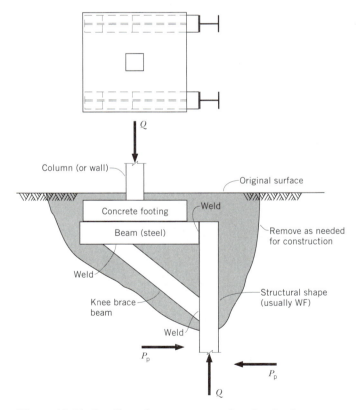

Figure 11.41 Cantilever-beam systems of underpinning.

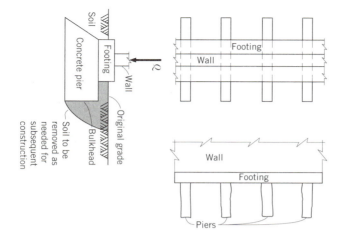

Figure 11.42 Underpinning via poured piers under the foundation.

corrosive elements to the vertical beam, and (3) increasing the surface area that generates the passive resistance. Indeed, pouring the concrete around the entire ''joint'' after the fabrication is a precautionary step in minimizing corrosive effects for all the steel.

In order to provide maximum contact with concrete footing and thereby minimize any stress concentration in the footing concrete, one may inject a cement grout or some equivalent admixture between the bottom of the footing and the steel (horizontal) beam.

Figure 11.42 shows the underpinning for a wall. This is sometimes a less expensive approach than a piling, and is particularly appealing when one anticipates an excavation close to the existing foundation.

A possible approach is to excavate the new excavation at a distance from the outside edge of the footing to the depth of excavation under the footing. For example, if the depth of excavation is 10 ft below the bottom of the existing footing, then the edge of the excavated cut should not be closer than 10 ft from the outside edge of the footing. This provides roughly 45° pressure configuration; 45° is generally safe for most soils. However, if the soil is particularly compressible or weak, a somewhat larger distance may be advisable. The next step consists of digging with a backhoe perpendicular to the wall with a relatively narrow bucket (say 18 to 24 in.), carefully so as to cause the least disturbance of the undersoil between the piers. The excavation will entail tunneling under the footing, extending beyond the footing for a distance adequate to provide a uniform bearing pressure for the new pier. The bearing should be at least as large but preferably larger than the area of the footing.

Subsequent to tunneling, the bottom of the excavation should be cleaned of any loose soil (via a hand hoe, etc.) in order to minimize potential settlement as the load is transmitted via this pier. Then, the concrete should be poured and vibrated under the existing footing in order to provide even support to the footing. A bulkhead or shoring is placed at the excavation end, at a point clear of the anticipated excavation line. For example, if the new excavation is to be within 3 ft from the outside edge of the existing footing, then the pier should not extend beyond 3 ft from the footer's edge. Typically, the piers are constructed in a staggered manner, with pier 1, 3, 5, poured and cured, followed by pier 2, 4, 6, and so on.

The spacing of these piers will be designed in the context of the anticipated load to be supported by each pier and the footing load.

Regardless of the type of underpinning systems, one may experience, and indeed expect and account for, some settlement that may be induced as a result of underpinning construction. For example, the new installation will deform and thus result in some settlement of the structure (e.g., piles may move somewhat, the contact between the footing and the underpinning system may permit some compressive movement, a drilled pier may compress the bearing soil). Settlements of $\frac{1}{4}$ in. or more are rather common even under the most careful conditions. Many of the conditions in connection with the underpinning make for difficult construction, and quality control is frequently also difficult to achieve with a total degree of confidence. Hence, it is important that the overall behavior of the supported structure be assessed in the context of such potential settlement.

Problems

11.1   State some of the (relative) advantages and disadvantages of: (a) Wood piles. (b) Steel piles. (c) Precast concrete piles.

11.2   Compare features of cast-in-place piles:
   (a) Constructed by a driven steel shell with concrete.
   (b) Auger-cast type.

11.3   A cast-in-place concrete pile, 10 m long, with a uniform diameter of 0.4 m, is cast in a steel shell driven in a sandy stratum whose properties are $\gamma = 17.6$ kN/m^3; $\phi = 28°$; $c = 0$. Determine the ultimate bearing capacity of a single pile.

11.4   Determine the ultimate bearing capacity of the pile in Problem 11.3 if the water table rises close to the surface and $\gamma_b = 8.6$ kN/m^3.

11.5   Determine the ultimate capacity of a concrete-filled pipe pile, 10 m long and 0.3 m in diameter, installed in a sandy stratum whose properties are $\gamma = 18.1$ kN/m^3; $\phi = 28°$; $c = 0$.

11.6   An auger-cast concrete pile, 13 m long and 0.4 m in diameter, is constructed in moist sandy stratum. $\gamma = 18.5$ kN/m^3; $\phi = 30°$; $c = 0$; average SPT number ≈ 17 blows/0.3 m. Determine the bearing capacity of a single pile:
   (a) Based on the general bearing capacity formula.
   (b) Estimated based on N.

11.7   A 0.3-m-diameter pipe was driven into a clay stratum to a depth of 15 m, and then filled with concrete. The clay properties are $\gamma = 19$ kN/m^3; $\phi = 18°$; $c = 24$ kN/m^2. Determine the ultimate bearing capacity of a single pile.

11.8   A single acting steam hammer weighs 18.8 kN and falls 0.6 m. Estimate the capacity of the pile, via the modified *Engineering News Record* formula, when

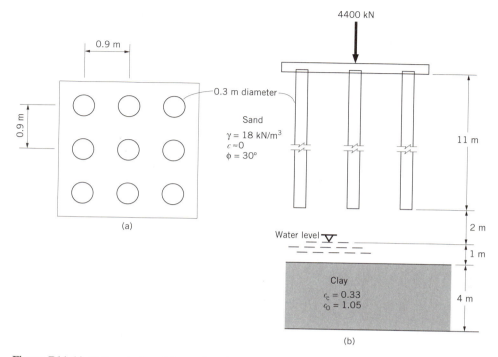

Figure P11.10 Estimate the ultimate bearing capacity of the group if the group was treated as an "equivalent" deep footing.

the resistance to penetration is approximately 17 mm per hammer blow. The pile is yellow pine, 10 m long and 0.4 in diameter. State any assumptions.

11.9   Determine the capacity of a steel pile 14 m long when the blow resistance is 10 mm per hammer blow. The pile is HP 30 × 1.49 (cm × kN/m); $A =$ 194 cm²; depth = 31.97 cm; $I_{zz} = 33,800$ cm⁴; $I_{yy} = 11,500$ cm⁴. State any assumptions.

11.10   A pile group of nine piles is arranged as shown in Fig. P11.10a and embedded as shown in Fig. P11.10b. Determine:

(a) The efficiency of the group.

(b) The settlement of the group (neglect the compressibility of the sand).

11.11   Estimate the ultimate bearing capacity of the group if the group was treated as an "equivalent" deep footing.

Refer to Fig. P11.12 from Problems 11.12 through 11.16. You are to assume the test data are typical for the entire site.

11.12   Based on SPT numbers shown on the boring logs, Fig. P11.12,

(a) Estimate the friction factor f_s (psf) for the points where the split-spoon samples were taken.

(b) Plot the values from part (a) over the 30-ft depth.

(c) Determine the average value of f_s for the 30-ft depth.

Remarks: Final water level taken 72 hours after the completion of drilling

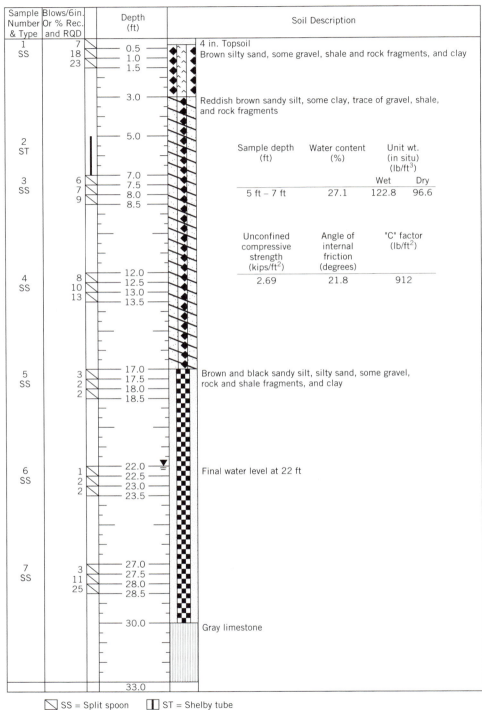

Sample Number & Type	Blows/6in. Or % Rec. and RQD
1 SS	7 18 23
2 ST	
3 SS	6 7 9
4 SS	8 10 13
5 SS	3 2 2
6 SS	1 2 2
7 SS	3 11 25

Depth (ft)

0.5
1.0
1.5
3.0
5.0
7.0
7.5
8.0
8.5
12.0
12.5
13.0
13.5
17.0
17.5
18.0
18.5
22.0
22.5
23.0
23.5
27.0
27.5
28.0
28.5
30.0
33.0

Soil Description

4 in. Topsoil
Brown silty sand, some gravel, shale and rock fragments, and clay

Reddish brown sandy silt, some clay, trace of gravel, shale, and rock fragments

Sample depth (ft)	Water content (%)	Unit wt. (in situ) (lb/ft^3)	
		Wet	Dry
5 ft – 7 ft	27.1	122.8	96.6

Unconfined compressive strength (kips/ft^2)	Angle of internal friction (degrees)	"C" factor (lb/ft^2)
2.69	21.8	912

Brown and black sandy silt, silty sand, some gravel, rock and shale fragments, and clay

Final water level at 22 ft

Gray limestone

☒ SS = Split spoon   ▯ ST = Shelby tube

Figure P11.12 Boring log and test data for Problems 11.12 through 11.16; computer-generated log (Courtesy, Geotechnical Computer Applications, Inc.)

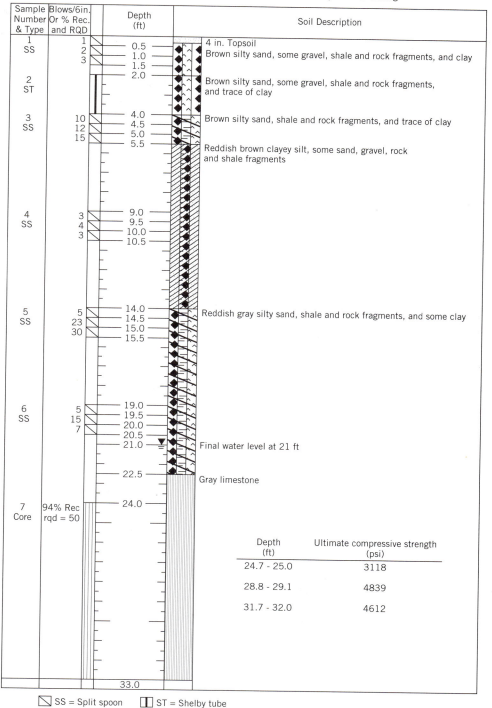

Remarks: Final water level taken 72 hours after the completion of drilling

Figure P11.17 Boring log and test data for Problem 11.17; computer-generated log (Courtesy, Geotechnical Computer Applications, Inc.)

11.13   Based on SPT numbers and test data (assumed to be representative of entire 30-ft depth), estimate the capacity of 30-ft long, auger-cast piles of diameters 16, 14, 12, and 10 in.

11.14   An auger-cast piling contractor gives a client a price of $15 per lineal foot, for 200, 16-in.-diameter, 30-ft-long piles. Options being considered are 234 piles, 14-in. diameter; 266 piles, 12-in. diameter. The piling contractor will reduce the price per foot by the amount of reduction in the cost of concrete mortar. Assuming the cost of concrete mortar to be $70 per cubic yard, determine the changes in cost for the 14-in.- and 12-in.-diameter piles, 30-ft long.

11.15   A group of four, 16-in.-diameter, 30-ft-long auger-cast piles are spaced 4 ft, c-c, and are capped with a square concrete cap. Estimate the combined capacity of the group.

11.16   (a) Estimate the capacity of a 42-ft-diameter, 30-ft-long straight shaft, drilled pier.

(b) What obstacles might you encounter in constructing a bell at the 30 ft depth?

(c) Assuming that a bell is feasible, what would be this pier capacity with an 8-ft bell at the 30-ft depth?

11.17   Referring to Fig. P11.17, assume the following allowable shear and normal stresses; $\tau = 160$ psi; $\sigma = 800$ psi. The concrete for the pier is 5000 psi. Determine the socket lengths for the following situations.

(a) $F = 2300$ kips; $D = 54$ in.; $E_r/E_c = 1$

(b) $F = 2300$ kips; $D = 48$ in.; $E_r/E_c = 2$

(c) $F = 2000$ kips; $D = 45$ in.; $E_r/E_c = 0.5$

(d) $F = 2000$ kips; $D = 42$ in.; $E_r/E_c = 1$

(e) $F = 1800$ kips; $D = 45$ in.; $E_r/E_c = 1.5$

(f) $F = 1800$ kips; $D = 42$ in.; $E_r/E_c = 2$

BIBLIOGRAPHY

1.   Alizadeh, M., and M. T. Savisson, "Lateral Load Tests on Piles—Arkansas River Project," *ASCE J. Geotech. Eng. Div.*, Sept. 1970.

2.   American Railway Engineering Association Committee, "Steel and Timber Pile Tests—West Atchafalaya Floodway," *Area Bull.* 489, 1950.

3.   Ashton, W. D., and P. H. Schwartz, "H-Bearing Piles in Limestone and Clay Shales," *ASCE J. Geotech. Eng. Div.*, July 1974.

4.   ASTM, "Standard Method of Testing Piles under Axial Compressive Load," *Annual Book of ASTM Standards,* p. 19, Designation D-1143-74, 1974.

5.   Baligh, M. M., V. Vivatrat, and H. Figi, "Downdrag on Bitumen-Coated Piles," *ASCE J. Geotech. Eng. Div.*, Nov. 1978.

6.   Barden, L., and M. F. Monckton, "Tests on Model Pile Groups in Soft and Stiff Clay," *Geotechnique,* vol. 20, no. 1, Mar. 1970.

7.   Barker, W. R., and L. C. Reese, "Load Carrying Characteristics of Drilled Shafts Constructed with the Aid of Drilling Fluids," Cent. Highw. Res., University of Texas at Austin, Austin, TX, Res. Rep. 89-9, Aug. 1970.

8. Begemann, H. K., "The Friction Jacket Cone as an Aid in Determining the Soil Profile," *6th Int. Conf. Soil Mech. Found. Eng.*, Montreal, Canada, vol. 1, 1965.

9. Berezantzev, V. G., et al. "Load Bearing Capacity and Deformation of Piled Foundations," *5th Int. Conf. Soil Mech. Found. Eng.*, Paris, France, vol. 2, 1961.

10. Bjerrum, L., I. J. Johannessen, and O. Eide, "Reduction of Negative Skin Friction on Steel Piles to Rock," *7th Int. Conf. Soil Mech. Found. Eng.*, Mexico, vol. 2, 1969.

11. Bowles, J. E., *Foundation Analysis and Design,* 2nd ed., McGraw-Hill, New York, 1977.

12. Bozozuk, J., "Downdrag Measurements on a 160-ft Floating Pipe Test Pile in Marine Clay," *Can. Geotech. J.,* vol. 9, 1972.

13. Broms, B. B., "Design of Laterally Loaded Piles," *ASCE J. Geotech. Eng. Div.,* May 1965.

14. Broms, B. B., and L. Hellman, "End Bearing and Skin Friction Resistance of Piles," *ASCE J. Geotech. Eng. Div.,* Mar. 1968.

15. Broms, B. B., "Settlements of Pile Groups," *5th Pan American Soil Mech. Conf.,* ASCE, vol. 3, 1972.

16. Broms, B. B., and P. Boman, "Lime Columns—A New Foundation Method," *ASCE J. Geotech. Eng. Div.,* Apr. 1979.

17. Butterfield, R., and P. K. Banerjee, "The Problem of Pile Group–Pile Cap Interaction," *Geotechnique,* vol. 21, no. 2, June 1971.

18. Caquot, A., and J. Kérisel, *Traité de Mécanique des Sols, Gauthier-Villars, Paris, France,* 1966.

19. Chellis, R. D., *Pile Foundations,* 2nd ed., McGraw-Hill, New York, 1961.

20. Chin, F. K., "Estimation of the Ultimate Load of Piles Not Carried to Failure," *2nd Southeast Asian Conf. Soil Eng.,* 1970.

21. Clark, J. C., and G. G. Meyerhof, "The Behavior of Piles Driven in Clay—Investigation of the Bearing Capacity Using Total and Effective Strength Parameters," *Can. Geotech. J.,* vol. 10, 1973.

22. Clemence, S. P., and W. F. Brumond, "Large-Scale Model Test of Drilled Pier in Sand," *ASCE J. Geotech. Eng. Div.,* June 1975.

23. Cole, K. W., and M. A. Stroud, "Rock Socket Piles at Coventry Point, Market Way, Coventry." *Piles in Weak Rock.* Instn. of Civil Engrs., London, England, 47–62, 1977.

24. Cooke, R. W., G. Price, and K. Tarr, "Jacked Piles in London Clay: Interaction and Group Behaviour under Working Conditions," *Geotechnique,* vol. 30, no. 2, 1980.

25. Coyle, H. M., and L. C. Reese, "Load Transfer for Axially Loaded Piles in Clay," *ASCE J. Soil Mech. Found. Div.,* Mar. 1966.

26. Cummings, A. E., G. O. Kerkhoff, and R. B. Peck, "Effect of Driving Piles into Soft Clay," *Trans. ASCE,* vol. 115, 1950.

27. D'Appolonia, E., and J. P. Romualdi, "Load Transfer in End-Bearing Steel H-Piles," *ASCE J. Geotech. Eng. Div.,* Mar. 1963.

28. D'Appolonia, E., and J. A. Hribar, "Load Transfer in a Step-Taper Pile," *ASCE J. Soil Mech. Found. Div.,* vol. 89, no. SM6, Nov. 1963.

29. Davisson, M. T., and J. R. Salley, "Model Study of Laterally Loaded Piles," *ASCE J. Geotech. Eng. Div.,* Sept. 1970.

30. De Beer, E. E., "Some Considerations Concerning the Point Bearing Capacity of Bored Piles," *Proc. Symp. Bearing Capacity of Piles,* Roorkee, India, 1964.

31. Desai, C. S., "Numerical Design Analysis for Piles in Sands," *ASCE J. Geotech. Eng. Div.,* June 1974.

32. Dos Santos, M. P., and N. A. Gomes, "Experiences with Piled Foundations of Maritime Structures in Portuguese East Africa," *4th Int. Conf. Soil Mech. Found. Eng.,* London, England, vol. 2, 1957.

33. Eide, O., J. N. Hutchinson, and A. Landva, "Short- and Long-Term Test Loading of a Friction Pile in Clay," *5th Int. Conf. Soil Mech. Found. Eng.,* Paris, France, vol. 2, 1961.

34. Ellison, R. D., E. D'Appolonia, and G. R. Thiers, "Load-Deformation for Bored Piles," *ASCE J. Soil Mech. Found. Div.*, vol. 97, no. SM4, Proc. Paper 8052, Apr. 1971.

35. Endo, M., et al., "Negative Skin Friction Acting on Steel Pipe-Piles in Clay," *7th Int. Conf. Soil Mech. Found. Eng.*, Mexico, vol. 2, 1969.

36. Fellinius, B. H., "Down-Drag on Piles in Clay Due to Negative Skin Friction," *Can. Geotech. J.*, vol. 9, Nov. 1972.

37. Fellinius, G. H., "Test Loading of Piles and New Proof Testing Procedure," *ASCE J. Geotech. Eng. Div.*, vol. 101, no. GT9, Proc. Paper 11551, Sept. 1975.

38. Freeman, C. F., D. Klajnerman, and G. D. Prasad, "Design of Deep Socketed Caissons into Shale Bedrock," *Can. Geotech. J.*, vol. 9, 1972.

39. Glos, G. H., and O. H. Briggs, Jr., "Rock Sockets in Soft Rock," *ASCE J. Geotech. Eng.*, vol. 109, no. 4, 1983.

40. Goodman, R. E., *Introduction to Rock Mechanics*, John Wiley & Sons, New York, 1980.

41. Horvath, R. G., and T. C. Kenny, "Shaft Resistance of Rock-Socketed Drilled Piers," *Symp. on Deep Foundations*, ASCE, 1979.

42. Horvath, R. G., W. A. Trow, and T. C. Kenny, "Results of Tests to Determine Shaft Resistance of Rock-Socketed Drilled Piers," *Proc. Int. Conf. on Struct. Foundation on Rock*, vol. 1, 1980.

43. Hagerty, D. J., and R. B. Peck, "Heave and Lateral Movements Due to Pile Driving," *ASCE J. Geotech. Eng. Div.*, vol. 97, Nov. 1971.

44. Housel, W. S., "Pile Load Capacity, Estimates and Test Results," *ASCE J. Soil Mech. Found. Div.*, vol. 92, no. SM 4, July 1966.

45. Ismael, N. F., and T. W. Klym, "Uplift and Bearing Capacity of Short Piers in Sand," *ASCE J. Geotech. Eng. Div.*, vol. 105, May 1979.

46. Jaime, A., M. P. Romo, and D. Resendiz, "Behavior of Friction Piles in Mexico City Clay," *ASCE J. Geotech. Eng. Div.*, vol. 116, June 1990.

47. Kay, J. N., "Safety Factor Evaluation for Single Piles in Sand," *ASCE J. Geotech. Eng. Div.*, vol. 102, Oct. 1976.

48. Kerisel, J., "Deep Foundations—Basic Experimental Facts," *Proc. Deep Foundations Conf.*, Sociedad Mexicana de Mecania de Suelos, Mexico City, Mexico, Dec. 1964.

49. Kim, J. B., and R. J. Brungraber, "Full-Scale Lateral Load Tests of Pile Groups," *ASCE J. Geotech. Eng. Div.*, vol. 102, Jan. 1976.

50. Koerner, R. M., and A. Partos, "Settlement of Building on Pile Foundation in Sand," *ASCE J. Geotech. Eng. Div.*, vol. 100, no. GT3, Mar. 1974.

51. Koutsoftas, D. C., "Caissons Socketed in Sound Mica Schist," *ASCE Geotech. Eng. Div.*, vol. 107, no. GT6, June 1981.

52. Kraft, Jr., L. M., "Performance of Axially Loaded Pipe Piles in Sand," *ASCE J. Geotech. Eng. Div.*, vol. 117, Feb. 1991.

53. Kuhlemeyer, R. L., "Static and Dynamic Laterally Loaded Floating Piles," *ASCE J. Geotech. Eng. Div.*, vol. 105, Feb. 1979.

54. Kuhlemeyer, R. L., "Vertical Vibration of Piles," *ASCE J. Geotech. Eng. Div.*, vol. 105, Feb. 1979.

55. Kulhawy, F. H., D. E. Kozera, and J. L. Withiam, "Uplift Testing of Model Drilled Shafts in Sand," *ASCE J. Geotech. Eng. Div.*, vol. 105, Jan. 1979.

56. Kuwabara, F., and H. G. Poulos, "Downdrag Forces in Group of Piles," *ASCE J. Geotech. Eng. Div.*, vol. 115, June 1989.

57. Ladanyi, B. Discussion of "Friction and End Bearing Tests on Bedrock for High Capacity Socket Design," by P. Rosenberg and N. L. Journeaux, *Can. Geotech. J.*, vol. 13, 1977.

58. Leung, C. F., and R. Radhakrishnan, "The Behaviour of a Pile-Raft Foundation in Weak Rock," *Proc. 11th Int. Conf. on Soil Mech. and Foundation Eng.*, vol. 4, 1985.

59. Lee, S. L., Y. K. Chow, G. P. Karunaratne, and K. Y. Wong, "Rational Wave Equation Model for Pile-Driving Analysis," *ASCE J. Geotech. Eng. Div.,* vol. 114, Mar. 1988.

60. Liao, S., and D. A. Sangrey, "Use of Piles as Isolation Barriers," *ASCE J. Geotech. Eng. Div.,* vol. 104, Sept. 1978.

61. Mansur, C. I., and J. A. Focht, "Pile Loading Tests, Morganza Floodway Control Structure," *ASCE J. Geotech. Eng. Div.,* vol. 79, paper no. 324, 1953.

62. Mansur, C. I., and A. H. Hunter, "Pile Tests—Arkansas River Project," *ASCE J. Soil Mech. Found. Div.,* vol. 96, no. SM5, 1970.

63. Mattes, N. S., and H. G. Poulos, "Settlement of a Single Compressible Pile," *ASCE J. Soil Mech. Found. Div.,* vol. 95, no. SM1, 1969.

64. McClelland, B., J. A. Focht, Jr., and W. J. Emrich, "Problems in Design and Installation of Heavily Loaded Pipe Piles," *Proc. ASCE Spec. Conf. on Civ. Eng., in the Oceans,* 1967.

65. McClelland, B., "Design of Deep Penetration Piles for Ocean Structures," *ASCE J. Geotech. Eng. Div.,* vol. 100, July 1974.

66. Meigh, A. C., "Some Driving and Loading Tests on Piles in Gravel and Chalk," *Proc. Conf. Behavior of Piles,* London, England, 1971.

67. Meyerhof, G. G., and L. J. Murdock, "An Investigation of the Bearing Capacity of Some Bored and Driven Piles in London Clay," *Geotechnique,* vol. 3, no. 7, 1953.

68. Meyerhof, G. G., "Penetration Tests and Bearing Capacity of Cohesionless Soils," *ASCE J. Soil Mech. Found. Div.,* vol. 82, no. SM1, Jan. 1956.

69. Meyerhof, G. G., "Compaction of Sands and Bearing Capacity of Piles," *ASCE J. Soil Mech. Found. Div.,* vol. 85, no. SM6, Proc. Paper 2292, Dec. 1959.

70. Meyerhof, G. G., "Bearing Capacity and Settlement of Pile Foundations," *ASCE J. Geotech. Eng. Div.,* vol. 102, no. GT3, Proc. Paper 11962, Mar. 1976.

71. Michigan State Highway Commission, "A Performance Investigation of Pile Driving Hammers and Piles," Lansing, MI, 1965.

72. Mindlin, R. D., "Force at a Point in the Interior of a Semi-Infinite Solid," *J. Am. Inst. Phys.,* May 1936.

73. Moorhouse, D. C., and J. V. Sheehan, "Predicting Safe Capacity of Pile Groups," *ASCE J. Civ. Eng.,* vol. 38, no. 10, Oct. 1968.

74. Moss, J. D. "A High Capacity Load Test for Deep Bored Piles," *Proc. 1st Australia-New Zealand Conf. on Geomech.,* vol. 1, 1971.

75. Neely, W. J., "Bearing Capacity of Auger-Cast Piles in Sand," *ASCE J. Geotech. Eng. Div.,* vol. 117, Feb. 1991.

76. Nonveiller, E., "Open Caissons for Deep Foundation," *ASCE J. Geotech. Eng. Div.,* vol. 113, May 1987.

77. Nordlund, R. L., "Bearing Capacity of Piles in Cohesionless Soils," *ASCE J. Soil Mech. Found. Div.,* vol. 89, no. SM3, May 1963.

78. Novak, M., and J. F. Howell, "Torsional Vibration of Pile Foundations," *ASCE J. Geotech. Eng. Div.,* vol. 103, Apr. 1977.

79. Novak, M., and J. F. Howell, "Dynamic Response of Pile Foundations in Torsion," *ASCE J. Geotech. Eng. Div.,* vol. 104, May 1978.

80. Nunes, A. J. C., and M. Vargas, "Computed Bearing Capacity of Piles in Residual Soil Compared with Laboratory and Load Tests," *3rd Int. Conf. Soil Mech. Found. Eng.,* Zurich, Switzerland, vol. 2, 1953.

81. Olsen, R. E., and K. S. Flaate, "Pile Driving Formulas for Friction Piles in Sand," *ASCE J. Soil Mech. Found. Div.,* vol. 93, no. SM6, Nov. 1967.

82. O'Neil, M. W., and L. C. Reese, "Behavior of Bored Piles in Beaumont Clay," *ASCE J. Soil Mech. Found. Div.,* vol. 98, no. SM2, 1972.

83. O'Neill, M. W., R. A. Hawkins, and J. M. Audibert, "Installation of Pile Group in Overconsolidated Clay," *ASCE J. Geotech. Eng. Div.,* vol. 108, Nov. 1982.

84. O'Neill, M. W., C. Vipulanandan, and D. Wong, "Laboratory Modeling of Vibro-Driven Piles," *ASCE J. Geotech Eng. Div.*, vol. 116, Aug. 1990.

85. O'Neill, M. W., R. A. Hawkins, and L. J. Mahar, "Load Transfer Mechanisms in Piles and Pile Groups," *ASCE J. Geotech. Eng. Div.*, vol. 109, no. 2, 1982.

86. Osterberg, J. O., and S. A. Gill, "Load Transfer Mechanism for Piers Socketed in Hard Soils or Rock," *Proc. 9th Canadian Symp. on Rock Mech.*, 1973.

87. Pells, P. J. N., et al. "Design Loadings for Foundations on Shale and Sandstone in the Sydney Region," *Res. Report No. R315*, School of Civ. Eng., Univ. of Sydney, Sydney, Australia, 1978.

88. Pitts, J. "A Review of Geology and Engineering Geology in Singapore." *Q. J. Eng. Geol.*, 1984.

89. Poulos, H. G., and E. H. Davis, *Pile Foundation Analysis and Design*, John Wiley & Sons, New York, 1980.

90. Poulos, H. G., and E. H. Davis, "The Settlement Behavior of Single Axially Loaded Incompressible Piles and Piers," *Geotechnique*, vol. 18, no. 3, Sept. 1968.

91. Poulos, H. G., "The Influence of a Rigid Pile Cap on the Settlement Behavior of an Axially Loaded Pile," *Civ. Eng. Trans., Inst. Eng. Aust.*, vol. CE 10, 1968.

92. Poulos, H. G., "Analysis of the Settlement of Pile Groups," *Geotechnique*, vol. 18, 1968.

93. Poulos, H. G., and N. S. Mattes, "The Analysis of Downdrag in End-Bearing Piles Due to Negative Friction," *7th Int. Conf. Soil Mech. Found. Eng.*, vol. 2, 1969.

94. Poulos, H. G., and N. S. Mattes, "The Behavior of Axially Loaded End-Bearing Piles," *Geotechnique*, vol. 19, 1969.

95. Poulos, H. G., "The Behavior of Laterally Loaded Piles: I. Single Piles," *ASCE J. Soil Mech. Found. Div.*, vol. 97, no. SM5, 1971.

96. Poulos, H. G., "The Behavior of Laterally Loaded Piles: II. Pile Groups," *ASCE J. Soil Mech. Found Div.*, vol. 97, 1971.

97. Poulos, H. G., "Lateral Load-Deflection Prediction for Pile Groups," *ASCE J. Geotech. Eng. Div.*, vol. 100, Jan. 1975.

98. Poulos, H. G., and E. H. Davis, "Prediction of Downdrag Forces in End-Bearing Piles," *ASCE J. Geotech. Eng. Div.*, vol. 101, Feb. 1975.

99. Poulos, H. G., "Settlement of Single Piles in Nonhomogeneous Soil," *ASCE J. Geotech. Eng. Div.*, vol. 105, May 1979.

100. Radhakrishnan, R., C. F. Leung, and R. V. Subrahmanyam, "Load Tests on Instrumented Large Diameter Bored Piles in Weak Rock," *Proc. 8th Southeast Asian Geotech. Conf.*, vol. 1, 1985.

101. Radhakrishnan, R., and C. E. Leung, 1985. "Load Transfer Behavior of Rock-Socketed Piles," *ASCE J. Geotech. Eng. Div.*, vol. 115, June 1989.

102. Ramiah, B. K., and L. S. Chickanagappa, "Stress Distribution around Batter Piles," *ASCE J. Geotech. Eng. Div.*, vol. 104, Feb. 1978.

103. Randolph, M. F., and C. P. Wroth, "Analysis of Deformation of Vertically Loading Piles," *ASCE J. Geotech. Eng. Div.*, vol. 104, Dec. 1978.

104. Reese, L. C., et al. "Generalized Analysis of Pile Foundations," *ASCE J. Soil Mech. Found. Div.*, vol. 96, no. SM1, Jan. 1970.

105. Reese, L. C., M. W. O'Neill, and F. T. Touma, "Bored Piles Installed by Slurry Displacement," *8th Int. Conf. Soil Mech. Found. Eng.*, Moscow, USSR, vol. 3, 1973.

106. Reese, L. C., F. T. Touma, and M. W. O'Neill, "Behavior of Drilled Piers under Axial Loading," *ASCE J. Geotech. Eng. Div.*, vol. 102, no. GT 5, Proc. Paper 12135, May 1976.

107. Reese, L. C., "Design and Construction of Drilled Shaft," *ASCE J. Geotech. Eng. Div.*, vol. 104, Jan. 1978.

108. Robinsky, E. I., and C. F. Morrison, "Sand Displacement and Compaction around Model Friction Piles," *Can. Geotech. J.*, vol. 1, 1964.

109.   Rosenberg, P., and N. L. Journeaux, "Friction and End Bearing Tests on Bedrock for High Capacity Socket Design," *Can. Geotech. J.,* vol. 13, 1976.

110.   Rowe, R. K., and H. H. Armitage, "A Design Method for Drilled Piers in Soft Rock." *Can. Geotech. J.,* vol. 24, 1987.

111.   Saul, W. E., "Static and Dynamic Analysis of Pile Foundations," *ASCE J. Soil Mech. Found. Div.,* vol. 94, no. ST5, May 1968.

112.   Scalan, R. H., and J. J. Tomko, "Dynamic Prediction of Pile Static Bearing Capacity," *ASCE J. Geotech. Eng. Div.,* vol. 95, Mar. 1969.

113.   Schlitt, H. G., "Group Pile Loads in Plastic Soils," *Proc. Highw. Res. Board,* Washington, D.C., vol. 31, 1951.

114.   Scott, R. F., "Freezing of Slurry around Piles in Permafrost," *ASCE J. Geotech. Eng. Div.,* vol. 85, Aug. 1959.

115.   Seed, H. B., and L. C. Reese, "The Action of Soft Clay along Friction Piles," *Trans. ASCE,* vol. 22, 1957.

116.   Sherman, W. C., "Instrumental Pile Tests in a Stiff Clay," *7th Int. Conf. Soil Mech. Found. Eng.,* Mexico, vol. 2, 1969.

117.   Singh, J. P., C. Donovann, and A. C. Jobsis, "Design of Machine Foundations on Piles," *ASCE J. Geotech. Eng. Div.,* vol. 103, Aug. 1977.

118.   Skempton, A. W., "The Bearing Capacity of Clays," *Proc. Build. Res. Cong.,* Div. 1, p. 111, London, England, 1951.

119.   Skempton, A. W., "Large Bored Piles—Summing Up," *Symp. Large Bored Pile,* Inst. of Civ. Eng. and Reinf. Conc. Assoc., London, England, Feb. 1966.

120.   Smith, J. E., "Pile Driving by the Wave Equation," *ASCE J. Soil Mech. Found Div.,* no. SM4, 1960.

121.   Stermac, A. G., K. G. Selby, and M. Devata, "Behavior of Various Piles in Stiff Clay," *7th Int. Conf. Soil Mech. Found. Eng.,* Mexico, vol. 2, 1969.

122.   Swedish Pile Commission, "Recommendations for Pile Driving Test and Routine Test Loading of Piles," Swed. Acad. Eng. Sci., Comm. on Pile Res., Rep. 11, Stockholm, Sweden, 1970.

123.   Tavenas, F. A., "Load Tests Results on Friction Piles in Sand," *Can. Geotech. J.,* vol. 8, no. 1, Feb. 1971.

124.   Thorburn, S., and R. S. MacVicar, "Pile Load Tests to Failure in the Clyde Alluvium," *Proc. ICE Conf. Behavior of Piles,* London, England, 1971.

125.   Thurman, A. G., "Discussion: Bearing Capacity of Piles in Cohesionless Soils," *ASCE J. Soil Mech. Found. Div.,* vol. 90, no. SM1, Jan. 1964.

126.   Tomlinson, M. J., *Pile Design and Construction Practice.* Pitman Publishing Ltd., London, England, 1975.

127.   Tomlinson, J. J., "Adhesion of Piles in Stiff Clays," Constr. Ind. Res. and Inform. Assoc., London, England, Rep. 26, Nov. 1970.

128.   Tomlinson, M. J., "Some Effects of Pile Driving on Skin Friction," *Proc. ICE Conf. Behavior of Piles,* London, England, 1971.

129.   Touma, F. T., and L. C. Reese, "Behavior of Bored Piles in Sand," *ASCE J. Geotech. Eng. Div.,* vol. 100, July 1974.

130.   Tsinker, G. P., "Performance of Jetted Anchor Piles with Widening," *ASCE J. Geotech. Eng. Div.,* vol. 103, Mar. 1977.

131.   Van Weele, A. F., "Negative Skin Friction on Pile Foundations in Holland," *Proc. Symp. Bearing Capacity of Piles,* Roorkee, India, 1964.

132.   Vesic, A. S., "Tests on Instrumental Piles, Ogeechee River Site," *ASCE J. Soil Mech. Found. Div.,* vol. 96, no. SM2, Proc. Paper 7170, Mar. 1970.

133.   Vesic, A. S., "Load Transfer in Pile-Soil Systems," *Proc. Conf. Design Installation of Pile Foundations,* Lehigh University, Bethlehem, PA, 1970.

134.   Vipulanandan, C., D. Wong, and M. W. O'Neill, "Behavior of Vibro-Driven Piles in Sand," *ASCE J. Geotech. Eng. Div.,* vol. 116, Aug. 1990.

135. Watt, W. G., P. J. Kurfurst, and Z. P. Zeman, "Comparison of Pile Load-Test Skin Friction Values and Laboratory Strength Tests," *Can. Geotech. J.,* vol. 6, 1969.

136. Wess, J. A., and R. S. Chamberlin, "Khazzan Dubai No. 1: Pile Design and Installation," *ASCE J. Geotech. Eng. Div.,* vol. 97, Oct. 1971.

137. Whitaker, T., "Experiments with Model Piles in Groups," *Geotechnique,* vol. 7, 1957.

138. Whitaker, T., "Some Experiments on Model Piled Foundations in Clay," *Proc. Symp. Pile Foundations,* Stockholm, Sweden, 1960.

139. Whitaker, T., and R. W. Cooke, "An Investigation of the Shaft and Base Resistance of Large Bored Piles in London Clay," *Proc. Symp. Large Bored Piles,* Inst. Civ. Eng., London, England, 1966.

140. Williams, A. F., and P. J. N. Pells, "Side Resistance Rock Sockets in Sandstone, Mudstone and Shale." *Can. Geotech. J.,* vol. 18, 1981.

141. Williams, A. F., I. W. Johnston, and I. B. Donald, "The Design of Socketed Piles in Weak Rock," *Proc. Int. Conf. on Struct. Found. on Rock,* vol. 1, 1980.

142. Woodward, R. J., R. Lundgren, and J. D. Boitano, "Pile Loading Tests in Stiff Clay," *5th Int. Conf. Soil Mech. Found. Eng.,* Paris, France, vol. 2, 1961.

143. Woodward, R. J., W. S. Gardner, and D. M. Greer, *Drilled Pier Foundations,* McGraw-Hill, New York, 1972.

144. Wooley, J. A., and L. C. Reese, "Behavior of an Axially Loaded Drilled Shaft under Sustained Loading," Cent. Highw. Res., University of Texas at Austin, Austin, TX, Res. Rep. 176-2, May 1974.

145. York, D. L., V. G. Miller, and N. F. Ismael, "Long-Term Load Transfer in End Bearing Pipe Piles," *Can. Geotech. J.,* vol. 12, 1975.

146. Zeevaert, L., "Compensated Friction—Pile Foundation to Reduce the Settlement of Buildings on Highly Compressible Volcanic Clay of Mexico City," *4th Int. Conf. Soil Mech. Found. Eng.,* London, England, vol. 2, 1957.

Appendix A

K_a and K_p Coefficients:
Rankine and Coulomb Equations

ACTIVE AND PASSIVE EARTH PRESSURES COEFFICIENTS K_a AND K_p

Coulomb's

$$P_a = \frac{H^2\gamma}{2}\left[\frac{\csc\beta\sin(\beta-\phi)}{\sqrt{\sin(\beta+\delta)}+\sqrt{\dfrac{\sin(\phi+\delta)\sin(\phi-i)}{\sin(\beta-i)}}}\right]^2 = \frac{H^2\gamma}{2}K_a$$

The corresponding passive thrust is expressed by

$$P_p = \frac{H^2\gamma}{2}\left[\frac{\csc\beta\sin(\beta+\phi)}{\sqrt{\sin(\beta-\delta)}-\sqrt{\dfrac{\sin(\phi+\delta)\sin(\phi+i)}{\sin(\beta-i)}}}\right]^2 = \frac{H^2\gamma}{2}K_p$$

Rankine's

For the active case:

$$P_a = \tfrac{1}{2}\gamma H^2 K_a = \tfrac{1}{2}\gamma H^2\left(\frac{\cos i - \sqrt{\cos^2 i - \cos^2\phi}}{\cos i + \sqrt{\cos^2 i - \cos^2\phi}}\right)\cos i$$

For the passive case:

$$P_p = \tfrac{1}{2}\gamma H^2 K_p = \tfrac{1}{2}\gamma H^2\left(\frac{\cos i + \sqrt{\cos^2 i - \cos^2\phi}}{\cos i - \sqrt{\cos^2 i - \cos^2\phi}}\right)\cos i$$

Notation

Phi = ϕ
Beta = β
Delta = δ
Alpha = α

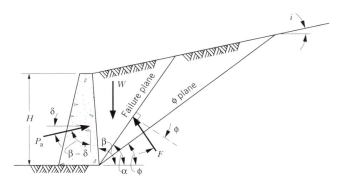

Figure A.1 Active and passive force resultants based on Coulomb's analysis.

Rankine Coefficients

i	0		5		10		15	
ϕ	K_a	K_p	K_a	K_p	K_a	K_p	K_a	K_p
24	0.422	2.371	0.428	2.318	0.449	2.159	0.492	1.895
26	0.390	2.561	0.396	2.507	0.413	2.346	0.448	2.083
28	0.361	2.770	0.366	2.715	0.380	2.551	0.409	2.284
30	0.333	3.000	0.337	2.943	0.350	2.775	0.373	2.502
32	0.307	3.255	0.311	3.196	0.321	3.022	0.341	2.740
34	0.283	3.537	0.286	3.476	0.294	3.295	0.311	3.002
36	0.260	3.852	0.262	3.788	0.270	3.598	0.288	3.293
38	0.238	4.204	0.240	4.136	0.246	3.936	0.258	3.615
40	0.217	4.599	0.219	4.527	0.225	4.316	0.235	3.977
42	0.198	5.045	0.200	4.968	0.204	4.744	0.213	4.383
44	0.180	5.550	0.181	5.468	0.186	5.228	0.193	4.842

Coulomb Coefficients ↓ $\beta = 90$
$\delta = 0$

i	0		5		10		15	
ϕ	K_a	K_p	K_a	K_p	K_a	K_p	K_a	K_p
24	0.422	2.371	0.448	2.709	0.482	3.093	0.528	3.545
26	0.390	2.561	0.414	2.943	0.443	3.385	0.482	3.913
28	0.361	2.770	0.382	3.203	0.407	3.713	0.440	4.331
30	0.333	3.000	0.352	3.492	0.374	4.080	0.402	4.807
32	0.307	3.255	0.323	3.815	0.343	4.496	0.367	5.352
34	0.283	3.537	0.297	4.177	0.314	4.968	0.334	5.981
36	0.260	3.852	0.272	4.585	0.286	5.507	0.304	6.710
38	0.238	4.204	0.249	5.046	0.261	6.125	0.276	7.563
40	0.217	4.599	0.227	5.572	0.238	6.841	0.251	8.570
42	0.198	5.045	0.206	6.173	0.216	7.673	0.227	9.768
44	0.180	5.550	0.187	6.864	0.195	8.650	0.205	11.209

$$\beta = 90$$
$$\delta = 14$$

i	0		5		10		15	
ϕ	K_a	K_p	K_a	K_p	K_a	K_p	K_a	K_p
24	0.378	3.553	0.406	4.357	0.442	5.382	0.493	6.747
26	0.352	3.916	0.376	4.840	0.407	6.042	0.449	7.682
28	0.326	4.325	0.348	5.393	0.374	6.813	0.409	8.797
30	0.303	4.790	0.321	6.032	0.344	7.720	0.374	10.143
32	0.280	5.321	0.296	6.775	0.316	8.797	0.341	11.785
34	0.259	5.932	0.273	7.644	0.290	10.088	0.311	13.816
36	0.239	6.638	0.251	8.670	0.266	11.652	0.284	16.368
38	0.220	7.460	0.230	9.892	0.243	13.571	0.259	19.630
40	0.202	8.426	0.211	11.364	0.222	15.960	0.235	23.888
42	0.185	9.569	0.193	13.156	0.202	18.984	0.213	29.590
44	0.168	10.938	0.176	15.369	0.184	22.885	0.193	37.459

$$\beta = 90$$
$$\delta = 16$$

i	0		5		10		15	
ϕ	K_a	K_p	K_a	K_p	K_a	K_p	K_a	K_p
24	0.375	3.792	0.403	4.706	0.440	5.894	0.491	7.513
26	0.349	4.195	0.373	5.250	0.404	6.652	0.447	8.611
28	0.324	4.652	0.345	5.878	0.372	7.545	0.408	9.936
30	0.300	5.174	0.319	6.609	0.342	8.605	0.372	11.555
32	0.278	5.775	0.295	7.464	0.315	9.876	0.340	13.557
34	0.257	6.469	0.272	8.474	0.289	11.417	0.310	16.073
36	0.237	7.279	0.250	9.678	0.265	13.309	0.283	19.291
38	0.218	8.230	0.229	11.128	0.242	15.665	0.258	23.494
40	0.201	9.356	0.210	12.895	0.221	18.647	0.234	29.123
42	0.184	10.704	0.192	15.076	0.201	22.497	0.213	36.894
44	0.168	12.335	0.175	17.812	0.183	27.581	0.192	48.035

$$\beta = 90$$
$$\delta = 18$$

i	0		5		10		15	
ϕ	K_a	K_p	K_a	K_p	K_a	K_p	K_a	K_p
24	0.372	4.057	0.401	5.099	0.438	6.482	0.490	8.413
26	0.346	4.507	0.371	5.716	0.403	7.359	0.446	9.717
28	0.322	5.020	0.344	6.434	0.371	8.401	0.407	11.309
30	0.299	5.609	0.318	7.274	0.341	9.651	0.371	13.280
32	0.277	6.292	0.293	8.267	0.313	11.168	0.339	15.758
34	0.256	7.088	0.270	9.451	0.288	13.030	0.309	18.928
36	0.236	8.022	0.249	10.878	0.264	15.349	0.282	23.070
38	0.218	9.130	0.229	12.616	0.241	18.287	0.257	28.620
40	0.200	10.456	0.209	14.764	0.220	22.080	0.234	36.284
42	0.183	12.061	0.191	17.458	0.201	27.092	0.212	47.276
44	0.167	14.028	0.174	20.900	0.183	33.897	0.192	63.815

$$\beta = 90$$
$$\delta = 20$$

i	0		5		10		15	
ϕ	K_a	K_p	K_a	K_p	K_a	K_p	K_a	K_p
24	0.370	4.354	0.399	5.545	0.437	7.163	0.490	9.485
26	0.345	4.857	0.370	6.249	0.402	8.186	0.446	11.049
28	0.320	5.436	0.342	7.074	0.370	9.414	0.406	12.986
30	0.297	6.105	0.316	8.049	0.340	10.903	0.371	15.423
32	0.276	6.886	0.292	9.212	0.313	12.733	0.338	18.541
34	0.255	7.804	0.270	10.613	0.287	15.014	0.309	22.617
36	0.235	8.892	0.248	12.321	0.263	17.903	0.282	28.081
38	0.217	10.194	0.228	14.433	0.241	21.636	0.257	35.629
40	0.199	11.772	0.209	17.083	0.220	26.569	0.234	46.459
42	0.183	13.705	0.191	20.468	0.201	33.270	0.212	62.760
44	0.167	16.111	0.174	24.885	0.183	42.686	0.192	88.892

$$\beta = 90$$
$$\delta = 22$$

i	0		5		10		15	
ϕ	K_a	K_p	K_a	K_p	K_a	K_p	K_a	K_p
24	0.369	4.686	0.398	6.056	0.436	7.960	0.491	10.774
26	0.343	5.253	0.369	6.864	0.401	9.164	0.446	12.676
28	0.319	5.910	0.341	7.820	0.369	10.625	0.406	15.068
30	0.296	6.675	0.316	8.960	0.340	12.421	0.371	18.130
32	0.275	7.574	0.292	10.334	0.312	14.659	0.338	22.136
34	0.254	8.641	0.269	12.011	0.287	17.497	0.309	27.507
36	0.235	9.919	0.248	14.083	0.263	21.164	0.282	34.930
38	0.217	11.466	0.228	16.685	0.241	26.013	0.257	45.584
40	0.199	13.364	0.209	20.011	0.220	32.602	0.234	61.627
42	0.183	15.726	0.191	24.352	0.201	41.864	0.212	87.356
44	0.167	18.714	0.174	30.161	0.183	55.447	0.192	132.367

$$\beta = 100$$
$$\delta = 0$$

i	0		5		10		15	
ϕ	K_a	K_p	K_a	K_p	K_a	K_p	K_a	K_p
24	0.494	2.089	0.527	2.364	0.569	2.667	0.626	3.007
26	0.463	2.226	0.493	2.532	0.531	2.872	0.580	3.259
28	0.434	2.374	0.462	2.714	0.495	3.096	0.537	3.537
30	0.407	2.535	0.431	2.912	0.461	3.343	0.498	3.846
32	0.380	2.710	0.403	3.129	0.429	3.615	0.461	4.190
34	0.355	2.901	0.375	3.368	0.399	3.916	0.427	4.575
36	0.332	3.109	0.350	3.630	0.370	4.251	0.395	5.007
38	0.309	3.336	0.325	3.920	0.343	4.624	0.365	5.496
40	0.287	3.587	0.302	4.242	0.318	5.043	0.337	6.050
42	0.267	3.862	0.280	4.600	0.294	5.514	0.311	6.683
44	0.247	4.167	0.259	5.000	0.271	6.047	0.286	7.410

$$\beta = 100$$
$$\delta = 14$$

i	0		5		10		15	
ϕ	K_a	K_p	K_a	K_p	K_a	K_p	K_a	K_p
24	0.455	2.898	0.492	3.478	0.539	4.174	0.605	5.036
26	0.429	3.136	0.462	3.784	0.503	4.574	0.558	5.568
28	0.403	3.399	0.433	4.126	0.469	5.025	0.516	6.177
30	0.379	3.691	0.405	4.508	0.437	5.537	0.478	6.879
32	0.356	4.014	0.379	4.938	0.407	6.121	0.443	7.692
34	0.334	4.376	0.355	5.425	0.380	6.791	0.410	8.644
36	0.313	4.781	0.332	5.978	0.353	7.565	0.380	9.764
38	0.293	5.237	0.309	6.609	0.329	8.465	0.352	11.097
40	0.273	5.754	0.288	7.336	0.305	9.520	0.325	12.699
42	0.255	6.341	0.268	8.176	0.283	10.767	0.301	14.647
44	0.237	7.015	0.249	9.157	0.262	12.257	0.277	17.047

$$\beta = 100$$
$$\delta = 16$$

i	0		5		10		15	
ϕ	K_a	K_p	K_a	K_p	K_a	K_p	K_a	K_p
24	0.453	3.050	0.491	3.695	0.539	4.480	0.606	5.467
26	0.427	3.311	0.460	4.033	0.502	4.926	0.559	6.070
28	0.402	3.599	0.432	4.411	0.468	5.433	0.516	6.765
30	0.378	3.919	0.404	4.837	0.437	6.011	0.478	7.571
32	0.355	4.277	0.379	5.318	0.407	6.674	0.443	8.513
34	0.333	4.678	0.354	5.865	0.379	7.441	0.410	9.624
36	0.312	5.129	0.331	6.491	0.353	8.332	0.380	10.946
38	0.292	5.641	0.309	7.210	0.328	9.378	0.352	12.534
40	0.273	6.223	0.288	8.043	0.305	10.615	0.326	14.467
42	0.254	6.890	0.268	9.015	0.283	12.092	0.301	16.849
44	0.237	7.658	0.249	10.159	0.262	13.877	0.278	19.831

$$\beta = 100$$
$$\delta = 18$$

i	0		5		10		15	
ϕ	K_a	K_p	K_a	K_p	K_a	K_p	K_a	K_p
24	0.452	3.216	0.490	3.933	0.539	4.820	0.608	5.955
26	0.426	3.501	0.460	4.307	0.502	5.321	0.560	6.642
28	0.401	3.818	0.431	4.728	0.468	5.892	0.517	7.439
30	0.377	4.171	0.404	5.204	0.437	6.548	0.479	8.372
32	0.354	4.567	0.378	5.745	0.407	7.306	0.443	9.471
34	0.333	5.014	0.354	6.363	0.380	8.189	0.411	10.780
36	0.312	5.520	0.331	7.075	0.353	9.223	0.381	12.354
38	0.292	6.095	0.309	7.899	0.329	10.448	0.353	14.268
40	0.273	6.755	0.288	8.861	0.306	11.911	0.326	16.630
42	0.255	7.516	0.268	9.993	0.284	13.680	0.302	19.586
44	0.237	8.400	0.249	11.339	0.263	15.844	0.279	23.356

$$\beta = 100$$
$$\delta = 20$$

i	0		5		10		15	
ϕ	K_a	K_p	K_a	K_p	K_a	K_p	K_a	K_p
24	0.452	3.397	0.490	4.196	0.540	5.201	0.611	6.510
26	0.425	3.710	0.460	4.612	0.503	5.765	0.562	7.298
28	0.401	4.059	0.431	5.081	0.469	6.413	0.519	8.219
30	0.377	4.450	0.404	5.616	0.438	7.162	0.480	9.306
32	0.354	4.891	0.379	6.227	0.408	8.034	0.445	10.600
34	0.333	5.391	0.354	6.930	0.380	9.057	0.412	12.157
36	0.312	5.959	0.331	7.744	0.354	10.268	0.382	14.052
38	0.292	6.611	0.310	8.695	0.330	11.715	0.354	16.389
40	0.273	7.363	0.289	9.815	0.306	13.465	0.327	19.317
42	0.255	8.237	0.269	11.146	0.284	15.607	0.303	23.052
44	0.238	9.260	0.250	12.743	0.264	18.266	0.280	27.915

$$\beta = 100$$
$$\delta = 22$$

i	0		5		10		15	
ϕ	K_a	K_p	K_a	K_p	K_a	K_p	K_a	K_p
24	0.452	3.596	0.491	4.489	0.542	5.630	0.614	7.146
26	0.426	3.940	0.461	4.952	0.505	6.269	0.565	8.056
28	0.401	4.326	0.432	5.479	0.471	7.008	0.522	9.129
30	0.377	4.761	0.405	6.081	0.439	7.868	0.483	10.407
32	0.355	5.254	0.380	6.775	0.409	8.878	0.447	11.945
34	0.333	5.815	0.355	7.579	0.382	10.074	0.414	13.817
36	0.313	6.458	0.332	8.518	0.356	11.504	0.384	16.128
38	0.293	7.200	0.310	9.624	0.331	13.233	0.355	19.024
40	0.274	8.063	0.290	10.938	0.308	15.351	0.329	22.718
42	0.256	9.073	0.270	12.517	0.286	17.980	0.304	27.532
44	0.239	10.267	0.251	14.436	0.265	21.301	0.281	33.961

$$\beta = 110$$
$$\delta = 0$$

i	0		5		10		15	
ϕ	K_a	K_p	K_a	K_p	K_a	K_p	K_a	K_p
24	0.584	1.939	0.628	2.188	0.682	2.453	0.755	2.740
26	0.554	2.043	0.594	2.315	0.643	2.607	0.706	2.927
28	0.526	2.155	0.562	2.451	0.606	2.773	0.662	3.131
30	0.498	2.274	0.532	2.597	0.572	2.952	0.621	3.352
32	0.471	2.402	0.502	2.754	0.539	3.147	0.583	3.593
34	0.446	2.539	0.474	2.924	0.507	3.358	0.546	3.857
36	0.421	2.687	0.447	3.108	0.477	3.589	0.512	4.147
38	0.398	2.846	0.422	3.308	0.449	3.840	0.480	4.466
40	0.375	3.018	0.397	3.525	0.421	4.116	0.449	4.820
42	0.353	3.204	0.373	3.762	0.395	4.419	0.420	5.212
44	0.332	3.406	0.350	4.021	0.370	4.754	0.393	5.650

$$\beta = 110$$
$$\delta = 14$$

i	0		5		10		15	
ϕ	K_a	K_p	K_a	K_p	K_a	K_p	K_a	K_p
24	0.554	2.532	0.604	3.008	0.667	3.555	0.755	4.198
26	0.527	2.703	0.572	3.222	0.629	3.827	0.704	4.548
28	0.501	2.887	0.542	3.456	0.593	4.128	0.658	4.938
30	0.476	3.088	0.514	3.713	0.559	4.460	0.616	5.374
32	0.452	3.306	0.487	3.995	0.527	4.829	0.577	5.864
34	0.430	3.544	0.461	4.306	0.497	5.240	0.541	6.417
36	0.407	3.806	0.436	4.651	0.469	5.701	0.508	7.044
38	0.386	4.093	0.412	5.034	0.441	6.219	0.476	7.760
40	0.365	4.411	0.389	5.461	0.415	6.805	0.447	8.582
42	0.345	4.763	0.367	5.941	0.391	7.471	0.419	9.531
44	0.326	5.154	0.346	6.481	0.367	8.232	0.392	10.637

$$\beta = 110$$
$$\delta = 16$$

i	0		5		10		15	
ϕ	K_a	K_p	K_a	K_p	K_a	K_p	K_a	K_p
24	0.553	2.638	0.605	3.158	0.670	3.763	0.761	4.483
26	0.527	2.822	0.573	3.391	0.631	4.061	0.708	4.871
28	0.501	3.021	0.543	3.646	0.595	4.392	0.662	5.305
30	0.477	3.238	0.515	3.926	0.561	4.759	0.619	5.792
32	0.453	3.476	0.488	4.236	0.529	5.168	0.580	6.343
34	0.430	3.736	0.462	4.579	0.499	5.627	0.544	6.967
36	0.408	4.022	0.437	4.960	0.470	6.143	0.510	7.680
38	0.387	4.339	0.413	5.386	0.443	6.726	0.479	8.499
40	0.366	4.689	0.390	5.863	0.417	7.389	0.449	9.446
42	0.346	5.079	0.368	6.401	0.392	8.148	0.421	10.548
44	0.327	5.515	0.347	7.010	0.369	9.022	0.394	11.844

$$\beta = 110$$
$$\delta = 18$$

i	0		5		10		15	
ϕ	K_a	K_p	K_a	K_p	K_a	K_p	K_a	K_p
24	0.554	2.751	0.607	3.319	0.674	3.989	0.767	4.796
26	0.528	2.950	0.575	3.573	0.635	4.317	0.714	5.227
28	0.502	3.166	0.545	3.852	0.598	4.682	0.667	5.712
30	0.478	3.402	0.517	4.159	0.564	5.089	0.624	6.259
32	0.454	3.661	0.490	4.500	0.532	5.544	0.584	6.881
34	0.431	3.945	0.464	4.879	0.501	6.058	0.548	7.590
36	0.409	4.260	0.439	5.303	0.473	6.638	0.514	8.405
38	0.388	4.608	0.415	5.777	0.445	7.298	0.482	9.348
40	0.368	4.996	0.392	6.313	0.420	8.053	0.452	10.446
42	0.348	5.429	0.370	6.919	0.395	8.923	0.424	11.737
44	0.329	5.916	0.349	7.610	0.371	9.932	0.397	13.268

$$\beta = 110$$
$$\delta = 20$$

i	0		5		10		15	
ϕ	K_a	K_p	K_a	K_p	K_a	K_p	K_a	K_p
24	0.566	2.873	0.610	3.494	0.679	4.236	0.775	5.143
26	0.529	3.087	0.578	3.771	0.639	4.598	0.721	5.624
28	0.504	3.322	0.548	4.077	0.602	5.002	0.673	6.168
30	0.479	3.579	0.519	4.415	0.568	5.454	0.629	6.785
32	0.456	3.862	0.492	4.791	0.535	5.964	0.589	7.490
34	0.433	4.174	0.466	5.212	0.505	6.541	0.552	8.300
36	0.411	4.521	0.441	5.684	0.476	7.197	0.518	9.237
38	0.390	4.906	0.417	6.215	0.449	7.948	0.486	10.330
40	0.370	5.337	0.394	6.818	0.423	8.813	0.456	11.614
42	0.350	5.821	0.372	7.505	0.398	9.816	0.427	13.138
44	0.331	6.367	0.351	8.293	0.374	10.990	0.400	14.965

$$\beta = 110$$
$$\delta = 22$$

i	0		5		10		15	
ϕ	K_a	K_p	K_a	K_p	K_a	K_p	K_a	K_p
24	0.559	3.004	0.614	3.684	0.685	4.507	0.785	5.528
26	0.532	3.237	0.582	3.987	0.645	4.907	0.729	6.067
28	0.507	3.492	0.552	4.323	0.607	5.356	0.680	6.680
30	0.482	3.773	0.523	4.696	0.572	5.862	0.636	7.379
32	0.458	4.083	0.496	5.114	0.540	6.434	0.595	8.184
34	0.436	4.426	0.469	5.582	0.509	7.086	0.558	9.115
36	0.414	4.809	0.444	6.110	0.480	7.831	0.523	10.200
38	0.393	5.236	0.420	6.708	0.452	8.690	0.491	11.477
40	0.372	5.717	0.397	7.390	0.426	9.687	0.460	12.992
42	0.352	6.259	0.375	8.173	0.401	10.854	0.432	14.809
44	0.333	6.876	0.354	9.077	0.378	12.230	0.404	17.013

$$\beta = 120$$
$$\delta = 0$$

i	0		5		10		15	
ϕ	K_a	K_p	K_a	K_p	K_a	K_p	K_a	K_p
24	0.705	1.891	0.765	2.138	0.838	2.394	0.936	2.664
26	0.675	1.974	0.731	2.239	0.798	2.517	0.884	2.814
28	0.647	2.062	0.699	2.347	0.760	2.648	0.837	2.973
30	0.619	2.155	0.667	2.460	0.724	2.788	0.793	3.144
32	0.592	2.253	0.637	2.581	0.689	2.937	0.752	3.328
34	0.566	2.357	0.608	2.710	0.656	3.096	0.712	3.525
36	0.540	2.467	0.579	2.848	0.624	3.268	0.675	3.738
38	0.516	2.585	0.552	2.995	0.593	3.452	0.640	3.968
40	0.492	2.710	0.526	3.152	0.563	3.650	0.606	4.218
42	0.469	2.844	0.500	3.322	0.535	3.864	0.574	4.490
44	0.446	2.987	0.475	3.504	0.507	4.096	0.543	4.787

$$\beta = 120$$
$$\delta = 14$$

i	0		5		10		15	
ϕ	K_a	K_p	K_a	K_p	K_a	K_p	K_a	K_p
24	0.687	2.350	0.759	2.787	0.849	3.273	0.973	3.823
26	0.659	2.480	0.726	2.948	0.807	3.476	0.915	4.079
28	0.633	2.618	0.694	3.122	0.768	3.695	0.862	4.359
30	0.608	2.766	0.664	3.309	0.731	3.933	0.814	4.664
32	0.583	2.925	0.635	3.511	0.696	4.193	0.770	5.000
34	0.559	3.095	0.607	3.730	0.662	4.476	0.730	5.370
36	0.536	3.279	0.580	3.968	0.631	4.787	0.691	5.779
38	0.513	3.477	0.554	4.227	0.601	5.128	0.655	6.233
40	0.491	3.692	0.529	4.510	0.571	5.504	0.621	6.739
42	0.469	3.926	0.505	4.821	0.544	5.921	0.588	7.305
44	0.449	4.180	0.481	5.162	0.517	6.384	0.557	7.942

$$\beta = 120$$
$$\delta = 16$$

i	0		5		10		15	
ϕ	K_a	K_p	K_a	K_p	K_a	K_p	K_a	K_p
24	0.690	2.428	0.764	2.899	0.857	3.428	0.987	4.035
26	0.662	2.566	0.730	3.072	0.814	3.648	0.926	4.314
28	0.636	2.715	0.699	3.259	0.774	3.886	0.872	4.619
30	0.610	2.874	0.668	3.462	0.737	4.145	0.823	4.955
32	0.586	3.044	0.639	3.681	0.702	4.429	0.779	5.325
34	0.562	3.228	0.611	3.919	0.668	4.739	0.737	5.734
36	0.539	3.427	0.584	4.178	0.636	5.081	0.698	6.189
38	0.516	3.642	0.558	4.461	0.606	5.458	0.662	6.695
40	0.494	3.876	0.533	4.772	0.577	5.875	0.627	7.263
42	0.473	4.131	0.509	5.114	0.549	6.338	0.595	7.901
44	0.452	4.409	0.485	5.491	0.522	6.856	0.563	8.623

$$\beta = 120$$
$$\delta = 18$$

i	0		5		10		15	
ϕ	K_a	K_p	K_a	K_p	K_a	K_p	K_a	K_p
24	0.694	2.510	0.770	3.017	0.867	3.594	1.002	4.262
26	0.666	2.658	0.737	3.204	0.823	3.832	0.939	4.568
28	0.640	2.817	0.704	3.406	0.783	4.092	0.884	4.903
30	0.614	2.988	0.674	3.626	0.745	4.375	0.834	5.273
32	0.590	3.172	0.644	3.863	0.709	4.685	0.789	5.682
34	0.566	3.371	0.616	4.123	0.675	5.026	0.746	6.136
36	0.543	3.586	0.589	4.406	0.643	5.403	0.707	6.643
38	0.520	3.820	0.563	4.717	0.612	5.820	0.670	7.210
40	0.498	4.075	0.538	5.059	0.583	6.284	0.635	7.849
42	0.477	4.354	0.513	5.436	0.554	6.802	0.602	8.572
44	0.456	4.659	0.490	5.854	0.527	7.383	0.570	9.395

$$\beta = 120$$
$$\delta = 20$$

i	0		5		10		15	
ϕ	K_a	K_p	K_a	K_p	K_a	K_p	K_a	K_p
24	0.700	2.597	0.779	3.144	0.879	3.773	1.020	4.509
26	0.672	2.756	0.744	3.346	0.834	4.031	0.955	4.844
28	0.645	2.927	0.712	3.564	0.793	4.314	0.898	5.213
30	0.620	3.111	0.681	3.802	0.754	4.624	0.847	5.621
32	0.595	3.310	0.651	4.061	0.718	4.965	0.800	6.075
34	0.571	3.525	0.623	4.344	0.683	5.341	0.757	6.581
36	0.548	3.759	0.596	4.654	0.651	5.757	0.717	7.148
38	0.525	4.013	0.569	4.996	0.620	6.221	0.679	7.787
40	0.503	4.292	0.544	5.373	0.590	6.738	0.644	8.509
42	0.482	4.597	0.519	5.791	0.561	7.319	0.610	9.332
44	0.461	4.934	0.495	6.257	0.534	7.974	0.578	10.275

$$\beta = 120$$
$$\delta = 22$$

i	0		5		10		15	
ϕ	K_a	K_p	K_a	K_p	K_a	K_p	K_a	K_p
24	0.707	2.689	0.789	3.279	0.893	3.965	1.040	4.777
26	0.679	2.860	0.754	3.498	0.847	4.247	0.973	5.145
28	0.652	3.044	0.721	3.735	0.805	4.556	0.914	5.552
30	0.626	3.243	0.689	3.993	0.765	4.896	0.861	6.005
32	0.601	3.458	0.659	4.275	0.728	5.271	0.813	6.510
34	0.577	3.691	0.631	4.585	0.693	5.687	0.769	7.076
36	0.554	3.945	0.603	4.926	0.660	6.149	0.728	7.714
38	0.531	4.223	0.576	5.302	0.628	6.666	0.690	8.435
40	0.509	4.528	0.551	5.720	0.598	7.245	0.654	9.257
42	0.487	4.864	0.526	6.185	0.570	7.900	0.620	10.199
44	0.466	5.236	0.502	6.705	0.542	8.642	0.587	11.286

Solution to Terzaghi's Consolidation Equation

$$C_v \frac{\partial^2 \mu}{\partial z^2} = \frac{\partial \mu}{\partial t}$$

Solution

$$\mu = \mu(z, t) \qquad \left.\begin{array}{l} \mu(0, t) = 0 \\ \mu(2H, t) = 0 \end{array}\right\} \qquad \text{Boundary conditions}$$

$$\mu(z, 0) = \mu_0 \qquad \text{Initial condition}$$

Symbols and Abbreviations

$$\frac{\partial^2 \mu}{\partial z^2} = Z''; \qquad \frac{\partial \mu}{\partial t} = T'$$

p.d.e. = partial differential, equation;   o.d.e. = ordinary differential equation.
Let $\mu = Z(z) \cdot T(t)$; separation of variable solution

$$\text{p.d.e.} \Rightarrow C_v Z' T = Z T'$$

or

$$\left(\frac{Z'}{z}\right) = \frac{T'}{C_v T} \equiv K = \text{separation constant}$$

p.d.e. reduces to a coupled system of o.d.e.

$$Z_1'' - KZ = 0 \qquad \text{and} \qquad T' - C_v KT = 0$$

In operator from

$$(D^2 - K)\{Z(z)\} = 0 \Rightarrow D = \pm \sqrt{K} = 0 \pm \sqrt{-K}i$$
$$(D - C_v K)\{T(t)\} = 0 \Rightarrow D = C_v K$$

$$\left.\begin{array}{ll}
K = 0: & \mu = ZT = \overline{C}_1 + \hat{C}_2 z \\
K > 0: & \mu = e^{C_v Kt}(\hat{C}_1 e^{\sqrt{K} \cdot z} + \hat{C}_2 e^{-\sqrt{K} \cdot z} \\
K < 0: & \mu = e^{C_v Kt}(\overline{C}_1 \sin \sqrt{-K}z + C_2 \cos \sqrt{-K}z)
\end{array}\right\} \begin{array}{l} \text{solution of} \\ \text{the p.d.e.} \end{array}$$

For a Fourier Series solution, assume $K < 0$. Thus, let $\mu = e^{C_v Kt}(C_1 \sin \sqrt{-K} \cdot z + C_2 \cos \sqrt{-K}z)$. Now to find C_1, C_2, and K, we apply the conditions:

$$\mu(0, t) = 0 \Rightarrow 0 = e^{C_v \cdot K \cdot t}C_2$$

Let $C_2 = 0$; therefore,

$$\mu = C_1 e^{C_v Kt} \sin \sqrt{-K}z \tag{a}$$

and

$$\mu(2H, t) = 0 \Rightarrow 0 = C_1 e^{C_v \cdot K \cdot t} \sin(\sqrt{-K} \cdot 2H) \tag{b}$$

For nontrivial solution, let $\sin \theta \equiv \sin \sqrt{-K} \cdot 2H) = 0$.

$$\theta = 2H\sqrt{-K} = n\pi; \qquad n = 1, 2, 3 \ldots$$

or

$$\sqrt{-K} = \frac{n\pi}{2H}$$

Squaring,

$$K = -\left(\frac{n\pi}{2H}\right)^2 < 0, \qquad \text{as assumed}$$

Therefore, from Eq. (a)

$$\mu = \mu(z, t) = C_1 e^{-C_v \left(\frac{n\pi}{2H}\right)^2 t} \sin\left(\frac{n\pi z}{2H}\right); \qquad n = 1, 2, 3 \ldots$$

Via "compounding" of solutions, we can also write

$$\mu = \mu(z, t) = \sum_{n=1}^{\infty} \left\{ C_1 e^{-C_v \left(\frac{n\pi}{2H}\right)t} \sin\left(\frac{n\pi z}{2H}\right) \right\}$$

For

$$\mu(z, 0) = f(z) = \mu_0 = \text{constant} \neq 0 \Rightarrow \mu_0 = \sum_{n=1}^{\infty} C_1 \sin\left(\frac{n\pi z}{2H}\right)$$

But this sum has the form of a sine Harmonic Fourier Series (Fourier Series in sines only):

$$\psi(z) = \sum_{n=1}^{\infty} l_n \sin\left(\frac{n\pi z}{L}\right); \ l_n = \frac{2}{L}\int_0^L \psi(z) \sin\left(\frac{n\pi z}{L}\right) dz$$

with

$$l_n = C_1 = \frac{2}{2H}\int_0^{2H} \mu_0 \sin\left(\frac{n\pi z}{2H}\right) dz$$

$$= \frac{1}{H}\cdot \mu \cdot \left(\frac{-2H}{n\pi}\right) \cos\left(\frac{n\pi z}{2H}\right)\Big]_{z=0}^{2H}$$

$$= -\frac{2\mu_0}{n\pi}[\cos(n\pi - 1)]$$

$$\begin{cases} 0; & n \text{ even} \\ \dfrac{4\mu}{n\pi}; & n \text{ odd.} \end{cases}$$

Therefore,

$$\mu = \mu(z, t) = \sum_{n=1,3,5,7}^{\infty} \left\{ \frac{4\mu_0}{n\pi} e^{-C_v\left(\frac{n\pi}{2H}\right)^2 t} \sin\left(\frac{n\pi z}{2H}\right) \right\}$$

Note: With the summation taken for *odd* positive integers, *n*, we could also write:

$$\mu = \sum_{m=0}^{\infty} \left\{ \frac{4\mu_0}{(2m+1)\pi} e^{-C_v\left(\frac{(2m+1)\pi}{2H}\right)^2 t} \sin\left(\frac{(2m+1)\pi z}{2H}\right) \right\}$$

Note the summation is taken on *m*; *m* = 0, 1, 2, 3. . . . Now let

$$M = \frac{\pi}{2}(2m+1) \qquad \text{and} \qquad \hat{T} = \frac{C_v}{H^2}\cdot t$$

Then the solution to the differential system becomes

$$\boxed{\mu = \mu(z, \hat{T}) = \sum_{m=0}^{\infty} \left\{ \frac{2\mu_0}{M}\cdot \sin\left(\frac{Mz}{H}\right)\cdot e^{-M^2\hat{T}} \right\}}$$

This is Eq. 9-8.

Boussinesq's Equation for Radial Stress

Boussinesq's equation for the radial stress (σ_r) can be expressed as

$$\sigma_r = \frac{Q}{2\pi} \left[\frac{3r^2 z}{(r^2 + z)^{5/2}} - \frac{1 - 2\mu}{r^2 + z^2 + z(r^2 + z^2)^{1/2}} \right]$$

where Q = the magnitude of the concentrated point load
     r = the radial distance from the concentrated point load
     z = the depth from the concentrated point load
     μ = Poisson's Ratio

The equation can be simplified in terms of (r/z):

$$\sigma_r = \frac{Q}{2\pi z^2} \left[\frac{3(r/z)^2}{[(r/z)^2 + 1]^{5/2}} - \frac{(1 - 2\mu)}{[(r/z)^2 + 1] + [(r/z)^2 + 1]^{1/2}} \right]$$

which can also be expressed as

$$\sigma_r = \frac{Q N_{BR}}{z^2}$$

Table C.1 Horizontal Stress Coefficients, N_{BR}, for a Case of a Concentrated Load (Boussinesq's Solution, $\sigma_r = QN_{BR}/z^2$)

r/z	$\mu = 0.0$	$\mu = 0.1$	$\mu = 0.2$	$\mu = 0.3$	$\mu = 0.4$	$\mu = 0.5$
0.00	0.0000	0.0000	0.0000	0.0000	0.0000	0.0000
0.05	0.0000	0.0000	0.0000	0.0000	0.0000	0.0012
0.10	0.0000	0.0000	0.0000	0.0000	0.0000	0.0047
0.15	0.0000	0.0000	0.0000	0.0000	0.0000	0.0102
0.20	0.0000	0.0000	0.0000	0.0000	0.0019	0.0173
0.25	0.0000	0.0000	0.0000	0.0000	0.0104	0.0256
0.30	0.0000	0.0000	0.0000	0.0048	0.0197	0.0346
0.35	0.0000	0.0000	0.0000	0.0146	0.0292	0.0438
0.40	0.0000	0.0000	0.0100	0.0243	0.0385	0.0527
0.45	0.0000	0.0056	0.0194	0.0333	0.0471	0.0610
0.50	0.0011	0.0146	0.0280	0.0414	0.0549	0.0683
0.55	0.0095	0.0225	0.0355	0.0485	0.0616	0.0746
0.60	0.0167	0.0293	0.0419	0.0545	0.0671	0.0797
0.65	0.0227	0.0349	0.0471	0.0592	0.0714	0.0836
0.70	0.0276	0.0394	0.0511	0.0628	0.0746	0.0863
0.75	0.0314	0.0427	0.0541	0.0654	0.0767	0.0880
0.80	0.0342	0.0451	0.0560	0.0669	0.0778	0.0887
0.85	0.0361	0.0466	0.0571	0.0676	0.0781	0.0886
0.90	0.0373	0.0474	0.0575	0.0676	0.0777	0.0877
0.95	0.0378	0.0475	0.0572	0.0669	0.0766	0.0863
1.00	0.0378	0.0471	0.0564	0.0658	0.0751	0.0844
1.05	0.0373	0.0463	0.0552	0.0642	0.0732	0.0821
1.10	0.0365	0.0451	0.0537	0.0623	0.0710	0.0796
1.15	0.0354	0.0437	0.0520	0.0603	0.0685	0.0768
1.20	0.0342	0.0421	0.0501	0.0580	0.0660	0.0739
1.25	0.0327	0.0404	0.0480	0.0557	0.0633	0.0710
1.30	0.0312	0.0386	0.0459	0.0533	0.0606	0.0680
1.35	0.0297	0.0367	0.0438	0.0509	0.0579	0.0650
1.40	0.0281	0.0349	0.0417	0.0485	0.0553	0.0621
1.45	0.0265	0.0330	0.0396	0.0461	0.0527	0.0592
1.50	0.0249	0.0312	0.0375	0.0438	0.0501	0.0564
1.55	0.0234	0.0295	0.0355	0.0416	0.0477	0.0537
1.60	0.0219	0.0277	0.0336	0.0394	0.0453	0.0511
1.65	0.0205	0.0261	0.0317	0.0374	0.0430	0.0486
1.70	0.0191	0.0245	0.0299	0.0354	0.0408	0.0462
1.75	0.0178	0.0230	0.0282	0.0335	0.0387	0.0440
1.80	0.0165	0.0216	0.0266	0.0317	0.0367	0.0418
1.85	0.0153	0.0202	0.0251	0.0300	0.0349	0.0397
1.90	0.0142	0.0189	0.0236	0.0284	0.0331	0.0378
1.95	0.0132	0.0177	0.0223	0.0268	0.0314	0.0359
2.00	0.0122	0.0166	0.0210	0.0254	0.0298	0.0342
2.05	0.0112	0.0155	0.0197	0.0240	0.0282	0.0325
2.10	0.0104	0.0145	0.0186	0.0227	0.0268	0.0309
2.15	0.0095	0.0135	0.0175	0.0215	0.0255	0.0294

Table C.1 (Cont'd) Horizontal Stress Coefficients, N_{BR}, for a Case of a Concentrated Load (Boussinesq's Solution, $\sigma_r = QN_{BR}/z^2$)

r/z	$\mu = 0.0$	$\mu = 0.1$	$\mu = 0.2$	$\mu = 0.3$	$\mu = 0.4$	$\mu = 0.5$
2.20	0.0088	0.0126	0.0165	0.0203	0.0242	0.0280
2.25	0.0080	0.0118	0.0155	0.0192	0.0230	0.0267
2.30	0.0074	0.0110	0.0146	0.0182	0.0218	0.0255
2.35	0.0067	0.0102	0.0137	0.0173	0.0208	0.0243
2.40	0.0061	0.0095	0.0129	0.0163	0.0197	0.0231
2.45	0.0056	0.0089	0.0122	0.0155	0.0188	0.0221
2.50	0.0051	0.0083	0.0115	0.0147	0.0179	0.0211
2.55	0.0046	0.0077	0.0108	0.0139	0.0170	0.0201
2.60	0.0041	0.0072	0.0102	0.0132	0.0162	0.0192
2.65	0.0037	0.0067	0.0096	0.0125	0.0155	0.0184
2.70	0.0033	0.0062	0.0090	0.0119	0.0147	0.0176
2.75	0.0030	0.0057	0.0085	0.0113	0.0141	0.0168
2.80	0.0026	0.0053	0.0080	0.0107	0.0134	0.0161
2.85	0.0023	0.0049	0.0076	0.0102	0.0128	0.0154
2.90	0.0020	0.0046	0.0071	0.0097	0.0122	0.0148
2.95	0.0018	0.0042	0.0067	0.0092	0.0117	0.0142
3.00	0.0015	0.0039	0.0063	0.0088	0.0112	0.0136
3.05	0.0013	0.0036	0.0060	0.0083	0.0107	0.0130
3.10	0.0010	0.0033	0.0056	0.0079	0.0102	0.0125
3.15	0.0008	0.0031	0.0053	0.0075	0.0098	0.0120
3.20	0.0006	0.0028	0.0050	0.0072	0.0094	0.0115
3.25	0.0005	0.0026	0.0047	0.0068	0.0090	0.0111
3.30	0.0003	0.0024	0.0044	0.0065	0.0086	0.0107
3.35	0.0001	0.0022	0.0042	0.0062	0.0082	0.0103
3.40	0.0000	0.0020	0.0039	0.0059	0.0079	0.0099
3.45	0.0000	0.0018	0.0037	0.0056	0.0076	0.0095
3.50	0.0000	0.0016	0.0035	0.0054	0.0073	0.0092
3.55	0.0000	0.0015	0.0033	0.0051	0.0070	0.0088
3.60	0.0000	0.0013	0.0031	0.0049	0.0067	0.0085
3.65	0.0000	0.0012	0.0029	0.0047	0.0064	0.0082
3.70	0.0000	0.0010	0.0027	0.0045	0.0062	0.0079
3.75	0.0000	0.0009	0.0026	0.0043	0.0059	0.0076
3.80	0.0000	0.0008	0.0024	0.0041	0.0057	0.0074
3.85	0.0000	0.0007	0.0023	0.0039	0.0055	0.0071
3.90	0.0000	0.0006	0.0021	0.0037	0.0053	0.0069
3.95	0.0000	0.0005	0.0020	0.0036	0.0051	0.0066
4.00	0.0000	0.0004	0.0019	0.0034	0.0049	0.0064
4.05	0.0000	0.0003	0.0018	0.0032	0.0047	0.0062

Table C.1 (Cont'd) Horizontal Stress Coefficients, N_{BR}, for a Case of a Concentrated Load (Boussinesq's Solution, $\sigma_r = QN_{BR}/z^2$)

r/z	$\mu = 0.0$	$\mu = 0.1$	$\mu = 0.2$	$\mu = 0.3$	$\mu = 0.4$	$\mu = 0.5$
4.10	0.0000	0.0002	0.0017	0.0031	0.0046	0.0060
4.15	0.0000	0.0001	0.0016	0.0030	0.0044	0.0058
4.20	0.0000	0.0001	0.0015	0.0028	0.0042	0.0056
4.25	0.0000	0.0000	0.0014	0.0027	0.0041	0.0054
4.30	0.0000	0.0000	0.0013	0.0026	0.0039	0.0053
4.35	0.0000	0.0000	0.0012	0.0025	0.0038	0.0051
4.40	0.0000	0.0000	0.0011	0.0024	0.0037	0.0049
4.45	0.0000	0.0000	0.0010	0.0023	0.0035	0.0048
4.50	0.0000	0.0000	0.0010	0.0022	0.0034	0.0046
4.55	0.0000	0.0000	0.0009	0.0021	0.0033	0.0045
4.60	0.0000	0.0000	0.0008	0.0020	0.0032	0.0044
4.65	0.0000	0.0000	0.0008	0.0019	0.0031	0.0042
4.70	0.0000	0.0000	0.0007	0.0018	0.0030	0.0041
4.75	0.0000	0.0000	0.0006	0.0018	0.0029	0.0040
4.80	0.0000	0.0000	0.0006	0.0017	0.0028	0.0039
4.85	0.0000	0.0000	0.0005	0.0016	0.0027	0.0038
4.90	0.0000	0.0000	0.0005	0.0015	0.0026	0.0037
4.95	0.0000	0.0000	0.0004	0.0015	0.0025	0.0036
5.00	0.0000	0.0000	0.0004	0.0014	0.0024	0.0035

Index

CONVERSION OF ENGLISH UNITS TO SI

Length

Multiply By ↓ To	From → Inches	Feet	Yards	Miles
mm	2.54×10^1	3.05×10^2	9.14×10^2	1.61×10^6
cm	2.54	3.05×10^1	9.14×10^1	1.61×10^5
m	2.54×10^{-2}	3.05×10^{-1}	9.14×10^{-1}	1.61×10^3
km	2.54×10^{-5}	3.05×10^{-4}	9.14×10^{-4}	1.61

Area

Multiply By ↓ To	From → Square Inches	Square Feet	Square Yards	Square Miles
mm^2	6.45×10^2	9.29×10^4	8.36×10^5	2.59×10^{12}
cm^2	6.45	9.29×10^2	8.36×10^3	2.59×10^{10}
m^2	6.45×10^{-4}	9.29×10^{-2}	8.36×10^{-1}	2.59×10^6
km^2	6.45×10^{-10}	9.29×10^{-8}	8.36×10^{-7}	2.59

Volume

Multiply By ↓ To	From → Cubic Inches	Cubic Feet	Cubic Yards	Quarts	Gallons
cm^2	1.64×10^1	2.83×10^4	7.65×10^5	9.46×10^2	3.78×10^3
liter	1.64×10^{-2}	2.83×10^1	7.65×10^2	9.46×10^{-1}	3.78×10^0
m^3	1.64×10^{-5}	2.83×10^{-2}	7.65×10^{-1}	9.46×10^4	3.78×10^{-3}

Force

Multiply By ↓ To	From → Ounces	Pounds	Kips	Tons (short)
dynes	2.780×10^4	4.448×10^5	4.448×10^8	8.897×10^8
grams	2.835×10^1	4.535×10^2	4.535×10^5	9.074×10^5
kilograms	2.835×10^{-2}	4.535×10^{-1}	4.535×10^2	9.074×10^2
newtons	2.780×10^{-1}	4.448	4.448×10^3	8.897×10^3
kilonewtons	2.780×10^{-4}	4.448×10^{-3}	4.448	8.897
tons (metric)	2.835×10^{-5}	4.535×10^{-4}	4.535×10^{-1}	9.074×10^{-1}